3000 8
St. Louis Commu

D0903057

STUDIES IN INDUSTRY AND SOCIETY

Philip B. Scranton, Series Editor

Published with the assistance of the Hagley Museum and Library

Related titles in the series:

David A. Hounshell, *From the American System to Mass Production, 1800–1932: The Development of Manufacturing Technology in the United States*

John K. Brown, *The Baldwin Locomotive Works, 1831–1915: A Study in American Industrial Practice*

Thomas R. Heinrich, *Ships for the Seven Seas: Philadelphia Shipbuilding in the Age of Industrial Capitalism*

Mark Aldrich, *Safety First: Technology, Labor, and Business in the Building of American Work Safety, 1870–1939*

The Carriage Trade

The Carriage Trade

Making Horse-Drawn Vehicles in America

Thomas A. Kinney

The Johns Hopkins University Press
Baltimore and London

This book has been brought to publication with the generous assistance of the Hagley Museum and Library.

Printed in the United States of America on acid-free paper
2 4 6 8 9 7 5 3 1

The Johns Hopkins University Press
2715 North Charles Street
Baltimore, Maryland 21218-4363
www.press.jhu.edu

Library of Congress Cataloging-in-Publication Data

Kinney, Thomas A., 1968–
The carriage trade : making horse-drawn vehicles in America / Thomas A. Kinney.
p. cm. — (Studies in industry and society)
Includes bibliographical references and index.
ISBN 0-8018-7946-9 (hardcover : alk. paper)
1. Carriage industry—United States—History. I. Title. II. Series.
HD9709.5.U62K56 2004
388.3′41′0973—dc22 2003025935

A catalog record for this book is available from the British Library.

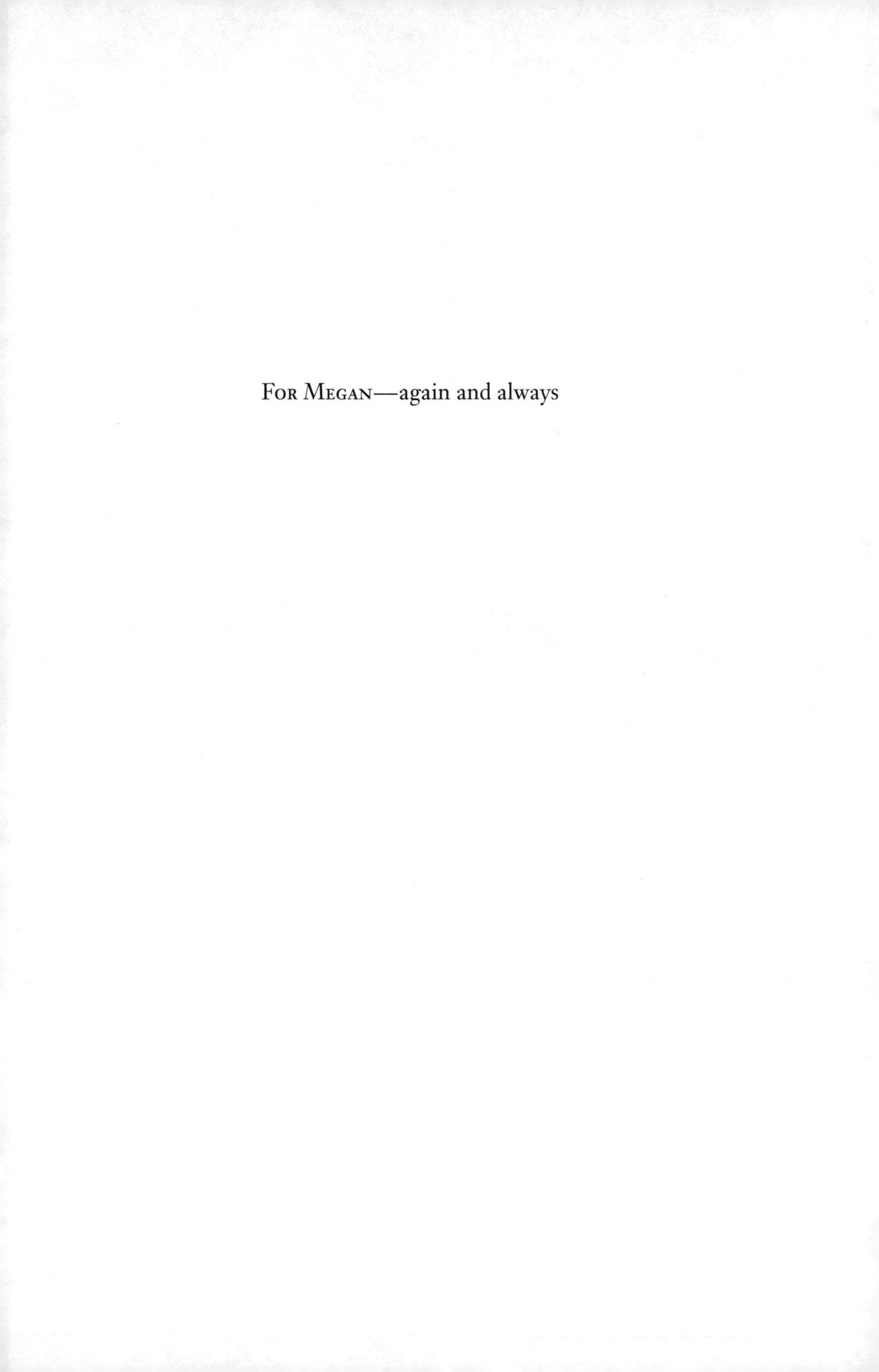

For Megan—again and always

Contents

Acknowledgments

Samuel Johnson, that crusty old English lexicographer, once averred that "a man will turn over half a library to make one book." In the darker moments of this project I would have thought this an understatement, but, even if only literal, the quip accurately describes the historian's plight. Indeed, in a decade's heavy tilling I have incurred many debts, not all of them scholarly, and I would like to acknowledge my many creditors here. Time and human memory being what they are, I fear omitting some key person, but I trust all will understand that my gratitude far outstrips my ability to call up every name and face. Likewise, though I relied on a great many sources of advice and information, I take full responsibility for the conclusions presented in this book.

Easily my greatest carriage-related debt is to Merri M. Ferrell, former curator of the carriage collection at the Long Island Museum of American Art, History & Carriages and one of the most knowledgeable experts on horse-drawn vehicles in the United States. Whether her missionary zeal for conveying the value of unrestored vehicles and original surfaces or her abiding fascination for the people who made them, Merri's contributions to this study are legion. That I made a valued friend along the way makes it that much sweeter. The Long Island Museum's Eva Greguski and Christa Zaros provided additional assistance with photographs, and I wish to thank them again here.

At the Smithsonian Institution's National Museum of American History, librarian James Roan provided invaluable reference assistance, while the Division of Transportation's Roger White lent office space, reference assistance, and a great deal of useful information. Curator Steven Lubar has always made himself available to answer questions, both before and after this book project. The Studebaker National Museum's Andy Beckman proved exceedingly helpful in steering me to some especially obscure but vital resources, and archivists Dr. and Mrs. Robert Denham provided numberless photocopies and invaluable research. Sandra Jacques of the Studebaker National Family Association answered my questions about Studebaker genealogy and aided my deciphering the lineage of this large and prolific family group. James Campbell of the New Haven Colony Historical Society gave me helpful background on a key Brewster manuscript, while Amesbury Public Library reference librarian Jen Haven happily fielded similar requests about other dead carriage makers.

Closer to home, the reference and circulation staff of Case Western Reserve University's Kelvin Smith Library handled various requests with cheerful dispatch, but I am especially indebted to the Cleveland Public Library's department of science and technology. Section head Jean Piety and her staff combine professional competence with unfailingly professional service, and they have measurably aided me in completing this project. I feel privileged to have such graceful assistance at hand. The reference desk and photo duplication staff of the Ohio Historical Society's library granted me access to rare copies of early trade journals, and I here extend heartfelt thanks to Duryea Kemp and Thomas Starbuck for their prompt and careful handling of numerous photocopy requests. I owe thanks as well to Christopher Duckworth, David Simmons, and the rest of the *Timeline* crew for a place to stow my baggage and for more than one lunch.

Fellow historians Carroll Pursell, Philip Scranton, Howard Segal, and Craig Semsel provided advice and feedback, and to me they epitomize the character of the Society for the History of Technology to which we all belong. SHOT is a vibrant group of scholars whose enthusiasm, accessibility, and warm fellowship shame much larger and better known historical organizations. I especially appreciate historian Carolyn Cooper both for her early interest in my work and subsequent suggestions, as well as Robert Gordon for sharing his voluminous knowledge of metals which helped clarify my descriptions of malleable iron.

I have benefited from the time and generosity of the staff of two fine old Cleveland firms, and Cleveland Hardware & Forging Company's James Hoban, Thomas A. Birkel, and Deborah Kubiak all have my gratitude for allowing me access to records and historic photographs. Cleveland Hardware's ancient rival, Eberhard Manufacturing Company, proved equally helpful, and I greatly enjoyed time spent with Brian Kay and Edward Hongosh poring over that firm's trove of papers, patents, and artifacts. Would that every manufacturing executive have such treasure in his office!

Rick Hossman, old family friend, and Dave Barker, Ohio Historical Society, both gave me expert photographic advice that materially improved my copy stand work. Mark L. Gardner graciously allowed me use of an image from his extensive photographic collection, as did Frank Frost, who shared of his assorted Studebaker ephemera. Cord Camera's Marie Simpson and Brian Miller provided tremendous assistance with traditional and digital imaging, while the Carriage Library of America's Susan Green provided photocopies during the early stages of this project. Don H. Berkebile, now long retired from the Smithsonian, provided useful insights, as did Donald H. Laing, associate professor of classics at Case Western Reserve University. My brother, Travis E. Kinney, drew on his architectural and design background to guide me in understanding carriage drafting and illustration. By curious coincidence he also came into possession of a surprisingly complete carriage blacksmith shop during the course of this book. Besides the enormous fun of identifying the various tools and machinery and the many hours of enjoyment we shared firing up the forge and making our simple projects at the anvil, the experience provided a tangible link with a past that I have tried to resurrect in these pages. I like to think that John Pillsbury, blacksmith, wagon maker, and original owner of the shop, would be proud.

The Johns Hopkins University Press has been an ideal partner in this enterprise, and I owe to history editor Robert J. Brugger my gratitude for his droll insight and highly seasoned editorial eye. More than that, I treasure his patience, and I hope this book has proven worth the wait. Series editor Philip Scranton, tall sage of so many conference book exhibits, never fails to come through with advice under pressure, and his helpfulness is matched only by his infectious good humor. I am pleased to know him. Thanks also to the Press's anonymous reader and to copyeditor Glenn Perkins, both of whom added enormously to the finished product. Melody Herr has assisted with everything from obscure formatting questions to educated guesses at Bob's whereabouts, and I deeply appreciate it all.

I owe an incalculable debt to my in-laws, who, besides their willingness to suffer the writing temperament, also tolerated a messy lumber pile, wood shavings on their rugs, and believed me when I promised not to install a forge in their basement. More importantly, they never ceased to listen, proffer wise advice, and show kindness and patience beyond all measure. I owe even more to Megan, my wife, best friend, and true partner in everything. She has given freely of her love, support, and advice, and even in the most trying times never stopped believing in me. "Thank you" seems hardly sufficient. My greatest debt of all, however, and one I can never begin to repay, can be summarized by a sentiment doubtless shared by James Brewster and Clement Studebaker alike: *Soli Deo Gloria.*

The Carriage Trade

Introduction

Remembering a Forgotten Industry

When the organizers of the Philadelphia Centennial Exposition arranged the layout of the buildings and grounds for the 1876 extravaganza, they put the American carriage industry in a wooden annex behind the glass-sheathed Main Exhibition Building. Sharing space with makers of furniture, stoves, household utensils, railroad equipment, and other goods "too bulky to occupy the limited and more valuable space in the Main Building," over sixty American carriage firms wedged nearly three hundred of their finest products into the display area. If they were pressed for room, they were at least near the action. The manufacturers of wagons for farm and business had to content themselves with a primitive shed on the edge of the grounds. Although there was grumbling in the trade journals, the slight was not intentional. Inundated with requests for exhibit space far exceeding all estimates, the Exposition committee had hastily erected dozens of auxiliary structures, all dwarfed by the cathedral-like main buildings.[1]

A change in venue would not have mattered much, for the American horse-drawn vehicle industry was not the only one hoping to curry public favor in Philadelphia that year. Indeed, it had stiff competition from every conceivable quarter. Visitors gawked at miniature models of the world's leading cities, strolled through humid greenhouses full of tropical plants, and sampled fare from eight major restaurants and a liberal scattering of popcorn, coffee, and soda water stands. They examined a Swedish schoolhouse, marveled at the discipline of the cadets at the West Point encampment, and attended a steady stream of events ranging from the Special Display of Melons and the Exhibit of Dogs to the International Cricket Matches and the Catholic Total Abstinence Society Parade. Compared to such exotic offerings, wagons and carriages must have seemed ordinary indeed.[2]

While visitors delighted in such eclectic variety, the Centennial Exposition's true purpose was to celebrate the American industrial revolution. A staggering array of consumer goods choked the aisles of Main and Machinery Halls: ornately painted safes, nickel-trimmed parlor stoves, and patented clothes wringers vied for attention with immense stalls of newly published books, bright bolts of cloth, and display cases crowded with glittering jewelry. It was impossible to get away from the machinery, but then nobody really wanted to. Freshly assembled examples of the latest woodworking machinery and machine tools fascinated the men nearly as much as the extensive sewing machine display captivated the women. Row upon row of operating looms, printing presses, and rivet headers noisily proclaimed the wonders of the industrial age. Nothing epitomized this better than "The Great Corliss Engine," the 2,500-horsepower steam engine driving all the line shafting in Machinery Hall. *Atlantic Monthly* editor William Dean Howells called it "an athlete of steel and iron" possessing "hoarded power that makes all tremble." Indeed, part of its appeal came from its sheer size, but more important was what it represented to thousands of daily admirers. For if Americans were enthusiastic over anything, it was machinery and its promise for their lives as individuals and as a nation.[3]

Celebrating American inventiveness was an unabashed goal of the event, so it comes as no surprise that machinery took center stage. It remains there still. Americans continue to be entranced by technology, and our fascination with the automobile is but one example. Even our scholarship reflects this in the annual crop of automotive history books that document every possible aspect of the subject—and rightly so, for they help us understand a pivotal American industrial development. Like the throngs of fair goers rushing past the buggies to get to the steam engines, however, most historians and their readers ignore the American horse-drawn vehicle industry on their way to the cars and trucks. The modest literature on the subject rarely goes beyond pictorial studies of carriage styles, reproductions of trade journals, and photographic collections celebrating wagons and working horses. In nearly every case the objects receive the attention; the industry that produced them remains virtually unknown.

This book proposes to change that. Rather than merely stroll the exhibit hall aisles and admire the design and decoration of the products, I want to step into the shops and factories that made the display possible and illuminate the processes and innovations that were developed there. For the story that the industry has to tell us today consists of far more than a simple account of what preceded the automobile. The rise and fall of the American wagon and carriage industry diverges significantly from the accepted pattern of American industrial development. Our understanding of industry depends entirely too much on the automobile, the steel mill, and the machine shop, just as it does on the vertically integrated corporation. Oak and ash, the horse-drawn vehicle, small shops, and legions of partnerships and family-run firms played equally important roles. As one of the nation's largest nineteenth-century industries, and one epitomizing this understudied alternative to mass production, the wagon and carriage industry supplies an ideal example.

Writing about his family's prosperous Ohio carriage firm, Ben Riker noted that "nearly every mechanic in the carriage industry nursed the idea of starting a shop of his own." Perhaps more surprising was the relative ease of turning that dream into reality. Moderate trade skills, quality tools, and self-discipline were the principal requirements, for, according to Riker, money mattered less than "energy, ambition and initiative." This sounds suspiciously like Horatio Alger idealism, but Riker knew of what he wrote. In towns and cities nationwide small shops turned out nearly every kind of horse-drawn vehicle through the end of the industry. Indeed, a typical medium-sized city often hosted them in impressive numbers: in 1891 Cleveland claimed more than seventy wagon and carriage manufacturers, the majority of them small shops.[4]

This seems hardly noteworthy until we remember that by the 1890s the trade had been industrialized for decades and featured a considerable number of large factories. As early as 1860, New Haven's G. & D. Cook & Company turned out a vehicle an hour; twenty years later the Studebaker Brothers Manufacturing Company averaged one hundred wagons daily, or one every six minutes. During the intervening years many other factories fashioned vast quantities of buggies, farm wagons, and pleasure carriages, while a few vied with Studebaker—largely in vain—for the right to advertise themselves as the world's biggest vehicle factory. How did small shops survive in an environment seemingly dominated by these sizeable and more productive factories? This paradox lies at the center of the American horse-drawn vehicle industry. Resolving it is the purpose of this book.[5]

Industrialization forever changed most forms of manufacturing over the course of the last two hundred years. This wholesale craft-to-industry transition created the diverse American manufacturing sector churning out the goods admired by the crowds during that long Philadelphia summer. Indeed, few areas of production remained exempt from industrialization. In many cases large mechanized factories replaced nearly all previous production formats. Consider the shoe industry: Colonial and antebellum one-room shops could in no way compete with ruthlessly efficient New England shoe factories. Shoe manufacture became effectively monopolized by large corporations that turned out enormous quantities of standardized, low-price footwear. Handmade shoes disappeared in all but the poorest and most remote regions of Appalachia and the South, and the village cobbler faded into the past. The conventional view of industrialization typically revolves around examples like this: the replacement of hand tools, craft methods, and small firms with machinery, mass production, and large corporations. Yet industrialization, which appears so monolithic at a distance, becomes far more textured when subject to detailed scrutiny. Some craft methods survived the introduction of high-speed machinery, skill remained a staple in the lives of many workmen, and small firms often stood their ground against outsized competitors. The wagon and carriage trade provided living proof.[6]

Historians of nineteenth-century business and technology have uncovered much that helps us understand this fundamental structural contradiction. Reacting to scholarly overemphasis on corporate capitalism and mass production, recent studies have revealed a preponderance of alternative manufacturing strategies. A groundbreaking examination of the Philadelphia textile industry demonstrated that huge New England textile mills exemplified by Lowell, Massachusetts, were not the only or even the most predominant organizational formats within textile manufacturing. Characterized by relatively small, family-owned firms and partnerships engaged in short-run, flexible production of rapidly changing goods, Philadelphian textile manufacture differed markedly from the Lowell model. Historians refer to their methods as "batch production." While mass production consists of sustained, quantity manufacture of a fixed product, batch production takes its name from short runs of products whose changing physical specification and episodic demand require technological and institutional flexibility. Batch production represents an aspect of what one historian has called "the other side of industrialization": noncorporate, small-scale production, which played a critical, if poorly understood, part in the history of American manufacturing.[7]

An expanded industrial typology embraces both bulk/mass and custom/batch production. Mass production relies on complex, highly mechanized processes for the creation of large quantities of standardized goods. While the automobile industry became its apotheosis, many other commodities exemplify mass production, including sewing machines, watches, bicycles, electric motors, light bulbs, and more. Bulk production involves similar but simplified, less machine-intensive methods resulting in equally large amounts of standardized goods in a range of sizes. Dimensional lumber, cement, nuts, bolts, simple castings, and flour all derive from bulk production. Both methods depend on routinized, mechanized processes to stock inventory and meet established market demands. Custom production creates individual goods to order. A locomotive built to client specifications qualifies as a custom product, as do bridges, turbines, some machine tools, and other goods tailored to individual requirements and specific orders. Batch production turns out small groups of similar or identical goods to order or in anticipation of demand. It shares some of mass production's efficiency but, like custom work, fills specific orders. Batch goods include locomotives, some machine tools and other heavy commodities, as well as looms, printing presses, coated papers, styled fabrics, and jewelry. Together referred to as flexible production, both custom and batch manufacture share the virtue of flexibility as a means of coping with fluctuating demand and the vagaries of fashion.[8]

Mass and batch production frequently coexisted in a single industry. The wagon and carriage trade featured this mixed format owing partly to parallel development of two primary factory types. One was the vehicle factory, which as its name suggests, manufactured complete horse-drawn vehicles. Ranging in size from owner-operated enterprises employing a dozen tradesmen, to plants with hundreds of workers and a complex administrative structure, vehicle factories formed the backbone of high-volume wagon and carriage production. The other was the specialty factory, which produced parts used in vehicle construction. Collectively known as the accessory industry, some individual firms specialized in one or two products while others fielded comprehensive lines of wagon and carriage parts in an impressive array of styles, sizes, and prices. These included relatively simple staple goods such as carriage bolts, body brackets, and shaft tips, as well as more complex components like fifth wheels, top assemblies, and dashboards. Together they revolutionized wagon and carriage building.[9]

Vehicle factories increasingly divided labor, mechanized processes, and simplified products in an effort to maximize production of standardized vehicles for middle-class markets. Numerous antebellum New England vehicle makers owed their prosperity to southerners who purchased inexpensive pleasure carriages. After war destroyed this trade, midwestern factories began meeting growing demand from the West for buggies and farm wagons. While skilled craftsmen still played important roles in such firms, particularly as department foremen or plant superintendents, even regular factory work often demanded considerable skill. Vehicle factories tended to restrict the degree of customization, but nearly all offered a range of standardized options. Clearly the degree to which a firm catered to customer-driven choices determined its level of adherence to mass production principles. Many large horse-drawn vehicle manufacturers qualified as mass producers while retaining at least a measure of batch production flexibility, further proof of the permeable border dividing the two production philosophies.

Specialty factories worked with a stricter bulk and mass production ethos. Wagon and carriage parts manufacturers fashioned huge quantities of goods as identical as they could possibly make them. Widespread demand from large and small vehicle manufacturers alike ensured a stable and often expanding market reached through retailers and direct sales. While stock production dominated the typical specialty manufacturer's business, many actively sought out custom orders. Since the most coveted orders emanated from the big midwestern buggy and farm wagon factories, filling them virtually mandated some form of mass production. Manufacture of basic items through relatively simple processes (malleable iron castings, carriage bolts, axle clips) fell into the bulk production category. Makers of more intricate goods (fifth wheels, assembled dashboards, wheels) qualified as mass producers. In both cases the need to meet large orders drove a series of production innovations that made the specialty factory the industry's interchangeability pioneer.

Small vehicle shops normally operated under a more relaxed division of labor better suited to respond to customer requirements. Most shops could capably manufacture virtually any horse-drawn vehicle: a typical year might see ice wagons, rockaways, farm carts, buggies, and buckboards all roll out the doors of the same small establishment. The necessity of vying with dozens of local competitors, however, suggested specialization as a prudent business strategy. Some concentrated on a single class of vehicles, perhaps the most common wagon variants or all pony vehicles. Shops catering to local farmers prospered with farm wagons, carts, and buggies, while their urban counterparts flourished on special-purpose delivery wagons and heavy drays. A few restricted themselves to a single vehicle type such as sprung delivery or ice wagons. Such operations accommodated single unit orders along with the coveted multiple vehicle jobs for fleet supply. Nevertheless, specialization never totally eclipsed flexibility; custom and batch methods remained even as highly specialized shops turned out the odd individual vehicle as necessary. Repair work, known as "jobbing," supplies an additional example. Whether simple tire resetting or more ambitious work like repainting, repairing remained a small shop staple.

The great irony of the small wagon and carriage shop's persistence lay in its enthusiastic adoption of ready-made parts. More than any other single factor, the easy availability of vehicle components allowed small enterprises to remain competitive. Every part purchased from the accessory industry obviated one more time-consuming production step and freed shop hands for assembly and finishing processes. Filling a custom order required ascertaining and obliging customer specifications, at least in matters of upholstery and finish. In any case the completed vehicle appeared on the market under the shop's name, to add or subtract to the proprietor's reputation according to the quality of the work. Wagon and carriage shops thus established a syncretic system built directly on custom and batch principles, yet reliant on mass production parts factories. Instead of obliterating small shops, factories proved the key to their survival.

Though ready-made parts became nearly universally applied in small shops and large factories alike, the degree of dependence varied greatly. Those lacking traditional wheelwrights or sophisticated special-purpose wheel-making machinery frequently outsourced their wheels, one of the earliest and most popular accessory industry products. Many firms extended the principle much further. A manufacturer could buy complete sets of wheels, axles, and springs and outfit them with shop-built bodies. Detroit's Easy Wagon Gear Company was one of many suppliers whose very name suggested labor saving possibilities. Even manufacturers inclined to refrain from using ready-made wooden components often succumbed to the lure of the malleable iron casting. Rather than laboriously hand forge the considerable quantity of metal hardware required of all vehicles, a wagon or carriage maker could buy malleable iron castings needing only minimal finishing prior to installation. Such advantages often won over the most hesitant carriage blacksmith or shop owner, who might be wooed by casting suppliers' thick, attractive catalogs full of the most useful parts.[10]

Wagon and carriage makers confronted a range of profound issues in determining their degree of reliance on ready-made parts. Heavy dependence on factory parts suggests that there was a diminished call for skilled craftsmen, though the reality proved more complex. One observer felt that "but little more comparatively is required of the builder than to place in the hands of his workmen the several parts to be joined together." As another put it: "They buy a body here, the wheels there, the axles in New Jersey, the springs in Connecticut; they clap them together by machine power, cover them with paint bought ready mixed, and finish them with decalcomanie coat-of-arms, which were painted in Germany."[11] A small shop owner-operator penned a vitriolic letter to *The Hub*, one of the leading wagon and carriage trade journals, in 1872 protesting widespread use of patent wheels in lieu of those made in the shop. Insisting on the illogic of this practice, given patent wheels' alleged inferiority, the letter embodies some of the attitudes and emotions aroused by the opportunities and dangers inherent in ready-made parts. These and other comments raise important questions about the quality and economy of the accessory industry's products and their impact on worker skill, but a full treatment will come later. Ready-made parts kept small wagon and carriage shops competitive with mass production rivals. Regardless of the finer details, this wed shop and factory in what had become a decidedly mixed format industry.[12]

Any work presuming to cut to the heart of a nationwide enterprise risks sacrificing detail for scope. I have sought balance through including descriptive surveys of general categories (shop, vehicle factory, specialty factory) and overarching processes (craft system, industrialization, industrial decline), supported by examples drawn from dozens of individuals and firms. Added to this are two chapter-length studies that extend scrutiny of these processes in the context of a single family or manufacturing concern. As some of the most prestigious American luxury carriage builders, New York City's Brewsters provide us with excellent examples of the highly touted eastern carriage manufacturers. Their long and successful run as custom builders, their sustained quality, much imitated style, and a reputation that lingers even today qualify them for inclusion. South Bend, Indiana's Studebaker Brothers Manufacturing Company occupies an equally unique position. Besides being one of the best known and fondly remembered American wagonmakers, the firm is an outstanding example of the wildly successful midwestern mass-production vehicle factory. Innovative advertising in conjunction with a well-made, purpose-specific product resulted in early brand-name prominence; the famed surname became identified as no other with the classic American farm wagon. As the only major wagon and carriage manufacturer achieving long-term success in automobile production, the firm's transitional years provide an unequaled glimpse into an industry struggling to define its place during a time of profound change.

Chapter 1 provides a general history of American horse-drawn vehicle manufacture through a survey of common wagon and carriage varieties, the trade's physical geography and westward progression, and the rise of a trade press and national trade organization. The focus shifts in chapter 2 to the small craft shop and examines craft practice in conjunction with specific examples from shops across the country. Chapter 3 opens with a detailed discussion of industrialization in general, moving on to explore its manifestation in wagon and carriage manufacture in particular. The accessory industry's contributions form the nucleus of chapter 4. Besides covering the origins of wagon and carriage parts production, the chapter features coverage of some of the nation's very largest specialty manufacturers. Chapters 5 and 6 are extended encounters with the same forces in the context of two leading firms, demonstrating how industrialization catapulted the Brewsters and Studebaker to the forefront of the American vehicle industry. Chapter 7 unites the foregoing chapters from the perspective of the wagon and carriage worker. Assessing the impact of industrialization on those actually building the vehicles requires careful attention to the changing details of their working world. I focus in particular on mechanization, rationalization, and ready-made parts, including individual and collective worker opinions to answer vital questions surrounding the fate of skill and the nature of labor in the industry. Chapter 8 explores the industry's response to the automobile, including attempts to compete with automobility, the reasons behind the industry's oft-noted but seldom explained failure to transition into automobile production, and the coping strategies successfully pioneered by a few fortunate exceptions. Using as a motif the final meeting of the Carriage Builders National Association in 1926, the epilogue briefly explores the death of an industry in the face of the stunning triumph of the automobile in America.

In the final analysis, the automobile accounts for the American wagon and carriage industry's forgotten status. From its faltering, smelly start in dozens of barns, basements, and homemade workshops, the internal combustion engine and the awkward mechanisms harnessed to it staggered into the light of day, frightening horses, upsetting carts, and annoying pedestrians. It has never left us. Impelled by incremental improvements and bequeathed to millions through the efforts of a Michigan farm boy named Henry Ford, the automobile revolutionized American life as no other form of personal transportation has done before or since. Americans today wrestle with its significance in matters ranging from traffic fatalities to fuel crises. We also wrestle with the automotive industry. The much-publicized strain and confusion engendered by Asian competition and its seemingly fresh concept of flexible, batch production, highlights the danger of forgetting the past. For we have in our own national history an example of a trade harnessing that very concept to become the mightiest industry of its kind in the world. That, to me, seems reason enough to remember it.

[1]

Rich Men's Vehicles at Poor Men's Prices

The Development of a Trade and Its Products

I happen'd to say, I thought it was [a] pity they had not been landed rather in Pennsylvania, as in that Country almost every Farmer had his Waggon.

Benjamin Franklin, 1788

The side-bar wagon is one of the few examples of a standard vehicle which is peculiarly identified with this country. For a quarter of a century it has changed very little in general shape, and is not likely to do so for some time to come. . . . The reason for this consistency is found in the practicality of the road wagon for its purpose, combining as it does lightness, strength, and a shape which is in keeping with the swinging stride of the trotter pure and simple.

Francis T. Underhill, 1897

We do not claim to be able to do wonders, but we do claim to be able to sell vehicles direct from the manufacturer to the consumer for less money than any other concern in existence." Thus proclaims the opening page of the Vehicle, Harness, & Saddlery Department of the 1897 Sears, Roebuck & Company catalog, the famed "wish book" of the late nineteenth century. Several pages of densely packed type and engravings follow, showcasing a impressive collection of horse-drawn vehicles. Emphasizing quality, uniformity, durability, and low prices, the Sears catalog offers ample testimony to the historian that by the turn of the century the horse-drawn vehicle had been thoroughly democratized. Neatly summarizing this transformation, the catalog copy reads "we are able to furnish a wealthy man's rig at a poor man's price, able to bring within reach of our many patrons such vehicles as have heretofore been in possession of the wealthy only."

Compare Sears's advertisements to those of eighteenth-century Philadelphia carriage manufacturer William Tod. Active during the years prior to the American Revolution, Tod had boasted in 1767 that he procured most of his materials from London where he himself had been "bred to the trade." The obvious point was that the English made the best carriages—and that it was going to cost the discriminating consumer to own one. There could be no doubt that Tod and his contemporaries catered to the well-heeled. The contrast between these two examples of merchandising goes beyond differences in the tone of their advertisements and their prospective clientele: in the intervening 130 years the entire American horse-drawn vehicle world had undergone massive changes. What separated William Tod from Sears, Roebuck, & Company was the transformation of a craft into an industry. Before examining that evolution, we should begin with an overview of American horse-drawn vehicle development, a sense of the evolving geography of the trade, and knowledge of the vibrant trade literature and manufacturer's organization that became an intrinsic part of the industry.[1]

Early Americans most commonly traveled by horseback or on foot. Each mode of transportation was suited to the terrain in that it required nothing more than space between the trees. Yet the necessity of conveying larger and heavier goods soon gave rise to the earliest American wheeled vehicles. The humble cart saw the most widespread initial use. Though later refinements brought it into the realm of passenger transport, it remained a port, farm, and factory work vehicle long after most people rode in four-wheeled carriages. With side stakes or a simple box body, the cart transported hay and manure as easily as it did garden produce or a farmer's family. Urban cousin of the farm cart, the dray dominated city haulage from the eighteenth century. Its design was ideal for transporting unwieldy items; the shafts extended rearward to form the frame, which sometimes supported floorboards and side stakes, terminating at the rear in a projecting loading ramp. Active seaports relied on these skeletal vehicles to shift cargo between ship and warehouse, while industrial districts employed them to move everything from ponderous castings to lumber. Though eventually supplemented by more stable and capacious four-wheeled trucks, drays and their often colorful drivers remained a vital part of freight transport until the advent of motor trucks.[2]

Gradual improvement in vehicle design and construction eventually resulted in more versatile four-wheeled wagons during the nineteenth century. Wagons freed horses from having to balance the load and offered far greater carrying capacity. Eighteenth-century Pennsylvania contributed the Conestoga, a sway-bellied, heavy freight wagon whose canvas top made it one of the earliest recognizably American work vehicles. Developed around 1750 by immigrants steeped in English and

German wagon design, the Conestoga dominated overland freighting in east-central America for the next hundred years. By the eve of the Civil War it had been pushed aside by other heavy wagons and largely supplanted by railroads. Lighter, simplified covered wagons rolled across the expanding American West over much of the nineteenth century, forever linking bow-topped freighters and hardy pioneers in the popular imagination. Likewise, four wheels replaced two on American farms, as carts gave way to farm wagons. In cities, four-wheeled drays began to replace the old two-wheeled versions; these and the truck, which featured a set of stiff leaf springs, were among the heaviest-duty of all common horse-drawn vehicles.[3]

The terrain-moderating effects of snow and ice long favored winter for heavy hauling, and in occupations such as logging the practice persisted well into the twentieth century. Given their innate simplicity, many sleds and sledges were "home built" by the user. Farmers employed crude work sledges even in warm weather to pull loads heavy enough to reduce spoked wooden wheels to kindling. Runnered vehicles also served as passenger transport in regions with consistent winter snowpack. Though eighteenth- and nineteenth-century Americans often used "sled," "sledge," and "sleigh" interchangeably, a sleigh technically referred to a vehicle designed to carry people. Farmers rode in wooden boxes mounted on runners called "bobs" or even on simple homemade sledges made from boards and iron strapping. Seventeenth-century Dutch settlers in what became New York made sleighs with runners fashioned from split saplings, and other simple variants existed. The cutter consisted of a gracefully curved, open wooden body with a bench seat for one or two people. Two specific styles easily outranked all others in popularity. Upstate New York carriage maker James Goold developed the swell-sided Albany cutter during the 1830s, around the same time that the Kimballs of Maine developed the square-bodied Portland cutter. Some builders turned out larger sleighs with extra seats, and a few built opulent closed-body models rivaling the finest carriages. Eventually firms designed convertible wagons or carriages, which in winter could substitute detachable runners for wheels. The virtues of simplicity and low cost ensured the sleigh's long reign while immortalizing winter pleasure driving in story and song. Snow made the rough places plain, and "speeding" became a popular and fashionable winter pastime.[4]

Workaday carts, wagons, and sleds defined early American road transport and increased in number and variety from the mid-eighteenth century. A slowly expanding road network in and outside of settlements fueled demand for more refined personal conveyances, vehicles generically known as carriages. The riding chair, popular in both eighteenth-century England and America, consisted of nothing more than an ordinary wooden chair mounted on a two-wheeled gear. These could often be found in the carriage houses of owners wealthy enough to own more elaborate equipages and were often favored for fair-weather pleasure use. An even more popular and better-known example in this category was the chaise ("shay" in New England), a two-wheeled vehicle with a folding top and thoroughbrace-suspended body. Popular in late Colonial America, its simple durability (which inspired Oliver Wendell Holmes's allegorical poem about the "One-Hoss Shay") kept it in general use until the rise of the four-wheeled buggy during the 1830s and 1840s. Numerous variants of the chaise existed, some of them with improbable names: the whiskey was simply a chaise stripped down for fast driving, the gig a slightly more sophisticated chaise with better suspension or a more elaborate folding top. Of the relatively small numbers of carriages found outside urban areas in the late eighteenth century, the majority were chaises.[5]

Though not as numerous as carts or chaises, coaches were a particular symbol of pride for those wealthy enough to afford them. A coach was a heavy, closed carriage consisting of a meticulously crafted body suspended on a substantial four-wheel gear and fronted by a raised box for the coachman. The body formed the ideal canvas for the coach maker's art, from the tight-fitting doors and silver-plated hardware to the glossy paint and family crest that frequently graced the exterior. In order to absorb some of the inevitable jarring, the body hung from leather thoroughbraces or steel C-springs, and the interior carpeting and upholstery provided a degree of soundproofing. The gear had to bear a great deal of weight, and the massive wheels and axles were connected by a beam called a perch, a feature more in common with heavy wagon construction than light carriage design. Traditionally the purview of European royalty, the coach's weighty solemnity made it ideal both for state occasions and as a badge of prestige. In America coaches were initially exclusive to Colonial governors, most of whom had theirs imported from England. The American Revolution and the development of a domestic, urban coach-making trade saw American-made coaches come into gradual favor outside the ranks of high government. Other carriages such as the chariot, a sort of abbreviated coach, became popular with some wealthy and powerful Americans, and by the 1760s and 1770s, four-wheeled carriages had joined the more numerous chairs and chaises in cities like New York (fig. 1.1).[6]

Those moderately skilled in basic woodworking might build a serviceable wagon body on a set of purchased wheels or hammer together a simple sled out of common lumber, but the knowledge needed to construct the flowing curves and intricate joinery of a coach far outstripped that of the carpenter or handy farmer. It took a carriage maker to make a carriage, and, initially, European craftsmen supplied the American colonies. English imports dominated the American pleasure vehicle market until the mid-eighteenth century, by which time a small handful of American carriage makers had become established. These native shops produced goods imitating traditional European styles; English and French designs exerted a long-standing influence. Europe also provided the craftsmanship, if not the craftsman himself: many early American carriage builders had emigrated from Europe or had been apprenticed to European master craftsmen. Eventually apprenticeships and more informal forms of training took place within the American trade, though European practice remained a leading model. Slow growth of better roads sustained the gradual evolution of American carriage building so that by 1791 Alexander Hamilton could list the trade as one having attained a "considerable degree of maturity." Carriage making had become an established urban business.[7]

Enterprisers like William Tod plied their trade in the leading American cities. Philadelphia and New York led in eighteenth-century carriage production, though British occupation retarded the latter for some time. Nicholas Bailey, one of New York's earliest carriage makers, made and repaired "Shaes and Coaches" in the 1740s and was joined by Irishman James Hallett in the 1750s ("at the sign of the Golden Wheel") and Samuel Lawrence in the 1760s. Even better known were the Deanes, also from Ireland and in business by the 1760s. Elkanah and William Deane billed themselves as "Coachmakers from Dublin" and boasted work "equal in quality, if not superior, to any imported from England." These men built, sold, and repaired all types of wheeled passenger vehicles but specialized in the highest grade work. They were part of an exclusive group, for as late as the 1780s fewer than a half dozen carriage makers worked in New York City. In Philadelphia such luminaries as Gouverneur Morris, John Adams, Alexander Hamilton, and Thomas Jefferson bought their carriages from Quarrier & Hunter. Alexander Quarrier, born and trained in Scotland, founded the firm in 1778, later taking his stepson, William Hunter, into partnership. The city contained a dozen coach makers in 1785, with Quarrier & Hunter and about three others supplying the bulk of the local trade. In Boston, Major Adino Paddock opened a shop in 1758 and went on to become one of the city's leading carriage manufacturers. Paddock built and repaired vehicles, a common combination throughout the horse-drawn vehicle era. All three cities also had blacksmiths and wheelwrights who busied themselves with the far more numerous carts and wagons.[8]

Indeed, urban growth greatly benefited those turning out work vehicles. In addition to carts and drays, more specialized equipages saw use in and between cities. Like the dray, the stagecoach owed its long-term popularity to a robust design ideally suited to its task. In fact, it changed little from its origin well before the Civil War into the early years of the twentieth century. Late-manufactured coaches retained many features long discarded in most other horse-drawn vehicle designs. Initially a rough but serviceable wagon with bench seats, stage wagons took their name from the successive stages of their routes. By the late 1820s more sophisticated closed-body designs began to appear, the two most famous taking their names from their places of origin. Accounts vary on which developed first, but the classic American stagecoach originated in Troy, New York, and Concord, New Hampshire, during the late 1820s. Troy evidently preceded Concord, but the latter's Abbot, Downing & Company went on to produce so many of them that the name became synonymous with the product. Though available in numerous variations, the basic design consisted of a rugged, round-bottom body suspended on thoroughbraces and mounted on sturdy gearing. The roof and the interior both accommodated passengers, and a specially designed rear platform carried the bulk of the luggage. Simple, seemingly crude, but undeniably robust and built for hard use, the Concord coach excelled at passenger, mail, and light freight transport in regions beyond the railroad. Of course it quickly became an icon of the American West, but many today are unaware of its equally pivotal role in the frontier regions of South America, South Africa, and Australia.[9]

As the Concord coach and its many imitations spread over the continental hinterland, other forms of public transport evolved for urban use. While never proliferating as they did in Europe, various forms of hired cabs gradually became part of daily traffic in the larger American cities. Long popular in London, the Hansom Cab had achieved limited use in New York City by the late nineteenth century. The design of this two wheeler evolved over the years, but its final and most familiar form consisted of a tall enclosed body that passengers entered through double front doors. These opened and closed via a mechanical linkage operated from the driver's perch at the rear. Less exclusive but much more common, the omnibus consisted of a long, enclosed wagon with bench seating for several passengers. Originating in early-nineteenth-century Paris, the concept quickly spread to other countries. London had them by 1829, and American Abraham Brower introduced them to New York City around the same time. John Stephenson's version followed in 1831 and soon became the prototypical American variant, making Stephenson's firm the nation's largest omnibus manufacturer.[10]

The wagon and carriage trade spread out to include shops in hundreds of small towns across the country. As the above examples suggest, however, the early-nineteenth-century carriage trade remained centered in the urban northeast, particularly in Massachusetts, Connecticut, New York, and Pennsylvania. Many towns and cities here achieved regional fame through the work of leading shops. Men such as James Goold (Albany, New York), Lewis Downing (Concord, New Hampshire), Robert D. Canfield (Newark, New Jersey), Jason Clapp (Pittsfield, Massachusetts), Thomas Goddard (Boston, Massachusetts), James Brewster (New Haven, Connecticut), and Joseph Sargent (Amesbury, Massachusetts) continued the tradition of fine carriage making even as they trained workmen who founded their own shops. Before 1850 numerous cities in several states had garnered well-earned reputations in the carriage trade thanks to reputable work, and the number of firms in each location often increased significantly. New Haven enjoyed typical growth, from nine carriage shops in 1811 to fifty-one in 1860. Boston, Worcester, Amesbury, and Amherst in Massachusetts trailed New Haven. Amherst had sizeable firms and a large export trade by the 1830s, but Amesbury garnered greater acclaim as a vehicle-building center. Amesbury shops and factories produced huge numbers of carriages, many of which were shipped west on open rail cars beginning in the 1850s. Named for the white sheeting protecting the vehicles, these "ghost trains" signified the national importance of the New England carriage trade. While not as numerous as their southern New England neighbors, northern New England wagon and carriage firms accounted for at least a portion of the regional output. Portland, Maine, had a small number of locally important shops, and Charles P. Kimball had already reaped acclaim for his well-crafted sleighs. Concord's Abbot, Downing & Company had an unparalleled reputation for its stagecoach, a product that defined overland passenger transport.[11]

Yet even Concord held no monopoly, for Troy led all of New York in stagecoach production. While Albany and some other upstate towns had long been home to influential carriage shops, New York City remained the single most important carriage making center in the Empire State and the nation. Having long since overcome the dampening effects of British occupation, early-nineteenth-century New York began a steady rise to prominence within the trade. Long arbiters of fashion in many other fields, New Yorkers established an early lead in carriage styling they would retain long after large midwestern firms dominated the low- and intermediate-price vehicle market. James Brewster and many other Connecticut carriage makers established New York City repositories in the 1820s, and Brewster's sons went on to become fixtures in New York's indigenous carriage trade.

Like New York, Pennsylvania hosted numerous vehicle tradesmen, ranging from rural Conestoga wagonmakers to builders of fine urban coaches. Similarly, it concentrated a significant percentage of its output in one major metropolitan region. Philadelphia lacked the reputation of New York, New Haven, or Amesbury but nevertheless produced a range of horse-drawn vehicles in large quantities. For most of the century Philadelphia wagon and carriage makers turned out more vehicles than any other eastern city save New York; from the 1850s and 1860s builders such as William D. Rogers and Charles M. Foulke produced carriages rivaling the best made anywhere. Having the same general combination of conditions so favorable to manufacturing, Philadelphia mirrored New York in the production of both high-grade coaches and carriages, as well as farm and commercial wagons. Philadelphia was not the only important mid-Atlantic city in the trade, however. Baltimore had a locally prestigious group of carriage builders doing a particularly remunerative business with residents of Washington, D.C., and other parts of the South. Newark, New Jersey had a similar-sized trade, and there were of course others.[12]

As it did in so many areas of American life, the Civil War brought change to wagon and carriage manufacture, especially in New England, where carriage makers had grown accustomed to steady southern sales. The disruption of war, inability to collect debts from the Confederate states, and subsequent collapse of traditional southern plantation culture combined to cut down a previously large and lucrative market. In 1874 a southern carriage dealer lamented that the supply of new carriages in the South was at a thirty-year low. Not that this made any difference, for in his estimate "not one in a hundred" could afford one anyway. Catastrophic market shifts sent New England's wagon and carriage industry into a slump from which it never really recovered. Horse-drawn vehicle building did continue in the region, and during the period 1859–64 the number of wagon and carriage makers there actually increased. Census data for 1860 and 1870 reveals growth in the number of wagon and carriage firms in Massachusetts and Connecticut, but these shops were no longer at the forefront of the industry. Eastern builders continued to set fashion trends and manufactured high-quality carriages for America's elite, but the center of large-volume horse-drawn vehicle manufacture had started its move west.[13]

The war loosened New England's stranglehold on the trade, creating an environment in which new regions could seize the lead. War meant opportunity: incessant government demands for baggage wagons and ambulances gave birth to new firms and new life to established ones, particularly those located in the industrializing Midwest. Cincinnati hardware dealer J. C. C. Hollinshade established a profitable business building government wagons, his success foreshadowing the mass-production buggy factories so prominent in postwar Cincinnati. Although several other individuals and firms similarly benefited from government work, one in particular owed a large measure of its success to federal contracts. Four enterprising brothers moved to northern Indiana to work iron and build wagons during the late 1850s, but lack of capital threatened to close their shop. One brother contributed a nest egg accumulated on a western business venture; however, the saving grace for the Studebaker brothers came in the form of Union ambulance and wagon contracts during the early 1860s. Flush with capital, which they shrewdly turned back into the South Bend firm, their Studebaker Brothers Manufacturing Company went on to become the nation's leading wagon manufacturer (fig. 1.2).[14]

The end of the conflict brought even more changes to the trade through resumption of prewar trends such as westward expansion. Momentous growth of towns and cities in the Old Northwest, as well as the establishment of new settlements west of the Mississippi, greatly increased demand for all types of horse-drawn vehicles. While the older eastern firms scrambled for market shares, established midwestern vehicle builders were already on hand, soon joined by many more during the late 1860s and 1870s. Midwestern vehicle builders enjoyed not only proximity to mushrooming regional markets but other benefits as well. One was the timely rise in popularity of a new carriage uniquely suited to the rapidly expanding rural market. Wagon builders had traditionally coped with wretched roads via heavy, rigid construction, but passenger-vehicle design took a different route. In order to lessen strain on the horse and extend the range of daily travel, American carriage manufacturers turned increasingly to lightweight, flexible construction. This trend manifested itself in various ways, but its greatest ideals were realized in a uniquely American vehicle that became a symbol of both Yankee ingenuity and the entire horse-drawn era.

> **Buggy**, *n.; pl.* Buggies. **1**. A light one horse two-wheeled vehicle. [Eng.] **2**. A light, four-wheeled vehicle, usually with one seat, and with or without a calash top. [U.S.][15]

As the definition suggests, the name "buggy" originated in late-eighteenth-century England where it denoted a small, single-person carriage. This did not become common American usage until well into the nineteenth century, and even then some makers preferred "road wagon" or "side bar wagon." Nomenclature aside, the buggy ranks as one of a handful of truly American horse-drawn vehicle designs. As early as the late eighteenth century, American carriages began shedding European design elements to take on what would become a characteristic lightness. The gigs, chaises, whiskeys, and coaches of the period 1750–1830 gave way during the 1840s, 1850s, and 1860s to a new series of carriages far better suited to American terrain and tastes. One of them was the pleasure wagon, a versatile, light vehicle developed in New England during the first quarter of the nineteenth century. A four-wheeled wagon with a rockered body somewhat reminiscent of the Conestoga, the pleasure wagon carried light freight and doubled for passenger use. Considered by some to be the inspiration for the later Concord wagon (a vehicle lighter than, but sharing the same gear as, the Concord coach) and a light farm wagon called a spring wagon, the pleasure wagon also presaged the buggy.[16]

The American buggy was a single-seated, shallow body mounted on a flexible, sprung gear, sporting four lightweight wheels, and was drawn by a single horse. Slender hardwood framing and thin basswood panels reduced body weight to an absolute minimum. While available in a variety of styles, these bodies invariably consisted of some form of tray with a center-mounted seat. The very lightest held a single person, but many buggies easily accommodated two, and some variants included sliding or removable "jump" seats for additional passengers. A folding top of cloth, leather, or patent leather stretched over wooden (later metal) bows completed the practical and lightweight body. The gear sported similar touches: instead of using heavy C-springs, primitive thoroughbraces, or no suspension at all, buggy builders hung their vehicle bodies from steel elliptical springs made available through improved metal production. Mounted in a variety of ways between body and gear the springs softened road shock and greatly reduced strain on the vehicle, its passengers, and the horse. Lithe, curved shafts replaced straight, heavy ones, while iron reinforcements trimmed weight and added strength. Front and rear wheels nearly equal in size improved stability and contributed to the overall lightness of the ensemble.[17]

The wheels themselves formed one of the most significant departures from convention. European and early American vehicle wheels consisted of a combination of heavy, dense hardwoods: split-resistant elm hubs, oak spokes, and ash felloes. Each wood possessed qualities ideally suited to its application, and their successful combination into a durable wheel was one of the masterpieces of woodworking craftsmanship. Strength came at the cost of weight, however; carriages intended to convey passengers over crude roads benefited enormously from a set of light yet strong wheels. American wagon and carriage manufacturers developed techniques to reduce wheel weight significantly without weakening the structure. Rather than relying on the traditional sawn and fitted felloes, manufacturers steam bent hickory to form a pair of rim halves, creating a far lighter and stronger loop of rock-hard, resilient wood. Hickory spokes possessed similar qualities, and while elm remained a favored hub material, strategically placed iron reinforcements helped hubs shed weight. Access to vast stands of excellent quality native hardwoods, along with innovative wheel-making machinery, gave birth to the lightweight American hickory carriage wheel during the 1840s and 1850s. Jared Maris, one of the trade's best-known timber experts, once said that "as to what God could have done in making better wood for light buggies than hickory, I do not try to know." Hickory wheels garnered a worldwide reputation and were a major ingredient in the buggy's success in the United States and abroad.[18]

The buggy made its appearance as a distinct vehicle type sometime during the 1830s. Early examples featured heavier construction and more primitive suspension, but the overall form had taken shape by this time. American buggies appeared at the Crystal Palace exhibition in 1851 in part because they were already becoming staple products of American carriage shops. Still, acclaim was not universal, and the British especially had reservations about this peculiarly American vehicle. Over an illustration of a buggy in the exhibition's official catalog appeared the polite yet faintly damning comment that "the body of the vehicle seems very light in its construction." London coach builder George Thrupp criticized the difficulty of entry and exit due to the tall wheels and objected to the "tremulous motion" caused by the springs but admitted their fitness for America's endemic bad roads. He also praised the vehicles for their low purchase price.[19]

This last feature contributed mightily to the buggy's popularity. A midcentury rise in the standard of living put horse-drawn vehicles into more barns and stables than ever before. One writer in an 1879 issue of *Scientific American* felt that general prosperity since the 1830s had cre-

TABLE 1.1. Carriage Production by Type, 1900

Buggies	513,565
Road wagons	147,126
Surreys and phaetons	116,615
Driving wagons	29,945
Carts, cars	24,994
Runabouts	22,747
Spiders, stanhopes, traps, and tallyhos	14,502
Buckboards	11,987
Pony and park wagons	9,869
Broughams and landaus	4,507
Mountain wagons	4,084
Sulkies and skeleton carts	3,853
Victorias, cabriolets, vis-a-vis	2,645
Coaches	588
Gigs	455
Total	907,482

Source: Adapted from James K. Dawes, "Carriages and Wagons," in U.S. Bureau of the Census, *Census Reports: Twelfth Census of the United States, 1900* (Washington, 1902), 10:310.

ated so many new carriage owners that it had in turn spawned "an industry which ranks among the first in scope and magnitude." The expanding vehicle market encouraged wagon and carriage builders to experiment with factory methods, and prices began a gradual descent. In the 1860s a well-made buggy still cost a tradesman several months' wages, or between $125.00 and $150.00, hardly a casual purchase. By century's end, however, the price had dropped to about a month's wages, thanks in large part to the economies of scale made possible by mechanized factory production. In 1902 Sears, Roebuck & Company sold open buggies for as little as $22.35, and for a mere pittance would ship one to the nearest rail depot. Downward spiraling prices soon enticed the vast rural market: farmers who once considered a spring-equipped passenger vehicle an unnecessary luxury found themselves purchasing at least one, sometimes more. In a sense, then, the cheap, factory-made buggy became to nineteenth-century Americans what the Model T

Ford would become to their children and grandchildren. Though surpassed in capacity by carts and wagons, and outclassed in style by broughams and cabriolets, no other vehicle so well symbolized the age of the horse-drawn vehicle in America.[20]

The evolution of buggy design flowed in part from radical manufacturing departures. Impelled by desire for competitive advantage and inspired by the well-known local textile mills, a few innovative southern New England carriage builders began applying factory methods to vehicle production. Experiments with the division of labor, mechanization, and standardization gradually yielded productive gains, and by the end of the 1850s at least one New Haven carriage manufacturer claimed to turn out a complete carriage every hour. Not surprisingly, such innovation quickly spread beyond the region; near the end of the 1850s a West Amesbury, Massachusetts, carriage maker relocated to Cincinnati and set the stage for dramatic developments in the industry. Carriage factories here made modest productive gains during the 1860s, but not until the economic backwash of the Civil War had subsided did the Queen City achieve its lasting reputation for low-priced buggies. Cincinnati and several other Ohio cities became synonymous with buggy manufacture during the 1870s, a decade that saw rationalized factory production transform the industry. By this time Ohio ranked third in American carriage production, and the Midwest began to overshadow New England and the east coast cities in the trade.[21]

The Midwest appealed to wagon and carriage manufacturers because of its cheap land, plentiful timber, and convenient rail access. Vehicle factories required substantial production buildings, powerhouses, lumber sheds, and the other structures so necessary to efficient industry. Escalating urban real estate prices and tales of cheap land beyond Pennsylvania naturally drew attention. As manufacturers of primarily wooden objects, wagon and carriage makers also took advantage of one of the region's more abundant raw materials. Eastern timber reserves were giving out from relentless overexploitation: by 1850 the center of U.S. timber production had shifted from Maine to New York, to Pennsylvania by 1860, and from 1870 until just after the turn of the century to the upper Midwest. In addition, rapidly expanding middle and far West populations made locating factories there a sensible move. Members of the vehicle industry expressed concern over the railroad's impact, yet the iron road proved a benison. Horse-drawn vehicle manufacturers shipped their goods via rail, railroad corporations used coaches and wagons for passenger and freight transfer, and the same conveyances continued to transport mail and consumer goods beyond even the most far-flung stations.[22]

As the wagon and carriage industry became increasingly midwestern, old-line eastern carriage centers continued to occupy an important place in the trade. Despite intense competition from Ohio, Illinois, and Michigan, both New York and Pennsylvania led in number of establishments during the 1870s and 1880s. Though Cincinnati, Chicago, and even San Francisco ranked in the top ten carriage-producing cities, New York and Philadelphia each boasted more than double the number of firms in the next highest cities during the 1870s. According to the 1880 census, San Francisco, Cincinnati, Providence, and Chicago each counted between forty and fifty firms, while Philadelphia had 110 and New York (including Brooklyn) a whopping 229. This would eventually change, and the totals themselves are only an approximation, but they nevertheless reveal the East's continuing importance even in the face of ever-larger midwestern wagon and carriage factories. Established eastern shops continued to exert a fashion influence that only seemed to strengthen with time. Inventor and automobile pioneer Hiram Maxim recalled his encounter with an early-twentieth-century New Haven carriage designer scornful of his clumsy drafts of an electric vehicle body: "Mr. Maxim! Do you want this carriage to look like a Western buggy maker's job, or do you want it to be a gentleman's carriage?" Such were the mores of eastern carriage firms, many of whom shared the New Haven designer's hot disdain for the "Western buggy school of architecture." While the old New England carriage cities like New Haven, Bridgeport, and Amesbury retained a place in the trade, New York remained the single most prestigious locale, in keeping with its lengthy traditions in publishing, the garment trade, and as a gateway to European culture and fashion (fig 1.3).[23]

Birth of a Trade Press

The rise of the midwestern carriage industry and the mass-produced carriage soon spawned an additional sign of industrial maturity. It began with the efforts of an eccentric Ohio inventor and carriage maker named Cyrus W. Saladee. Though trained by his father as a traditional carriage builder, Saladee engaged in many other occupations that gave vent to his penchant for invention. The majority of his patents pertained to horse-drawn vehicles, but they also included improved doorbells, modified playing card markings, and linen-finished paper collars. By the 1850s Saladee had moved from his native southern Ohio to Columbus, where he established the first American wagon and carriage trade publication in 1853. An annual in book format, *The Coach-Maker's Guide* consisted of approximately 150 pages of technical articles and news items, concluded by several pages of "fashion plates," side elevation line drawings of carriage styles. Saladee adopted a monthly format and a new title, *The Coach-Maker's Illustrated Monthly Magazine*, in 1855. Besides running the same kind of trade news and advertisements, he began incorporating more original matter, like his series on the advanced style of European carriage drafting called the French Rule. In applying geometric principles to carriage draftsmanship, nineteenth-century Parisian carriage builders had found a way to raise the craft above reliance on patterns and rule-of-thumb to vastly increased levels of accuracy and detail. The method proved enormously influential on both sides of the Atlantic, and Saladee's was an early voice urging its American adoption.[24]

By the mid-1850s Saladee appears to have realized the limitations of his location, and he founded a New York branch office as a remedy. Somewhere in his travels he had met John W. Britton, one of the partners in New York's Brewster & Company. Well acquainted with shops in New York and southern New England, Britton recommended a fellow New York carriage maker named Ezra M. Stratton. Stratton learned the trade in his native Connecticut during the 1820s and had operated his own New York City shop since 1836. Located just a block from Brewster's Broome Street factory, Stratton supplemented his own work by crating Brewster carriages for shipment. A curious combination of practical tradesman and amateur scholar, Stratton was an autodidact who overcame his limited education by aggressive after-hours study of literature, history, and foreign languages. The combination of craft knowledge and literary competence perhaps struck Saladee as ideal, but Britton's endorsement sealed the matter. Stratton became assistant editor around 1855, initially combining his new duties with work in his Elizabeth Street shop.

In two years' time *The Coach-Maker's Illustrated Monthly Magazine* had virtually ground to a halt. One reason was financial: antebellum wagon and carriage manufacturers hardly constituted a cohesive national body, and the editors found it difficult even to identify their audience by name. They laboriously compiled their prized mailing list by personal acquaintance and word of mouth. The second reason concerned a bitter disagreement between the two editors. Talented tradesmen with ambitions outside the shop, Saladee and Stratton were both intensely opinionated, highly sensitive, and easily offended. All this and a marked age disparity made for a potentially volatile combination, and it was perhaps inevitable that there would be a falling out.[25]

The break came in 1858 when Stratton split off to establish *The New York Coach-Maker's Magazine*. According to an account written fifty years later, Stratton took advantage of Saladee's absence on a European trip to found a rival journal that drove his former employer out of business. Journalistic rumor had it that Stratton absconded with Saladee's hard-won mailing list ("And Ezra did covet his neighbor's goods, even his ox, his ass, and his *subscription list* . . ."), a charge Stratton, interestingly, never refuted. He probably felt justified, given Saladee's effective abandonment of the enterprise. Abruptly leaving Columbus in 1858 or 1859, Saladee's whereabouts remained unknown until Stratton received a clipping from an 1860 issue of a Galveston, Texas, newspaper concerning newly arrived plantation owner Colonel C. W. Saladee. Stratton preferred to tell his readers that the former editor had fled creditors, though it appears that he instead had to take over a southeast Texas plantation left to him by his deceased father. Little beyond his actual presence there has ever been verified, and Saladee's intent toward his original publishing venture is impossible to ascertain. Stratton certainly seemed to relish poking fun at his former employer, with frequent references to his "'old friend' beyond the Alleghenies" and cartoons portraying Saladee as a ranting, top-hatted freak.[26]

The New York Coach-Maker's Magazine was the second major carriage trade journal, though some readers clearly thought the new editor a bit freakish in his own right. Stratton included "Literary and Social" aspects of the craft alongside instructional material, items such as his carefully researched articles on the horse-drawn vehicle's ancient history. Strange promises such as "a prominent feature of Volume nine will be a series of articles on Egyptian history, copiously illustrated" elicited snickering references to "The Egyptian Magazine" and its "monthly collection of dog-carts, Egyptian and Assyrian carts and processions." Besides reflecting the editor's antiquarian interests, however, such material also acted as filler. Early technical journalists were voices howling in the wilderness, and it took time to develop a truly useful literature. A general educational strain also accounted for the diverse content, for many early trade journal editors saw their role as elevating the education and status of American tradesmen. Despite its peculiarities, *The New York Coach-Maker's Magazine* established the basic content formula followed by nearly all subsequent wagon and carriage trade journals. This included a mix of news, anecdote, and technical articles divided into appropriate departments; vehicle fashion plates; foreign and domestic carriage news; patent and invention notices; and a steadily increasing amount of advertising.[27]

Stratton continued to publish *The New York Coach-Maker's Magazine* while remaining at least a nominal carriage maker through the end of the Civil War. In late 1866 or early 1867 he left his old trade for good and entered the coal business, a lucrative enterprise he retained until his death. He also began dealing in household and carriage hardware, some of the latter appearing for sale in his journal. Despite all this he still managed to find time for an astonishing range of scholarly pursuits. Besides establishing a small museum of Near Eastern artifacts and models, Stratton mastered several foreign languages and used them in his increasingly impressive historical research. Illustrated articles on vehicle history continued to appear in his journal until its sale in 1871, and he subsequently gathered them into a book published in 1878. However odd it seemed to his contemporaries, Stratton's painstaking research and impressive historical scope served him well, for *The World on Wheels* remains an authoritative work on the ancient history of the horse-drawn vehicle. He stayed in close contract with the New York carriage trade and wrote articles for the trade press until his death in 1883.[28]

Cyrus Saladee continued his peripatetic existence both in and outside the trade. His trail of patents shows him back in central Ohio by 1864, where he busied himself inventing everything from padlocks to plow points. In 1870 he moved to Saint Catharines, Ontario, where he created numerous innovations widely used in buggy construction. Returning to the United States in 1873, he spent the next twenty years moving around the East and Midwest, working, investing, and amassing a total of nearly two hundred patents. The breadth of his creative output, however, was matched only by the magnitude of his financial incompetence. In addition to a tendency to invest in failing business ventures, he repeatedly sold his patents for cash to fund his increasingly profligate lifestyle. It all ended in the northern Illinois manufacturing town of Freeport. Attempting to cure himself of cholera, the brilliant inventor and founder of the carriage trade literature died from a self-administered morphine overdose in the fall of 1894.[29]

The journal that purchased *The New York Coach-Maker's Magazine* in 1871 was itself only two years old. *The Hub* originated as a simple company advertising sheet produced by an employee of the Boston varnish manufacturer Valentine & Company. Company president Lawson Valentine soon invited his talented brother-in-law George W. W. Houghton to edit an improved version proclaiming the virtues of a new wood putty aimed at wagon and carriage makers. He proved an ideal choice. A bright and energetic nineteen-year-old from Cambridge, Massachusetts, Houghton had literary talents that likely attracted Valentine's attention. The young man accepted the offer and the first issue of *The Hub* (besides the obvious carriage-related connection, the title referred to Oliver Wendell Holmes's quip about Boston being the center of the universe) came out in the spring of 1869. Valentine & Company moved to New York City the following year, where the publication quickly outgrew its original purpose. The transformation from advertising sheet to trade journal may have been a result of Houghton's ambition, though Valentine's funding certainly helped. Houghton soon obtained the benefit of talented New York tradesmen, as well as correspondents from other parts of the country, and acquisition of Stratton's journal helped propel *The Hub* to the forefront of the carriage trade press.[30]

Houghton shared Stratton's literary bent, but he exploited it more adroitly through biographies of eminent wagon and carriage men, profiles of old and respected firms, and the occasional piece on the American history of the horse-drawn vehicle. These items interested his readership and wed the journal to an industry appreciative of its own rich past. Houghton seems also to have established a better balance in appealing to both shop owners and workmen. Saladee and Stratton had catered to proprietors in a trade still dominated by the small craft shop, but the advent of the carriage factory and the increasing separation of owners and workers created new realities that a younger man was perhaps better equipped to handle. Houghton managed to publish a journal as relevant to the man on the shop floor as to the man in the front office. The wealth of practical "how-to" articles spoke to the carriage worker, and the proliferation of group subscriptions ("clubs") by workers suggests he successfully bid for their attention.[31]

The Hub's main rival started life as a union publication. An extended slump in vehicle sales in 1861 and 1862 resulted in widespread wage cuts and layoffs, which in turn generated discontent among carriage workers. In an attempt to alleviate these harsh new realties, a group of them met and established the Coach-Makers' International Union in late 1864 or early 1865. In 1865 the union authorized an official publication named *The Coach-Makers' International Journal* and recommended as editor a young New Jersey carriage maker named Isaac D. Ware. Ware knew the trade intimately. The product of a traditional apprenticeship during the 1840s, he had plied his craft in Ohio, Indiana, and New Jersey during the 1850s before coming to Philadelphia and becoming the union's secretary. The new publication consisted of the usual mix of fashion plates, articles, work hints, and miscellaneous space fillers. Despite having amassed a respectable subscription list, it suffered financial problems stemming from union instability. A late 1860s economic upswing sent the organization into decline, and Ware purchased the publication in 1866, eventually changing its name to *The Carriage Monthly*. He expanded his advertising base and filled the pages with detailed articles on body making, trimming, and carriage drafting. The journal made particularly important strides in the realm of illustration; by the 1870s trade journal illustrations began reflecting the growing sophistication of carriage draftsmanship and lithography. Ware and *The Carriage Monthly* helped popularize a pair of techniques contributing significantly to this trend.[32]

The first was the scale drawing, referred to in the trade as a "working draft." Sharing the fashion plate's flat elevation, it added refinements of enormous use to carriage workmen. The most important consisted of rendering the illustration to scale (usually three-quarters of an inch to the foot) and including key dimensions. These changes transformed published carriage illustrations from general stylistic models to detailed construction guides, a distinction that would eventually have a profound effect on the trade. Ware began turning out illustrations in sizeable numbers during the 1880s. The second innovation, also a descendant of the fashion plate, was the perspective drawing. It portrayed the carriage as it appeared to the human eye, providing visual depth absent from scale drawings. Perspective drawings combined practical accuracy with aesthetic appeal, and they became indispensable to the trade journals and trade catalogs alike. *The Carriage Monthly* began featuring them around 1878, and they were quickly adopted by other journals and product catalogs. As a popularizer of these illustrations and proprietor of an increasingly capacious publishing company (his sons having formally joined him in 1881), Ware shrewdly exploited this new demand. Sale of illustrations to other publications formed a lucrative sideline, and close scrutiny of wagon and carriage engravings from the 1880–1900 period reveal "Ware Bros., Phila." lurking in a corner of a significant percentage of them.[33]

While *The Hub* and *The Carriage Monthly* achieved definitive leadership by the 1870s, they were by no means the only carriage trade publications. *The Harness and Carriage Journal* also boasted a pre–Civil War founding date (1857), and several others entered circulation during the following decades. Chief among these were *The Blacksmith and Wheelwright* (1880), catering to metalworkers, wagonmakers, and repair shops; *The Coach Painter* (1880), devoted to carriage painting; *The Spokesman*

(1884), a general interest publication aimed at the wagon, carriage, and horse markets; *The Wagon Maker* (1886), focused on heavy farm and delivery vehicle manufacture; and *Varnish* (1888), another painting publication. In addition, several other newspapers and journals dealt with vehicle and farm implement sales, horse and harness matters, and more. While some failed within a few years, others achieved impressive circulation and publication that sometimes ran into the twentieth century. Nevertheless, *The Hub* and *The Carriage Monthly* easily surpassed the rest as the leading journals in the industry and were some of the finest of all American trade publications.[34]

Fraternal Ties

Though thousands of wagon and carriage builders practiced their craft across the United States by the end of the Civil War, the average shop owner knew little of fellow tradesmen beyond his immediate vicinity. The nascent trade press dispelled a degree of this isolation, yet Saladee's difficulty in even obtaining the names of potential subscribers illustrates the limited effects of such early efforts. Part of the appeal of a national organization devoted to horse-drawn vehicle manufacture lay in the opportunity for fellowship, but fear also played a part. The explosive success of the carriage factory had generated growing unease among small wagon and carriage makers since the 1850s, and the specter of overproduction haunted the heads of the very largest firms. Mechanization, decline in apprenticeships, loss of southern and export markets, and labor unrest all contributed to a growing sense of malaise. By the end of the 1860s these concerns coalesced into a movement to create a national trade association. Such a body would ideally aid in reducing the isolation of individual manufacturers, eradicate destructive price wars, determine the best factory and machine practices, prevent overproduction, and perhaps extract federal trade and tariff concessions. Obstacles existed, however, particularly in the extraordinary degree of jealousy within the trade and collective suspicion of any overarching body. Generated in part by cultural distrust of centralized authority and small manufacturers' fear of domination by large competitors, it also owed something to traditional trade secrecy. Labor proved an even more delicate matter. Adoption of the factory system and machinery by segments of the trade in the 1850s and 1860s, combined with the hardships of the post–Civil War economic slump, created tension between labor and management in a trade unaccustomed to sharp labor disputes. Already nervous about the outcome of these trends, some proprietors equated a national association with trade unionism—or worse.[35]

While many carriage makers did profess at least mild interest, the decisive impetus came from the northeast, home of prominent carriage manufacturers like the Brewsters, Charles P. Kimball, and Abbot, Downing & Company. Many had been in business for decades; some exerted considerable financial and political influence; and a few had already dabbled in collective action. Early in 1868 over two dozen northeastern carriage manufacturers, most members of the New England Manufacturers' Convention, met in an attempt to persuade the federal government to protect domestic carriage manufacturers from imported vehicles. While the effort failed, the attempt bequeathed valuable experience. Others, like John W. Britton, had experimented independently. Britton pioneered an innovative profit-sharing plan at Brewster & Company in the late 1860s; though it failed to completely prevent labor strife, it did reveal Britton's willingness to innovate, foreshadowing his later service to the industry at large. Britton became one of the prime movers behind a national carriage makers group, joined by fellow Brewster & Company employee Charles J. Richter, Maine carriage builder Charles Porter Kimball, and *Hub* editor George Houghton. Houghton, Ware, and *Harness and Carriage Journal*'s editor William N. Fitzgerald lent their support to the project, all three publications running copies of pivotal correspondence and announcements. Such collaboration engendered a long-standing partnership between trade press and trade organization that would endure for the life of the industry.[36]

During the summer of 1872 Houghton composed a brief rationale for a national association of carriage makers and circulated it among interested individuals and firms to solicit suggestions, responses, and signatures. By mid-August he had received an encouraging number of replies, and he mailed a revised copy to about six hundred wagon and carriage makers across the country. The pamphlet proposed a tentative fall convention in order to "become better acquainted with one another, and form a more perfect union of sentiment and action." Eager to assuage the fears of strike-wary shop and factory owners, Houghton assured them that the purpose was "not to agitate, or even to touch upon, the question of labor and its reward." This version closed with the signatures of forty-three proprietors of wagon and carriage firms from Maine to Missouri, with an emphasis on the old eastern centers: Massachusetts (six), New Hampshire (five), New York (five), the District of Columbia (four), Pennsylvania (three), Delaware (three), and Indiana (three). The eleven other participating states had one or two signatures each. Some signatories owned modest works, but others headed concerns that had practically become household names. By early September sufficient response convinced its authors to proceed, and they called for assembly in New York City on Tuesday, November 19, 1872.[37]

Participants trickled into the St. Nicholas Hotel the preceding day and the following morning saw them gathered in the reading room beneath a portrait of Albany's James Goold, one of the nation's oldest active carriage makers. Prevented by ill health from attending, Goold had generously loaned the freshly painted canvas. John W. Britton welcomed the conventioneers on behalf of the city of New York and nominated Clement Studebaker as temporary chairman. After electing temporary officers the men heard a speech by newly elected convention president Charles P. Kimball. Opening with an outline of the long history of European and American horse-drawn vehicles, he turned next to American carriage manufacture and the challenges facing the trade. He closed by exhorting his audience to consider a national association, which he confidently predicted would be of great value to them all. They agreed, for they wasted no time in arranging the details, and a committee of organization proposed a permanent association to be named the "Carriage-Builders' National Association." Besides stipulating the number and tenure of officers, the rules required each member to pay a ten dollar initiation fee, made all carriage makers eligible for membership, and recommended an annual convention. They elected Kimball president, Britton treasurer, and Richter secretary, with several others becoming vice presidents each representing a state. The remainder of the meeting, which spilled over into a second day, concerned organizational matters punctuated by brief discussions of the advisability of uniform track, cost estimation, and several other subjects. Before adjournment on Wednesday, November 20, they arranged to meet again in New York the following year. When the weary delegates filed out of the hotel, the Carriage Builders National Association had ninety-nine members.[38]

Its first years proved troubled. Overcoming the suspicion and resistance of their fellow tradesmen required more effort than initially planned, and the Panic of 1873 only exacerbated the problem. Indeed, economic depression, apathetic membership, and unpaid dues precluded an 1875 meeting altogether, but the slowly rallying economy and some key internal changes saw the CBNA roar to renewed life by decade's end. Reduced initiation fees and annual dues along with establishment of a morale-boosting annual banquet helped, as did carriage makers' gradual acceptance of the benefits of membership. Expanding membership eligibility also helped. Whereas only makers of complete horse-drawn vehicles were eligible for "active" (voting) membership, "honorary" ("associate," or nonvoting) membership became open to anyone deemed eligible by the officers. At first primarily consisting of members of the trade press, this category soon became dominated by the accessory industry. This reflected an ironic structural shift: ready-made parts not only transformed the work of building wagons and carriages, but the companies producing them assumed size, output, and stature often exceeding that of the vehicle makers. While active members invariably governed the CBNA, honorary members provided vital financial backing, as well as leadership in planning and hosting the annual conventions. Inclusion of the accessory trade proved absolutely vital to the organization's continued solvency.[39]

Though largely honoring the ban on labor matters, the association found a great deal worthy of its notice. It amassed trade statistics, established uniform vehicle and storage guarantees, lobbied (largely unsuccessfully) for protective trade legislation, monitored railroad freight rates, and served as a discussion forum on the many confusing issues facing the industry. These ranged from factory production, mechanization, and the perils of overproduction to ready-made parts, mass-produced buggies, and the continued role of the small shop.

While the CBNA would one day look back with satisfaction at many achievements, its leaders pointed with pride to one in particular. With the assistance of George Houghton and the trade press, the CBNA was the driving force behind the establishment of a technical school for carriage builders. Though developed with the goal of comprehensive trade instruction, the primary motivation came from continued trade interest in advanced French carriage drafting. The CBNA's advocacy of a technical school revolved around the perceived need to promote the French Rule through actual instruction, and several established European institutions provided the model. Inspired by the imposing exhibit of European carriages at the Paris exhibition of that year, the 1878 CBNA convention appointed a committee to investigate the possibilities for expanding drafting instruction. Despite widespread interest, efforts seemed forever stalled in the discussion stage. The honorary membership again provided vital money: a $1,000 donation the following year by Lawson Valentine anchored the fund drive, and George Houghton's relentless persuasion began to have an impact. The committee reports, funding, and trade press advocacy finally brought action in 1880. In November a specially chosen committee met in New York and created the Technical School for Carriage Draftsmen and Mechanics, which convened its first class on December 6, 1880.[40]

Located initially at New York's Metropolitan Museum of Art Schools, the Technical School featured instruction by John D. Gribbon, an Irish-born draftsman employed by Brewster & Company. Guest lectures by skilled carriage makers on subjects including timber selection, wheels and axles, body painting, and the history of the horse-drawn vehicle supplemented the curriculum. Although it changed locations and experienced fluctuating levels of support for its first few years, the school grew until it became one of the CBNA's shining achievements. Moving beyond its initial evening-class format, the institution eventually added day classes and a corresponding department, with draftsmanship still the primary emphasis throughout. Gribbon spearheaded a scholarship fund to send the most promising graduate to the Dupont School in Paris, there to be exposed for several months to the work of some of the finest Parisian coach builders. The Paris scholarship became a much vied-for prize, enriching recipients and school alike, as the award obligated the winner to a year's teaching on his return. The winner of the prize for the 1884–85 season, Andrew F. Johnson, went on to become the school's longest-lasting instructor and the one who ended up bridging the gap between the carriage and automobile industries.[41]

While ostensibly an effort to enhance the industry's competitiveness with European imports, the Technical School provided a vital service through disseminating the design and drafting techniques necessary for industrialized production. Mechanical drawing had long been central to a handful of American industries that required the control detailed plans conferred. Drafting techniques saw only limited application outside bridge and locomotive manufacture until the 1860s and 1870s, when makers of smaller goods began exploring improved drafting to increase production significantly. American wagon and carriage makers followed suit, and the Technical School's emphasis on projection, rendering, and scale drawing enabled these young men to move the industry toward rationalized production during the late 1870s and 1880s. Technical School graduates filtered out into the carriage industry, bringing knowledge of new vehicle styles and much-needed skills into the drafting departments of hundreds of firms. While better drafting by no means completely accounted for the shift from craft to industry, it formed a vital ingredient—and one of the CBNA's most important legacies.[42]

Of course the CBNA was not the only group devoted to the trade. Even before 1872 a handful of smaller organizations had already formed, mostly to regulate prices and production, though a few concerned themselves with labor. A freshly minted group of Wilmington, Delaware, carriage makers sent a representative to the 1872 carriage makers' convention, perhaps influencing the structure of the larger

TABLE 1.2. Growth of Wagon and Carriage Industry, 1849–1909

	Number of Firms	Number Employed	Wages ($)	Cost of Materials ($)	Value of Products ($)
1849	1,822	14,040	4,268,904	3,955,689	11,073,630
1859	7,222[a]	37,102	13,417,816	11,898,282	35,552,842
1869	11,847[a]	54,928	21,272,730	22,787,341	65,362,837
1879	3,841	45,394	18,988,615	30,597,086	64,951,617
1889	4,572	56,525	28,972,401	46,022,769	102,680,341
1899	6,204	58,425	27,578,046	53,723,311	113,234,590
1904	4,956	60,722	30,878,229	61,215,228	125,332,976
1909	4,870	52,540	29,621,148	63,890,422	125,366,912

Source: Adapted from U.S. Bureau of the Census, "Vehicles for Land Transportation," *Census Reports: The Thirteenth Census of the United States, 1910* (Washington, 1913), 8:473.
[a]Figures for 1859 and 1869 include many small repair shops not included in later tabulations under this heading, which explains the seeming peak in firm number.

group. Other cities developed similar associations, which were joined eventually by other new regional and national organizations. They included the National Wagon Manufacturers' Association (1879), the Southern Carriage Builders' Association (1879), the National Association of Carriage Hardware Manufacturers (1884), the Wheel Manufacturers' National Association (1886), the Carriage, Harness and Accessory Traveling Men's Association (1891), and several others. Some lasted only a few years; others succumbed to trade fluctuations; and at least a few managed to accommodate the automobile. Yet the Carriage Builders' National Association came first, served as a blueprint for others, and easily retained its place as the premier national body in the American horse-drawn vehicle industry.[43]

By the late nineteenth century that industry had reached impressive proportions. In 1900 nearly 67,000 men at over 7,500 firms produced more than 1.5 million horse-drawn vehicles valued at $121,537,276. New York, Pennsylvania, and Ohio led in number of firms, while Ohio claimed the largest share of product value and laborers employed. The Midwest continued to predominate the manufacturing trade: over 60 percent of the nation's carriages poured out of factories in Ohio, Michigan, and Indiana, while over half the annual farm wagon production came from Indiana, Wisconsin, Illinois, and Michigan. The older eastern centers like New York City continued to produce respectable numbers of vehicles, and of course they retained a firm grip on the reins of fashion. Finally, the South, once a market for northern carriages, showed signs of impressive growth in its own wagon and carriage making sector, particularly in North Carolina, Tennessee, and Virginia.[44]

The trade was very different from what it had once been just a hundred years earlier. From its humble beginnings in the American colonies as a craft dependent on European design, materials, and skills, horse-drawn vehicle manufacture grew with the nation. As American vehicle builders advanced in numbers and sophistication, they began shedding the heavy vestiges of European design to create uniquely American vehicles like the buggy. Manufacturing innovations and geographical expansion marked the trade's growth, as did the rise of a trade literature and a national trade organization. All this marked the passage of the craft into a mature and lucrative industry, one that all Europe watched with amazement, chagrin, and, one suspects, a little envy. Indeed an enormous amount of change separated William Tod from Sears, Roebuck, & Company. Yet for all the importance of surveying this historical terrain, knowing where buggies came from, who founded *The Hub*, and which state led in production, these facts serve only as a basis for answering more essential questions about that vital period of time between Tod and Sears, cart and buggy, craft and industry.

[2]

Knights of the Draw Knife

The Craft Origins of Horse-Drawn Vehicle Manufacture

They were friends, as only a craftsman can be, with timber and iron. The grain of the wood told secrets to them.

GEORGE STURT, 1923

The hard labor imposed by getting everything "out of the rough" at that period can scarcely be realized in these days of saw-mills and steam-power, from which nearly every thing comes to our hand ready dressed.

EZRA STRATTON, 1859

ON THE AFTERNOON of the first day of his apprenticeship to a southern Connecticut carriage maker, a skinny sixteen-year-old farm boy found himself standing in his employer's wood shop. Someone cleared him a spot at a bench, handed him a worn-out draw knife, and set him to making wooden wedges. So it was that in the spring of 1824 young Ezra Stratton first entered the domain of the traditional American carriage maker. He began his training in a time-honored fashion, for instruction in the use of this staple hand tool formed the first lesson for countless new apprentices. At the time Stratton would doubtless have been vexed to know that both his simple tool and menial task would occupy hallowed places in the memories of later generations, but they did. During the first years of the twentieth century Colonel James Sprague, carriage-top manufacturer and former wagonmaker, wrote a brief account of his experiences in a pre–Civil War Ohio wagon shop. Waxing eloquent over a generation of traditional craftsmen, Sprague referred to his father and contemporaries as "grand old knights of the draw knife, spoke shave, and forge."

Struggling with an unfamiliar tool under the cool gaze of alien coworkers, young Stratton considered himself more a serf than a knight. Despite his initial forebodings, however, he too eventually exulted in the satisfaction and pride that came from mastering a set of complex skills. While never so elegiac as Sprague, Stratton nevertheless recognized that his apprenticeship enabled him to become an independent carriage maker. This was, after all, how the system was supposed to work, and he knew it. Indeed, anyone applying for entry to the ranks of trade had no alternative, for the craft system had fulfilled this role since Colonial days. While both Stratton and Sprague would live to witness the demise of the arrangement, we must begin as they did, with the traditional craft apprenticeship.[1]

An early book on the subject of the Colonial American craftsman begins with the reminder that the word "craft" "suggested a trade or occupation, the 'Art and Mistery' of which was acquired only after a long period of tutelage by a master craftsman." The master craftsman might be a family member, in which case the instruction formed an integral part of family life. In the absence of such kinship ties, or in order to pursue a different line of work, a trade education outside the family consisted of a formal apprenticeship, for centuries the primary means of transmitting technological expertise. Once thought to have begun in the Middle Ages, the craft system appears to have originated in the centuries before Christ. It reached its apex in late Medieval Europe, particularly in England. More than just a social practice, apprenticeship received legal sanction in many countries as a way not only of creating a skilled work force but also of controlling young men. Apprentices served a fixed term under a master craftsman in the trade they wished to learn, from three to five years in Germany and France to as many as seven years in England. The working relationship between apprentice and master was controlled by a legally binding contract called an indenture. This document codified the process by which the apprentice would become a fully trained master craftsman. His goal was to learn the trade completely by the end of the indenture and afterward seek paid employment as a journeyman. When he had saved enough money he would establish his own shop, finally becoming his own master and completing the cycle. The crowning achievement of a newly established master craftsman was the creation of a supreme example of his handiwork, the masterpiece. This also served as an entry into his guild, a regulatory body comprised of master craftsmen from the trade. The guild controlled the trade by restricting the number of apprentices, which in turn prevented an excess of master craftsmen, protected craft knowledge and quality, and ensured high prices.[2]

Time endowed the system with a series of rituals and traditions, giving it a sometimes rigid formality. It did not remain an entirely static institution even in Europe, though the overall form remained essentially unchanged for centuries. Americans, living in a world long on natural resources and short on labor, radically simplified the system. Greater distances between population centers, a perpetual shortage of skilled labor, a largely agricultural economy, and an underdeveloped legal system all worked against guild formation. Without their regulatory function anyone claiming to be a master craftsman could take on apprentices. Those possessing even the rudiments of a vital craft became valuable to new settlements regardless of their exact trade status. Lack of guild oversight meant that apprentices unhappy with conditions in one shop could simply run away and find work in another or present themselves as journeymen in a different town, an extremely difficult feat to carry off in the tightly regulated world of the European trades. A modified and less structured version of apprenticeship remained standard in most early American trades, wagon and carriage building included.[3]

Trade apprenticeships typically coincided with early adolescence, the time in a boy's life when he had acquired most of his schooling and attained sufficient physical development to commence the long and often arduous training. While the specific figures varied with time, place, and trade, ages twelve to fourteen were the most common, with age sixteen gradually becoming the norm. Of course exceptions occurred; boys as young as ten might become apprentices, though this was more likely within a family setting or when economic hardship dictated abandoning school for work. Although a late adolescent might still be eligible for instruction, custom favored an earlier start, often for good reasons. Given an apprenticeship's relatively long duration, younger boys were at a decided advantage by being able to attain journeyman status ahead of their older counterparts. It also had to do with widespread belief in the efficacy of trade knowledge imparted to a mind and body still developing. Not only were younger boys more likely to be attentive and to bear fewer preconceived notions, they also stood a much greater chance of acquiring the subtle muscle memory and eye-hand coordination so central to the hand crafts.[4]

The timing had little to do with the maturing of the boy's own volition, for the choice of trade often lay with his parents. Ideally they would take into account the child's aptitudes and desires; however, other considerations frequently carried more weight—whether family tradition, an opportune opening at a nearby shop, or the vicarious ambitions of one or both parents. This latter factor proved decisive in Ezra Stratton's case. Impressed by the rare sight of a fine carriage passing through their small coastal Connecticut town, his mother became fixated with carriage making as the ideal occupation for her son. Her motives appear to have been as much social as pecuniary; while she was well aware of a local carriage maker's prosperity, she felt strongly that this was also a highly respectable trade. Stratton professed an early interest in printing but eventually became a reluctant convert in the face of relentless maternal pressure. Young James Brewster had a different experience. While his mother had hopes for his higher education, she allowed him to make his own decision. Surprisingly, given his own deep love of books, Brewster chose to learn a trade. Fatherless from early childhood and anxious to avoid the financial insecurity plaguing his mother, he opted for the potential profit and independence of a trade and became apprenticed to a carriage maker at age sixteen.[5]

Location frequently formed the chief consideration in choosing a master craftsman. Parents often sent their boys to a shop in the same town or in the next nearest settlement. Stratton had to travel to the nearby tidewater community of Saugatuck (now Westport) in order to learn the trade; Platt & Townsend were the only substantial carriage makers within a reasonable distance of the family farm. Reputation also played a part. Since the quality of a master craftsman's work and his community standing would naturally affect the apprentice's chances in the trade, the decision warranted careful scrutiny. This factor may have weighed heavily in Brewster's choice. Connecticut born and raised, he could doubtless have found an opening somewhere in his home state, but instead he settled on a shop in the western Massachusetts town of Northampton. Colonel Charles Chapman had achieved regional fame as both militia leader and carriage maker, and his reputation likely played a part in Brewster's decision.[6]

However they made their choices, those entering apprenticeships had first to reach an agreement with their prospective masters. Apprenticeship was traditionally regulated by a written document called an indenture. A social and civil contract between the boy's parents and the master craftsman, the indenture bound the apprentice and master to a series of obligations designed to protect the rights of each. Used mostly in apprenticeships carried on outside the immediate family, written indentures were common in the European and American trades through the late eighteenth century, though by then the practice was on the wane in England. It lasted longer in America, persisting into the nineteenth century in some trades. James Brewster almost certainly entered his apprenticeship under a written indenture in 1804, and while Ezra Stratton commenced in 1825 with an oral agreement, complications soon impelled him to sign a document outlining renegotiated terms.[7]

An 1819 indenture between a Pennsylvania wagonmaker and his apprentice typifies the genre. In language richly redolent of ancient custom, the document announced that by consent of his father, the boy "his said master faithfully shall serve, his secrets keep, his lawful commands everywhere gladly obey." In homage less to the master's power than to legitimate parental concerns, it further stipulated that the apprentice would not fornicate, marry, gamble, engage in commerce, absent himself without permission, "nor haunt ale houses, taverns, and playhouses." In return his master would teach him "the art and mystery of a wagonmaker" and supply him with "sufficient meat, drink, wearing apparel, lodging and washing fitting for an apprentice." He further agreed to permit the apprentice six days annual leave during harvest, and that at the end of his term would give him, according to long lost calculations, the oddly specific sum of twenty-six dollars and sixty-six cents. Notarized, signed by all concerned, and in some cases deposited with the local town or city government, written indentures like this provided a concrete set of guidelines to govern the working relationship.[8]

Most indentures terminated when the youth "attained his majority," which was almost invariably age twenty-one. The indenture's actual length, however, depended on the age of initiation. English apprenticeships, nominally the American model, had long stipulated seven years. The pressures of labor scarcity, industrialization, and a gradual increase in schooling shortened English apprenticeships by the mid-eighteenth century. The same was true in America. For example, mid-eighteenth century apprenticeships in Philadelphia ranged anywhere from as little as two years to as many as ten, with the average falling between four and seven years. During the early 1770s apprenticeships in that city's carriage shops averaged roughly the same length. In the early nineteenth century we can compare the seven-year terms of both Ezra Stratton and James Brewster and the three-years-and-three-month term of Marylander Horatio Price, whose indenture is described above. While a difference in the age of commencement may explain the disparity, other factors likely played a part as well.[9]

Regardless of its length or the manner of agreement, the apprenticeship's purpose was to turn out a fully competent craftsman. Here we come up against the durable myth of the traditional carriage maker producing complete horse-drawn vehicles alone and unaided. To do so would require sufficient tools, techniques, and trade knowledge to give him equal facility in the four fundamental processes—woodworking, metalworking, painting, and trimming. Many wagon and carriage makers were indeed competent in more than one. Recalling his craft training in an 1850s country shop, Hoosier carriage maker Charles Eckhart explained that he became a thoroughly practical woodworker, painter, and trimmer capable of doing "one just about as well as the other." Even so, this had become increasingly rare by the 1850s, for in all save the smallest rural shops a number of factors favored specialization. The range of intricate skills necessary for carriage building demanded the talents of several specialized craftsmen, not a workman briefly acquainted with each. Tradesmen devoted to one branch generally produced better work, and shops dividing the process along the four main divisions often turned out higher quality vehicles and more of them. "In the modern carriage," wrote Isaac Ware in 1867, "there is less wood and less iron, but more head-work and hand work, although quite as strong and durable if not more so." In looking over the evolution of carriage design and construction since the 1830s, he noted vast improvements in quality and durability, a development that further dissuaded carriage makers from attempting to master the entire art. Restriction to a single branch made for better carriages, and it hardly harmed the craftsman, who found himself hard pressed to keep pace with the latest developments in his particular branch of the trade.[10]

It is impossible to assign a specific date to this shift, in part because a degree of specialization had been common from the beginning. The making of wheels, arguably the most complex vehicle component, had been a separate trade from antiquity, and the wheelwright was able to remain distinct from the carriage maker. So could the village blacksmith, who might iron a horse-drawn vehicle as part of his regular line of work. While eighteenth- and early-nineteenth-century wheelwrights and blacksmiths frequently built entire farm carts and wagons, such cross-trade work declined with demand for more elaborate carriages. Although thousands of carriage builders had mastered the intricacies of tight joinery and flowing curves necessary to build carriage bodies and thousands more could cut, forge, and weld the myriad iron reinforcements for both gear and body, these workers increasingly formed two separate groups by the nineteenth century. Each specialty simply required too much time for a tradesman to master more than one.[11]

Eighteenth- and early-nineteenth-century carriage makers were far more likely than their successors to train apprentices in more than one department; as masters of small craft shops, they valued versatility and interdependence. Other reasons for this diversity in training included the type of vehicles produced, local availability of craftsmen in the ancillary trades, and the firm's location. Rural wagonmakers far from the demands of the urban market were more likely to train workmen capable of a broad range of tasks, while the owners of city shops demanded specialization. This distinction had always existed and became more pronounced with time, though exceptions always occurred. Ezra Stratton's master owned a rural carriage shop, but his near-exclusive sales to New York City made work there more like that in an urban firm. Stratton began his apprenticeship with the understanding that he would learn carriage woodworking, which in his case leaned heavily toward carriage body design and construction. Though he did learn to fabricate the other wooden components, the end of his apprenticeship saw him become a skilled body maker.[12]

Even given the logic of specialization, apprentices had to be on hand to help with nearly anything. They were traditionally at the beck and call of not only the master but also his journeymen and any older apprentices anywhere in the shop. Stratton attributed this practice to the lingering ideal of the "jack of all trades" carriage maker and, in later years, agreed that the practice was still useful for those destined to work in small country shops. Yet it also reflected the boy's position on the craft ladder: the lower the rung, the greater the chance that the day's work might include tasks hardly related to one's fledgling specialization. In fact, it might have nothing to do with the trade at all—hence Stratton's disgust on finding that his first morning's work consisted of repairing the stone walls around his master's two-acre field. Although he spent the afternoon learning the art and mystery of wedge-making, the lunch hour had revealed the hopelessness of trying to evade nontrade tasks. Noticing the glee of a seventeen-year-old apprentice at the table, Stratton quickly divined the cause: "the cow and horse, the pig and woodpile no longer claimed *his* attention . . . for his initiatory year of 'chores' had expired." Stratton's, as revealed by the exercise in dry masonry, had only just begun. It continued the next morning when his boss roused him at dawn for the morning chores. His frustrations increased with the onset of cold weather and the necessity of rising at 3 A.M. to kindle the shop fires. Stumbling around in the frosty predawn darkness, the groggy and incredulous boy began to discover the reality of craft apprenticeship. While prone to somewhat extended lamentation about this in later years, Stratton at least recognized that his was no unique case; apprentices a hundred years earlier complained of similar treatment, and he would hardly be the last.[13]

Nontrade tasks reflected the fundamentally mixed nature of work in the traditional craft shop. In a time and place when many workshops were extensions of the master's home, the line between habitation and business was blurred at best. While some masters certainly took advantage of their apprenticed labor, the arrangement did have the virtue of job security. An indenture's legal and ethical obligation prevented a master from simply discharging an apprentice during hard times. Apprentices formed an integral part of the wider domestic economy, which also made them a form of "human credit" with multiple uses. When business slackened and the shop became embarrassingly idle, masters turned their apprentices and journeymen toward any of the thousand-and-one daily, seasonal, and yearly tasks found on all rural homesteads. They might be rented out to a busy fellow craftsman, or to a neighbor who wanted a little extra help harvesting his corn. The decline of traditional apprenticeship saw the decline of such duties. Those apprentices who remained in the shop no longer lived in the master's home, and he no longer felt compelled to care for them outside the workplace. When work slowed to a trickle, the nature of supply and demand in the cash economy dictated laying off workers, not inviting them home to hoe turnips.[14]

The wise young apprentice submitted to such tasks in order to rise above them. In trying to ease the way for new carriage apprentices many years later, Stratton advised them to "*never refuse to turn the grindstone!*" Such willingness paid off in the form of trade instruction. Neophyte woodworkers began with lowly wedge fabrication, thereby practicing with a fundamental tool (the draw knife) and producing a simple object (a wedge) that had uses ranging from centering hub boxes to leveling work benches. Beginner's exercises eventually gave way to more complicated tasks. For woodworkers like Stratton this might include mending broken wheels, fabricating gear parts, and sawing out poles and shafts. Blacksmithing apprentices moved beyond fire tending and bellows pumping to forging simple hardware and installing the more rudimentary gear and body reinforcements. Beginning painters graduated from the endless tedium of color grinding and body filling to learning basic brush strokes and the application of filler, paint, and varnish to wheels and gears. Apprentices in the trimming department continued to expand their repertoire of stitching techniques to embrace more complicated fastening and decorative work, as well as learning to upholster seat cushions, cover dashes, and install head liners.[15]

An apprentice's education consisted of the same general learning process regardless of his chosen specialty. It began with observation, ranging from simple daily exposure to the shop environment to detailed scrutiny of specific techniques. The mere act of watching bequeathed valuable information about the selection and care of tools, their proper application, how to avoid mistakes, and how to correct errors that did occur. This led naturally to the next step, actual assistance. Combining the observer's knowledge-absorption with the worker's material assistance, "helper" status transformed onlooker into participant. While the apprentice's low skill level limited his ability, his simple availability was itself often indispensable to his shop mates. Practice came with the gradual assumption of a complete task without the aid of a more experienced coworker. Extremely simple jobs like wedge and dowel making, bolt heading, or carpet tacking might only be demonstrated once before the apprentice continued alone. More complex work like spoke setting, tire welding, or wheel striping required additional demonstration and supervised practice. When deemed sufficiently skilled to be trusted with a particular task, the apprentice then began performing it completely unaided and as often as the work demanded.

Woodworkers necessarily had to spend a great deal of time in stock preparation, consisting primarily of hand planing rough-sawn planks into smooth, dimensional boards. Hours of what Stratton called "shoving the jack plane" preceded any actual joinery, and apprentices had to master this tedious but essential step before moving on to more complex and rewarding tasks. Indeed, Stratton complained of having to spend most of the first two years of his apprenticeship at "dressing stuff up," but all competent woodworkers had first to master this essential step. Even when an employer used an apprentice as a sort of human-powered planer, the practice developed the eye and muscle memory so essential to such tasks. Platt also assigned Stratton to elemental repairs like replacing broken spokes and sawing out felloes and other curved parts, this performed with a frame saw. As James Sprague recalled it: "Did you ever saw out felloes with a hand saw? If you have, you know how to pity a saw mill." This too passed as the apprentice worked his way up to more complicated work. Platt eventually entrusted Stratton with wagon shaft installation; despite initial bungling he soon mastered it and moved on to axle fabrication. As he successfully grappled with each successive task, he found, somewhat to his surprise, a gradual swelling of pride. "Who that has pursued his calling," he later wrote, "has not felt a peculiar sensation, as step by step he made advancement?"[16]

In this way the apprentice progressed from ignorance to competence in each of the many tasks comprising the wagon and carriage maker's work. Yet any number of contingencies might delay the process: friction between apprentice and journeyman, excessively slow business, or perpetual consignment to menial chores. Down the years craft apprentices complained of plights ranging from inadequate food and shelter to lack of trade instruction and even downright cruelty. Besides deploring his monotonous pork diet, Stratton took serious issue with the clause in his indenture charging his master with providing religious and ethical instruction. Platt evidently manifested no overt spiritual inclinations, and if Stratton's acerbic account is correct, he also proved ethically deficient. Of course, masters often had their own just causes for complaint. These might only stem from the minor irritants of youthful inattention and misbehavior, or they could come from more serious matters such as intoxication, theft, vandalism, or desertion. As a young master craftsman James Brewster caught two of his apprentices stealing watermelons from a neighbor's garden. Informing them of the principle of crime and its punishment, he "chastised the two delinquents in the open shop," and petty theft ceased among his apprentices.[17]

The range of potential problems encompassed the gamut of human nature, and the long list of proscribed behaviors on a typical indenture form reveal the more common ones. The layers of incrementally added customs, byproduct of the institution's great antiquity, could themselves generate additional conflict. Both Brewster and Stratton quickly discovered the prominent role alcohol played in many nineteenth-century craft shop rituals. New apprentices and journeymen often found their first duty consisted of treating the entire shop to a round of drinks, or "paying one's footing." Stratton considered this merely one more obstacle to overcome, but Brewster saw it as something else entirely. His temperate upbringing had instilled in him an aversion to alcohol bordering on visceral terror. When informed that his chores included regular trips to the local "grog shop" on behalf of his shop mates, the serious young man feared the worst. Indeed, his refusal to imbibe provoked the oldest apprentice to force him to stand on a workbench, informing the rest of the shop that this boy was a "no-souled, blue-skinned fellow." Determined to stick to his principles, Brewster asked him how it could possibly be fair to have to partake of something he loathed. Seizing what must have been a pregnant pause, he offered the abashed apprentice his pocket money as proof that morality, not greed, informed his stance. Yielding to this combination of logic, integrity, and fierce Yankee stubbornness, the apprentice softened, excused him from drinking, and called him "a general." The nickname stuck, and so General Brewster survived the first great challenge of his apprenticeship to the carriage trade.[18]

From Jour to Master

For all its limitations and potential hazards, apprenticeship succeeded in transmitting craft knowledge in the carriage trade just as it had in harness making, silversmithing, joinery, and other trades. The apprentice's realization that his was a temporary station formed an important ingredient in its success. Taking the next step meant becoming a journeyman. Though apprentices could only do so upon mastery of the craft, the actual event signifying the transformation was the apprentice's twenty-first birthday and his legal status as a fully independent adult. As the name suggests, a journeyman traveled from place to place plying his newly learned trade wherever opportunity beckoned. His goal was to gain additional experience while accumulating sufficient funds to establish a shop and become a master craftsman. Journeying was by no means mandatory, for many "jours" continued working at the same old bench in the same old shop, some even becoming partners with their former masters. However, local economic slumps, strained working relationships, dissatisfaction with the community, or simple youthful wanderlust often compelled the carriage jour to embrace the road. Exposure to different craftsmen and other shops added new techniques to his repertoire and increased the likelihood of his own eventual success.

Becoming a jour, then, signified a double achievement: mastery of an important set of skills and the privilege of being paid to exercise them. While skill naturally played the central role in the enterprise, the whole point of obtaining it was essentially economic. The ability to command a cash wage marked an important transition in the life of a fledgling tradesman; it meant the opportunity to edge out into the turbulent stream of capitalist enterprise, a significant step toward the overarching goal of independence that lay at the very heart of the system. Apprentices did not traditionally earn wages because their social contract revolved around the exchange of labor for craft knowledge; money had no place in the transaction. The regular earning and saving aimed at financing a new shop began with commencement of the journeyman phase. Yet the growth of a cash economy during the first quarter of the nineteenth century put increasing pressure on masters to pay their apprentices to prevent them from running off to factory jobs. This had the short-term effect of attracting and retraining apprentices, but it created more serious problems in the long run. It certainly added a greater element of instability in that apprentices were now willing to shop around for better wages, a practice formerly restricted to journeymen. It also increased the chances of friction between journeymen saving for the future and apprentices who found themselves now able to do the same.[19]

Advancement to journeyman status marked an important milestone, one often celebrated with some kind of party. Orchestrated by the master craftsman or fellow jours, the event supplied all hands with one of the episodic breaks punctuating traditional craft shop work. In many cases it also served as the jour's farewell party. In 1809 Colonel Chapman held an "entertainment" for twenty-one-year-old James Brewster, ironically hoisting a tankard to the young man's health and future success. Alcohol's central role in this tradition was yet another legacy of the hard-drinking eighteenth century. After spending most of his last year of apprenticeship working with a newly hired New York body maker, Ezra Stratton experienced roughly the same thing, though he called it a "freedom frolic." According to custom he footed the refreshment bill for the entire shop and any interested townsmen. Perhaps anticipating the outcome, Stratton took early leave. Indeed, the celebrants' subsequent excesses ended up making this the last "frolic" ever held at Platt & Townsend.[20]

The jour expected more than just a memorable headache on his release, for most indentures specified that the master supply him with "freedom dues." Usually a modest sum of money, the dues might instead consist of a suit of new clothes or a set of basic tools. Some journeymen journeyed no farther than back to their accustomed workbenches, for masters often hired new jours whose talents promised additional profit for the shop. At times this combination of enticing wages and familiar surroundings outweighed the lure of the unknown; Stratton put in several months of remunerative labor this way before moving along to another shop. Sometimes such arrangements led to master craftsman status via partnership. Daniel Platt's partner, Charles Townsend, began as a journeyman and ended up marrying Platt's daughter, becoming in a single stroke both a husband and a full-fledged carriage maker. Even before the end of his apprenticeship, Brewster fielded Chapman's offer of partnership. Though he turned it down, owing to his unspoken concern about Chapman's drinking, Brewster appreciated the gesture and departed on amicable terms.[21]

The new jour's first concern was often to obtain or augment his own set of tools. The best plan was gradual acquisition during the apprenticeship, but if his freedom dues did not include tools the jour would soon purchase more. Stratton's eagerness to visit New York City derived in part from the opportunity to shop for woodworking tools with money his mother had secretly saved for the purpose. As a body maker he relied on the same basic hand tools common to most woodworking tradesmen. Blacksmiths might flesh out their kit to include at least a representative sample of the most basic hammers, tongs, and punches. Journeymen painters and trimmers traveled lightest, their tools consisting of modest assortments of varnish and paint brushes, striping pencils, tack hammers, needles, and stuffing irons. New skills and experiences complemented these purchases, and many jours traveled from shop to shop to obtain such vital commodities. Rural-trained jours often sought out city work, lured by the dual benefits of higher wages and the latest trade knowledge. Important carriage-making centers like Boston, New Haven, New York, and Philadelphia enticed generations of eager young jours fresh out of country shops. Both Brewster and Stratton aimed for New York, and while the former never went further than New Haven, the latter spent the rest of his life as a downtown New Yorker.[22]

The "tramping jour" occupied a significant place in the craft world not only for what he sought to acquire but also for what he could bequeath. Masters hired them for their labor as well as their knowledge. In a heavily fashion-driven trade like carriage making, a jour familiar with the latest vehicle styles was a capital asset for a shop owner seeking a competitive edge. This was particularly true in the case of city-trained jours seeking work in rural locations. As early as the late eighteenth century, country shops wishing to tap into the growing urban carriage market eagerly hired city jours for their fashion savvy. Stratton's master shipped most of his output down Long Island Sound to New York City, and on at least two occasions he hired city jours to work and instruct the others in the latest urban styles. Not all country workmen appreciated the implied condescension. Stratton recalled the resentment he and his shopmates felt toward a New York woodworking jour, expressing mock relief that they "would not be necessitated to go to France to perfect themselves . . . when they had the *ne plus ultra* in their very midst." Despite the jour's arrogance and strange city manners, Stratton came to admire his swift efficiency: it took him only three days to turn out a complete paneled gig body from a pile of rough planks.[23]

Sometimes jours brought problems with their expertise. Though Platt's woodworker proved his mettle at the bench, incessant bragging about his family's superior wealth and social status quickly soured his fellow tradesmen. The end came with his disastrous attempt to attract the local ladies by driving a tandem-harnessed wagon about town; the resulting storm of ridicule drove him from the community. Some jours suffered from other maladies. One day Platt arrived back at the shop with a genuine "New York trimmer." He was an outstanding addition to the trim shop, but intoxication rendered him useless about 50 percent of the time. Accustomed to sleeping off his binges at work, the jour often awakened to find himself pinned to his bench by a nailed-down cowhide, trussed up hand and foot, or with a freshly blackened face. These expedients failed to separate him from the shop or his habits. In a desperate attempt to retain an otherwise valuable hand, Townsend told Stratton to fetch an eel from the nearby Saugatuck River. Slipping it into the trimmer's jug in hopes that the resulting sickness would cure him, his shop mates watched in silent amazement as he pulled from the jug, became violently ill—and then celebrated his subsequent recovery with another binge. Clearly the man had become a hardened alcoholic requiring more assistance than any eel could provide.

Stratton claimed the frequency of city jours in country shops came from the inability of some of them to refrain from heavy drinking. Unable to hold on to their jobs, they fled to outlying areas where people were unacquainted with their problems. This may have been the case with men like the trimmer, but other factors also propelled jours out into the countryside. Some left the city to sell their services where scarcity might command better wages, while others departed in the face of the intense competition and market saturation that diminished their opportunity to rise above their rank. Relocation frequently offered broader horizons and the opportunity to start a successful business. As industrialization gradually transformed jours from intermediate protocapitalists to permanent wage earners, some sought out areas where mechanization had made fewer inroads into the trade. These disruptions and discontinuities ensured that the tramping jour fulfilled the most ancient and vital role of spreading the trade to other areas.[24]

Ultimately, of course, few tradesmen wanted to remain "tramping jours" indefinitely. A large part of the rationale for traveling came from the opportunity to discover a location best suited to the final transition from journeyman to full-fledged master craftsman. By the time a jour reached this point, the salient difference between the two lay more in the realm of finances than trade skill. Experienced jours probably knew nearly as much about wagon and carriage building as most master craftsmen; what the master had that the jour lacked was his own business. If the jour exercised due economy, he had been saving toward this end for some time. James Brewster claimed that limiting his personal expenses as a new journeyman enabled him to accumulate a significant sum of money. Jours most likely to become their own masters were the ones like Brewster who saved sufficient funds to turn the dream into reality. Shortcuts were possible, such as marrying the boss's daughter and becoming a partner. Sons trained by their fathers often became members of the firm automatically, once they achieved both the necessary competence and legal age. While jours choosing these paths to advancement sometimes put up money to cement the deal, it was much easier than starting entirely from scratch.[25]

Still, many chose to go out on their own. Brewster's decision to do so stemmed not only from his concern about Chapman's drinking but also from his ambition to be his own boss. He knew, however, that ambition alone would not suffice. Buying, building, or renting a shop required money, and this was just the beginning. It took additional funds to outfit the premises, for even the well-equipped jour lacked the more ponderous tools of his trade. The would-be entrepreneur had to purchase such indispensables as grindstones, vises, tuyere irons, bellows, anvils, mullers, and more. The image of the craft shop bereft of all save bench and tool chest is decidedly false, for it took a surprising number of larger tools and fixtures to get things done in the absence of machinery. Although some items might be purchased later, others were absolute necessities in even a minimally equipped establishment. Blacksmithing could no more proceed without an anvil than could woodwork without a bench, painting without a color grinder, or trimming without a cutting table. Even when items like workbenches and saw trestles were customarily shop-fabricated, store-bought wood, nails, and screws were still necessary for their production. Enterprising young men might trade services for some of them, but in the end one truth became self-evident: it took capital to become a capitalist.[26]

There was also the matter of one's particular strengths and weaknesses in the trade. Even apprentices subject to equal amounts of training in all four branches often found themselves better suited by facility or simple preference to some more than others. Though consigned for a time to help the shop's exuberant and kindly blacksmith, young Stratton resented this distraction from his chosen specialty. Any multiple specialization that did occur frequently appears to have maintained a fundamental divide between wood- and metalwork. Charles Eckhart, apprenticed to a carriage maker in the late 1850s, learned woodworking, painting, and trimming, but not ironwork, and his experience was not unique. Some of the reason was the growing complexity of body making and carriage ironing, the two primary construction processes; these were arguably harder to combine than were the finishing processes of painting and trimming. The additional expense and bulk of the apparatus may also have played a part, especially since most communities already had blacksmiths, many willing and able to iron a vehicle.[27]

Specialization presented a problem for the jour determined to have his own shop but not quite willing or able to do it all. One option involved tapping into the local trade network. Tradesmen within the same community often exchanged or hired one another's services. A carpenter might repair a blacksmith's roof in exchange for horse shoeing, a joiner hire a shoemaker to fabricate a leather writing surface for a desk, or a house painter help a boatbuilder paint a new skiff. This kind of interchange was as old as the trades themselves, and wagon and carriage builders had long availed themselves of it. Many new carriage builders performed all the woodworking but relied on the local blacksmith for ironing. Some might also farm out the painting, but many woodworkers also painted and trimmed their own vehicles, even if the final result fell short of a truly expert job. The other option for new carriage builders lacking certain tools and skills was to find another having both. He might be a jour skilled in the necessary specialties or a fellow struggling craftsman willing to enter into partnership. The latter was especially preferable if he also brought sufficient capital to buy the remaining items necessary to equip the shop.[28]

Whether embarking alone or in conjunction with another, the new carriage builder also had to consider the matter of location. Some chose to remain in the city or town where they had served their apprenticeship, particularly if it seemed able to support an additional shop, while others chose similarly familiar surroundings. Stratton returned to his boyhood village in part for sentimental reasons but also because the community wanted a carriage maker of its own. Unexpected opportunities sometimes played a part. Brewster would have ended up in New York City in the fall of 1809 had the stagecoach not lingered in New Haven. As it turned out, he stayed on and established that city's reputation for fine carriages. Once a particular community was chosen, there still remained the question of where to set up shop. For the average cash-strapped new carriage maker, this decision hinged on the availability of suitable rental property. Though obviously a much more expensive option, those able to build their own shops spared themselves the frustrations of having to adapt to some ill-suited space, the fate of many new carriage builders occupying a general-purpose building or store front. Erecting a new building permitted a structure ideally suited to the complicated task of building horse-drawn vehicles.[29]

Most new craftsmen could not afford this luxury and settled for renting. Here the lone tradesman relying partially on outwork had an advantage of sorts. A savvy renter might only require sufficient space for a work bench and an open area for a vehicle under construction, requirements easily met by many available rented spaces. The urban carriage maker might locate in a store-front shop in the business district, an old storehouse by a busy waterway, or a small building on the edge of town. The rural carriage builder enjoyed fewer options but could also obtain larger quarters at a reduced price. One carriage maker recalled the common antebellum practice of renting a small building, usually a former dwelling, for use as a first shop. All of these arrangements had the advantage of reasonable price and widespread availability, and they served as a starting point for many carriage makers. As the craftsman gained more apprentices, hired more jours, and generally expanded his operation, he required not only more space but space of a certain type.[30]

The nature of the work favored shops arranged in specific ways. While there was a great deal of variation in other matters, most wagon and carriage shops shared two vital characteristics: space and subdivision. The size of a typical horse-drawn vehicle mandated more space than silversmiths, cobblers, and other tradesmen required. Ground floor vehicle access was imperative for even the smallest establishment, and few carriage makers ever felt they had sufficient interior space. The vehicles themselves used up an alarming amount of it, likewise their parts and accessories. Bulky, awkward items like shafts and poles, folding tops, wheels, and other impedimenta occupied a great deal of room. Similarly, while sufficient quantities of cloth, leather, paint, fasteners, and the like might occupy corners and closets, iron and lumber required dedicated storage space. Owners of small shops soon found themselves exploring the virtues of expansion or relocation in order to cope with an increasingly cluttered and inefficient work environment. The second characteristic, subdivision, had to do with the conflicting requirements of the various construction processes. Wagon and carriage makers found it impossible to confine their work to the single, unified work spaces of other trades because their own consisted of four distinct branches.

Woodworking, blacksmithing, painting, and trimming each had specific, and often conflicting, shop requirements. Blacksmithing called for a well-ventilated area with solid, vibration-free footing, sturdy storage racks and bins, and easy access to the outside for fresh air and for processes customarily undertaken outdoors, such as tire heating. Woodworking required a large, wide room with wall space sufficient for workbenches and tool chests, adequate natural light, and an unobstructed central work area supplied with trestles and saw horses. The varied requirements of priming, rough sanding, wet sanding, painting, varnishing, and drying, combined with the unceasing war against airborne impurities, meant painting required a particularly large space. Similarly concerned with cleanliness, trimmers sometimes shared this space, provided they steered clear of wet paint and varnish. Whether doubling up with the painters or woodworkers or occupying their own room, trimmers required space for a workbench or two, a cutting table, cloth and leather storage, and sufficient clearance to maneuver around vehicles being trimmed. The inadvisability of situating everything in a single room becomes readily apparent when considering the effects of soot on new varnish, sparks on dry wood shavings, or primer dust on clean upholstery. Even the earliest Colonial American carriage shops featured at least rudimentary compartmentalization, and this helps account for the persistence of the divisions throughout the life of the trade.[31]

Such subdivision took various forms. Those occupying small shops frequently walled off portions of the interior, erected additions, or even connected several extant structures. A portion of the "old shop" at Walborn & Riker typified this approach. Old even before the Civil War, one former occupant remembered it consisted of three small buildings "haphazardly joined, with step ups and down for uneven floor levels, and cut up into dozens of rooms." While hardly lending itself to architectural harmony, the vernacular tradition of incremental additions served carriage makers admirably. At other times separate structures made more sense. Rural shops often had the advantage of larger lots, and many country carriage builders divided their operation between several buildings. Daniel Platt situated blacksmithing in one building, trimming and harness making in another, and woodworking and painting in a third. Even urban carriage builders often boasted multiple structures. Many of the more prosperous city shops consisted of several buildings grouped around a central courtyard that served as a wagon and lumber yard, receiving and shipping dock, outside work area, and more. Another option consisted of "vertical expansion." The addition of upper floors or the erection of a new, multistory building made especially good use of often limited space, and the two- and three-story vehicle shop became a common sight in both city and country.[32]

Whatever the specific building format, a typical pattern of subdivision emerged. Provided the building had one, the basement usually housed the blacksmithing department. Not only did this make use of a space too dirty for painting and too damp for woodworking, it had other virtues particularly suited to blacksmithing. The hard-packed dirt floor provided vibration-free anvil footing (blacksmiths in wood-floored shops often sunk their anvil posts through the flooring into the ground below) for powerful, perfectly dead hammer blows. Basements provided ample and inexpensive support for enormously heavy masonry forges, stock racks, and other ponderous fixtures, while greatly lessening danger from fire. They also permitted gravity-loaded coal bins and isolated much of the noise and vibration from the rest of the building. Though it offered reduced ventilation, clamminess, and the possibility of upper-floor soot contamination, many carriage makers preferred this arrangement. The Spragues' antebellum shop followed this custom, as did hundreds of others. Where prevented by architecture, carriage makers installed ground floor forges as the next best choice.[33]

Wood shops frequently occupied the first floor, though upper-story locations were also quite common. Because simultaneous vehicle building and repair, and the attendant manipulation of large quantities of lumber, was common, the wood shop frequently demanded the largest space. The necessity of allotting a central area to work in progress, as well as the desirability of maximum natural light for men at their benches, mandated placement of the vise-equipped workbenches beneath windows. These outside walls also supported shelving for supplies and the wall-mounted wooden tool chests that sometimes supplemented the traditional free-standing chests interspersed with the benches. Additional implements and fixtures, such as grindstones, lathes (great wheel, spring-pole, or treadle), saw horses, hewing benches and chopping blocks, took up the remaining room. If the shop produced its own wheels there might be a wheel bench or pit, perhaps a hand-propelled spoke tenoner, or a few other fixtures devoted to wheel making. Sometimes this equipage edged out into the central space, dominated by partially constructed carriage bodies, wheels and gears awaiting ironing, and other vehicles in various stages of completion (fig. 2.1).[34]

The frequent placement of paint shops in the upper stories underscores the necessity of isolating freshly painted work from the smoke and dust of the other operations. Ideally consisting of carefully sealed floors and walls, paint shops had interior arrangements roughly similar to the wood shop: tools, supplies, and fixtures along the perimeter and a central area for standing work. Wall-mounted supply shelves supported containers of pigment, oil and other raw materials, while work benches held color grinding and mixing apparatus, rags, the workers' individual brushes, striping pencils, and paint cups. A screened-off area or separate room contained the rottenstone, water buckets, and rags used to rub down preliminary coats of body filler and rough stuff, while painting and varnishing took place in the center of the floor. Side windows cast the light necessary for good finish application, though skylights were even better. Moreover, while indoor drying usually gave the best results, many shops found space limitations forced them to use the courtyard, an adjacent flat roof, or a balcony for drying work. Trimmers often benefited from a work area handy to these places, as they normally received a vehicle only after it had been painted. Supplies of cloth, carpeting, leather, and fabrics occupied overhead racks or closet space, while shelves, work benches, and tool boxes contained the myriad tacks, brads, needles, thread, shears, and awls of the craft. Work benches or tables acted as cutting surfaces, and trimmers worked in, on, and around vehicles being trimmed.[35]

From Concept to Reality

Regardless of the specific kind of building or its exact floor plan, the design and construction of wagons and carriages within followed an established pattern. The first step was the choice of vehicle, a decision informed by the builder's market knowledge. Before any iron glowed or shavings curled, the vehicle had first to take shape in his mind's eye. Given that market considerations had already settled the matter of basic type, how then to fix the many details entering into its design? Tradition often took the lead here, as a considerable portion of the knowledge the builder had absorbed as an apprentice consisted of the acceptable design parameters of various vehicle types. Why had a C-spring to assume its particular shape? Because it always had, at least in the memory of the carriage maker, and likely the customer's as well. Yet few wagon or carriage types remained totally unchanged. Some of the more basic designs had a certain timelessness; a farm cart was a farm cart, for example, whether transporting a Colonial family to church or hauling manure on a Kansas dairy farm. Even here, though, there were incremental changes, primarily in new materials or modified construction. Sometimes these two factors produced immediately noticeable differences. An 1845 buggy looked very different from an 1895 buggy in part because the latter's spindly patent wheels gave it a lightness impossible to achieve with traditional, "keg-hubbed" versions. Nothing reflected this design flux better than the babble of names applied to vehicle styles. The trade tacitly recognized the confusion and made various attempts to standardize nomenclature, though these efforts met with only limited success. Fashion seldom fits into neatly organized categories.[36]

Although he might possess a crystal-clear mental image of his goal, even the most hidebound traditionalist resorted to some kind of written plans, if only rough pencil sketches on scrap wood. These not only clarified the details for the master, they also steered his apprentices and jours through the more difficult portions of the work. Occasionally a book or newspaper illustration provided a rough guide, and starting in the 1850s the fledgling trade press produced a trickle of fashion plates for this purpose. By the 1860s the trickle had become a flood, as editors

Ezra Stratton and Isaac Ware vied for the lead in publishing plates in their journals and in posters hawked to their readers. Most editors bowed to fashion in promoting their plates as the latest New York, English, or French designs sure to generate profit for any builder wise enough to follow their lead. The editors' claims aside, fashion plates did inspire countless carriage builders to attempt work they might have overlooked and close sales they might have failed to make.[37]

While the trade press eventually published detailed working drawings, the pre-1860 carriage builder had to confront the problem with his own pencils, paper, and wits. Those who tried often managed with simple side elevations, perhaps adorned by a few key measurements. Advanced carriage makers might render more detailed and exact scale drawings containing additional information. Another option consisted of full-sized blackboard drawings, again invariably side elevations. Blackboards permitted easy changes while drawings were portable, and some firms used both. City shops catering to the capricious whims of wealthy clients frequently opted for blackboards for their realistic scale and ease of emendation. Blackboard drawings also lent material aid to workmen by permitting comparison of patterns and finished parts to a full-scale master drawing. In either case, reliance on side elevations reflected the English practice followed by so many American carriage builders. Though they lacked specifics, they served admirably as general models for skilled workmen, who used trade knowledge to fill in the details. Doing so invariably called for patterns. Commonly made from thin wood, these served as exact and durable guides to construction, especially of wooden gear and body parts. Oft-used sets acquired dense coverings of scrawled dimensions that reminded one observer of Egyptian hieroglyphics. Most shops accumulated vast quantities of them, hanging from nails or slipped up over exposed rafters in haphazard piles. Such simple draftsmanship and rudimentary methods served many small manufacturers long after their larger competitors had adopted more accurate and sophisticated practices.[38]

Whether working from a blackboard, patterns, a trade journal illustration, or an idea in his head, the carriage builder could commence building a particular vehicle in a number of ways. A lone craftsman might start with wheel and gear fabrication, sending these to the local blacksmith for ironing when he turned to the body. Larger shops might begin various portions of the vehicle simultaneously, though without truly exact drawings and construction methods, this could only be performed to a degree. Cutting and fitting occupied a prominent place in the traditional carriage shop and necessitated sequential construction methods. As befitting an object composed chiefly of wood, horse-drawn vehicles came to life in the wood shop. The popular image of the traditional woodworking process, whether that of carpenter, joiner, or carriage maker, usually begins in the forest where the craftsman felled his own trees. Indeed, particularly in rural locations, carriage makers often procured some of their lumber this way, either off their own land or that of a customer. Cutting one's own timber had its advantages: the carriage maker could select prime quality growth, pick out trees with bends lending themselves to curved parts, and exercise special care in felling, bucking, and yarding. Charles Eckhart recalled that he and his shop mates customarily felled their own hardwood trees for hubs and spokes, even though most city carriage builders were by then buying theirs ready made.[39]

Beginning directly at the stump had its disadvantages, however. Not only did it presume handy access to suitable timber, it also required a significant investment of time and effort that not all craftsmen felt willing or able to make. Logs could not become lumber without sawing, and the amount of time required to pit saw even a small softwood log into planks persuaded many craftsman to contact the local lumber mill. The abundance of waterpower along the eastern American seaboard encouraged sawmill erection from a very early date, and craftsmen patronized them in great numbers. One could also supply the sawyer with wood; James Sprague's father obtained most of his timber in trade from local farmers, much of it as logs delivered to a nearby sawmill. Another option that became increasingly common was purchase of rough-sawn lumber from the sawyer. City carriage makers usually had no other recourse, but even rural shop owners preferred the enormous savings of time and effort in purchasing lumber rather than logs.[40]

Like most other woodworking tradesmen carriage makers normally bought large quantities of lumber well in advance. Much of it consisted of green wood, which required at least one or two years of seasoning before use. Outdoor lumber piles or semi-enclosed sheds formed an essential feature of most shops. Eckhart's employer piled his boards and billets alongside the building, an exceedingly common practice. While kiln-drying would become a more widespread option after the middle of the nineteenth century, slow air drying remained the preferred means of lumber seasoning throughout the carriage era. Many wagon and carriage makers and other woodworking tradesmen insisted that this produced lumber with fewer checks, end-splits, twisting and cupping than did even the most carefully operated lumber kiln. This rationale extended to other forms of wood. Many carriage makers turned out large quantities of hubs from green wood, it being easier to work in that state, then stacked them in available corners or in overhead spaces to season. Green-rived spoke blanks received similar treatment, with finishing postponed until sufficient drying had taken place.[41]

Air-cured lumber fresh from the pile had still to be planed, and working rough planks down to smooth, finished boards required hours of labor with the jack plane. This is what Stratton referred to when recalling the difficulties of "working from the rough": all joinery had to be preceded by lengthy periods of stock preparation. Even though the earliest rural communities often had sawmills, power planers did not exist, so carriage builders began by attacking vast quantities of shaggy planking with their arsenal of hand planes. Eckhart routinely "dressed down" thick planks up to thirty inches wide using nothing but a standard wooden jack plane. He and his contemporaries had a great deal of hand planing to do on the large quantities of hard-and softwoods used in making horse-drawn vehicles. Though the list might vary according to the builder's fancy and local availability, a handful of dominant woods passed across the benches of most carriage makers. These included oak, hickory, ash, beech, maple, basswood, and poplar. Generally speaking, the hardwoods formed the weight-supporting structural members such as spokes, rims, gears, and body framing, while the softer woods served as body panels, moldings, and other parts benefiting from lightness.[42]

Woodworkers used these materials to construct wheels, gear, and body. Those trained in the most traditional methods built all three, a practice rural shops perpetuated longest. If the shop manufactured its own wheels, the woodworker began with seasoned oak or elm which had been dried "in the round." If it had not already been done, the men then bored out the center of the hubs, turned their outer surfaces on the great wheel lathe, and chopped out the spoke mortises. After shaping the spokes from previously seasoned split blanks, a woodworker drove them into the snug-fitting hub mortises with a heavy sledge or mallet while another sawed out felloes from oak or ash planks. Steam-bent two piece rims eventually replaced the sawn felloes, but this was uncommon before the last half of the nineteenth century. Once the felloes were ready, the woodworker fit them onto the spokes with their ends abutting to form the rim. This completed the woodworkers' portion of wheel construction; the next major step called for the skilled hands of the blacksmith (fig. 2.2).[43]

An iron or steel tire greatly increased the durability of the wheel by not only protecting the rim from impact but also binding the separate components into a durable whole. The smith began by measuring wheel circumference with a small hand wheel called a traveler. Counting the revolutions and adding sufficient length for the weld and shrinkage, he then marked off the required length of iron or steel strapping. After cutting this off he then bent it to a rough circle by hand or by wrapping it around a large bolt of wood or a worn-out grindstone kept handy for this purpose. Exceptionally thick stock might even require heating beforehand. The smith then brought the ends to white heat in the forge and quickly hammered them together on the anvil to form a solid, welded hoop. He and a helper heated the tire in a large circular fire outside, dropping it into place over the wheel on the ground or on a low stone platform. The smith made adjustments with a hammer while apprentices or wood shop men applied buckets of water. The shrinking tire compressed the wheel into a slightly concave profile ("dish") intended to brace it against lateral thrusts. Sanding off burned wood and painting completed the process (fig. 2.3).[44]

Gear and body making required the usual range of woodworking hand tools, though as with cabinetmaking, carriage building called for the finer versions. The basic woodworking tools included rough shaping implements like the axe and hatchet. Controlled splitting came with the froe, the same tool used to rive (split) shingles and clapboards. Rip and crosscut hand saws, as well as the wooden-framed frame (bow) saw for curved work, were staples of basic shaping. Standard surfacing tools included jack, smooth, joiner, and block planes, but carriage makers' tool chests brimmed with more specialized planes for cutting rabbets, dados, and moldings. Mortising, paring, hollowing, and carving required a wide range of chisels, gouges, and mallets, while brace and bits took care of boring. Fastening depended on hammers and nails, screwdrivers and screws, plus the animal-hide glue and double-boiler glue pot found in every wood shop. Few of the carriage maker's hand tools were exclusive to his trade, the notable exceptions being a handful of rabbet and dado (sometimes called coach- or body maker's) planes designed for use on body panels and other curved surfaces. The others consisted of patented brace-mounted spoke tenoners, drawknives or hand routers for body work, and of course the spoke shave, useful for many tasks wholly unrelated to wheels.[45]

Through sawing stock on saw horses, shaping parts and gluing components at the bench, and assembling vehicles in the central work area, woodworkers transformed their hardwood stock into wagon and carriage gears. They punctuated their work with occasional visits to the shop's blacksmith for iron reinforcements, attachment of springs and axles, and any other steps requiring the attention of the metalworker. As the work progressed, more and more of it consisted of adding parts to the increasingly complete gear resting on trestles. Body work called for ripping out, planing, and joining hardwood frame members; sawing, shaping and fitting of whitewood body panels and molding; and making and installing flooring, doors, and seats. As with gear construction, body fabrication centered around saw horse and bench, with assembly on centrally located body trestles and with episodic trips to the blacksmith. Finer in nature than gear construction, body work required a keen artistic sense, working knowledge of draftsmanship and elemental mathematics, and woodworking acumen at least equaling that of the cabinetmaker. While English practice often separated body and gear making, American carriage woodworkers frequently built both, with apprentices and lesser skilled journeymen assigned to the rougher preliminary work. Formal division of carriage woodworking usually fell between wheel making and gear and body construction.[46]

No matter how well executed, the solidity of gear and body depended on more than snug joinery; few horse-drawn vehicles could long maintain structural integrity without strategically placed iron reinforcements. Wheels required tires and sometimes tire bolts, hub bands, and a few other fittings; gears needed a host of rods, brackets, braces, and bolts; bodies required lesser quantities of corner brackets, stay irons, and the like. Bodies also needed dash and toe rails, door hinges and handles, shaft or body steps, lamp brackets, and other metal fittings. Their manufacture and installation fell to the blacksmith, and since many parts were easier to install before the woodwork was complete, woodworker and blacksmith had to constantly coordinate their efforts. As with the other branches of the trade, blacksmithing would eventually be impacted by ready-made parts, but the early carriage smiths hand forged nearly everything from a limited variety of iron bar, sheet, and rod stock. The smith might forge some of the more standard items in advance, while making others only as required, but in any case the necessity of hand forging and fitting each part was extremely time-consuming. Eckhart remembered some ironing jobs taking as long as two weeks. Forging and tempering springs took time and great skill: success required first-rate steel, an advanced knowledge of metals, and the ability to formulate and follow a durable design. While many smiths did make them, springs were also one of the earliest carriage parts available ready made. Regardless of the part in question, the smith worked primarily from his forge, anvil, and leg vise, moving to the nearby vehicle for installation. He "burned on" some parts, and while this could weaken the adjacent wood, if done properly it ensured a tight fit with only minor, and easily sanded, surface damage.[47]

Once woodworkers and smiths had built the vehicle, painters and trimmers proceeded to finish it. Painting required more time than all the other steps combined, taking weeks to finish a single carriage. Before Cleveland firms like the Sherwin-Williams Company invented and popularized ready-mixed paints in the 1880s, paint shops ground their own colors on a stone slab using a pestle-like implement called a muller. This required an enormous amount of time and monotonous motion to yield sufficiently fine pigments and so proved an ideal outlet for the apprentice's allegedly boundless energy. As Sprague remembered it, "the muller got lopsided and the stone wore out, but the boy goes on forever." Hand grinding likely seemed endless because it sometimes took as long as two hours to prepare a batch of pigment this way. Good results were indeed possible, but uneven quality ingredients and limited chemical knowledge lent an air of mystery to the process. Eckhart insisted it was guesswork: "We did not know whether the paint was going to dry in twenty-four hours or in six days." Varnish worked the same way. Though ready-made varnish had been available far earlier than paint, many early shops still preferred to make their own. Stratton took his turn tending the simmering kettle, as did the other young apprentices at Platt & Townsend.[48]

While paint shop apprentices understandably loathed such work, surface preparation and drying occupied the majority of the time required to create a deep, glossy carriage finish. Painting demanded a dust-free location with sufficient space to dry several jobs simultaneously. The process began with a complete, unpainted vehicle said to be "in the white." The bare wood surface required a great deal of preparatory work. Primer coats composed of various mixtures of lead ground with linseed oil filled the pores and sealed the surface; sanding between each coat promoted adhesion and created a progressively smoother finish. Painters followed this with a coarser mixture of lead-based primer called "rough stuff," followed by vigorous rubbing with pumice stone or a similar abrasive. Apprentices applied putty to nail holes, cuts, and gouges, using assorted shop-made concoctions or, in later years, ready-made fillers. The monotony of base coat work owed a great deal to the repeated sanding and rubbing; small wonder that painters assigned it to apprentices or jours at every opportunity.[49]

Having laboriously built up a sufficiently smooth surface, the carriage painter began mixing the pigments his apprentice had so carefully ground. Color mixing was an art that called for the eye of an expert. The exact proportions of pigments and oils required for a particular color stood as one of the carriage painter's trade secrets. Depending on the expected degree of finish, the painter applied anywhere from one to several color coats before moving on to striping. Using thin, fine-haired brushes called striping pens, he applied various styles of decorative stripes to wheels, gear, and body. These and the underlying layers had to be fully cured before proceeding to the varnish room, what one instructor called the "sacred temple of the painter's hopes." Any reverence a painter attached to this step reflected its importance in creating high gloss finishes. Multiple varnish coats interspersed with sanding completed a process that had commenced several weeks earlier. Opinions differed concerning the drying process. Many manufacturers simply let jobs sit on flat roofs or outdoor platforms, though doing so risked finishes marred by dust, leaves, or insects. Other carriage painters preferred indoor drying. Tight-joined or plastered walls, puttied windows, and muslin hanging on walls and ceilings made for an uncomfortably humid work environment, but they also made for clear, unspoiled surfaces.[50]

Trimmers shared painters' concerns about cleanliness, though dust was less of a concern than cloth-staining smoke and soot. Trimmers had the cleanest job, one requiring artistic application of widely varied materials. Selecting from their stock of cloth, leather, leatherette, glue, nails, tacks, cord, and thread, they put the finishing touches on the mechanically complete vehicle in the form of seat cushions, interior carpet, head lining, folding top, dash cover, and more. Early trimmers hand stitched everything; not only did this mean dozens of linear feet of painstaking hand work, it also often included additional purely decorative touches. On some vehicles trimmers used contrasting thread to stitch decorative patterns in the upholstery. These might utilize shop-made designs or patterns called "stitching plates," which appeared in trade journals. Though a truly well-painted carriage was, in one editor's opinion "never improved by being crowded full of white stitching," trade journals continued to publish stitching plates to meet continued demand. Upholstering seats and laying carpet required thread and tacks, as did the attachment of various bits of leather to parts of the gear. Evenly stretching material over top bows took time and expertise, proving that the best trimmers had aesthetics and hand skills rivaling those of a good tailor.[51]

Work flow through the shop bore little of the linear quality of rationalized factory production. A vehicle under construction might make several trips between woodworking and blacksmith departments as each added incrementally to the job. The same vehicle could shuttle between trimming and paint shops in a similar manner, though usually to a lesser degree. In any case, if design changes or mistakes intervened, the job would be rolled back for the requisite adjustments. In some cases a craftsman went to the vehicle in another department, particularly if only to complete a minor task, as with a blacksmith fitting a newly forged body brace. Department placement within a building normally reflected space constraints and architectural limits more than efficient work flow. While craftsmen could take the stairs to an upper floor, vehicles required other arrangements. Hand-hoisted and later steam and electrically powered freight elevators eventually met this need in the larger firms, but most early shops used ramps. Owners located these outside buildings, positioned to terminate in large, upper-story doors or outside balconies wide enough to accommodate vehicles. Simple and cost effective, ramps and platforms remained a small shop feature long after the advent of elevators in big city enterprises.[52]

Modest output, departmental interplay, and scant task specialization characterized a manufacturing approach known as custom or batch production. Each buggy or surrey, every cart or wagon, passed through the various departments several times, just as it passed more than once through the hands of the shop's skilled craftsmen. Obviously everyone benefited from a steady stream of work, so all but the tiniest one-man shops contained multiple vehicles under construction. A typical shop on any given day might feature a buggy being built for a town doctor, a light express wagon for a local merchant, a heavy wagon for a nearby farmer, and perhaps a coach for a wealthy client. Fleet work occupied the shop sequentially or in batches if a builder was fortunate enough to land these desirable contracts. Such work-to-order jobs were small-shop staples, as were vehicles made ahead to meet anticipated demands. Of a type and pattern designed to appeal to a known market segment, these goods underwent a similar construction process. Though several jobs of the same type would then be commenced, each still reached completion in the same manner as custom work. Whether building to order or for stock, custom work and batch fabrication characterized the production format of the typical wagon and carriage craft shop.

Life and Labor in the Small Shop

The nature of craft shop work had little in common with the staccato regimen of the assembly line. Each man generally exercised the full range of his craft skills over the course of an average work week. Each occupied his own bench or shop space, owned most of the tools he used, and utilized skills first learned as an apprentice. The craftsman's work featured variety and pride of accomplishment, but it also meant laboring for remarkably long periods of time. Late-eighteenth- and early-nineteenth-century craftsmen generally worked a six day week, with a daily average of anywhere from twelve to sixteen hours. The longest days meant returning to the shop after supper to remain as late as 9 P.M. Daily hours often depended on the season, with shorter workdays in summer and longer in winter being common, though there were certainly variations. This venerable tradition likely originated as a concession to the demands of warm-weather farm work. In many regions the beginning and ending of the extended-hour period coincided with the first day of fall and spring, though a political event like election day might also mark its commencement. In either case it meant sixteen-hour days for as much as six months of the year. Both James Brewster and Ezra Stratton apprenticed under this regimen, Brewster in the first decade of the nineteenth century and Stratton in the late 1820s. Stratton deeply resented the four additional evening hours, especially as his master periodically attempted to extend them as late as 10 P.M.[53]

The seasonal schedule contrasts sharply with the wagon and carriage trade's annual business cycle, which had become well established by the antebellum period. Warm weather sparked demand and created the "spring rush," making it the busiest time in most firms; winter brought reduced hours and temporary unemployment. While the exact timing of the transition between these opposing seasonal patterns has yet to be explained, both are well documented. The older winter evening work was likely nothing more than a simple carryover from the agricultural daily and seasonal hours kept by the majority of Americans. It may also have owed something to attempts by early manufacturers to extract maximum labor from their employees, but even so the overarching nineteenth-century trend was toward shorter hours anyway. Even the most ardent capitalist found that mechanical workdays profitably extended only so far. This realization stemmed partly from growing awareness of the limits of human endurance in industrial work, which dictated against keeping the same hours as farmers. Not that farm work required any less exertion; rather, the kinds of exertion differed markedly. Many trades required much finer motor skills and intense concentration that could not continue indefinitely. Mechanization's early inroads in the textile industry also suggested that shorter hours could prove as productive and even more so than sixteen hours of handwork every day. Labor scarcity in some areas evidently assisted some craftsmen in their bid for higher wages and shorter hours, gains that spilled over into other trades. All these factors, increasingly abetted by various reform movements, trades union agitation, and ongoing newspaper dialogue, contributed to the moderation of manufacturing work hours starting in the 1820s. Stratton recalled shorter hours in the carriage trade by this time, and while the trend proceeded unevenly, it became more pronounced everywhere as the decades wore on.[54]

Just as they exercised unofficial but tangible leadership in matters of vehicle style, a handful of prestigious urban carriage builders set the standard in working hours. James Brewster and a few others led the American carriage trade in adopting the twelve-hour day during the early 1830s and the ten-hour day just a few years later, changes that ended the old seasonal differential. Around 1836 a joint agreement between Brewster, his business partner, and their employees, apprentices, and jours settled on a ten-hour daily standard. Maintenance of the previous wage scale prevented discord and encouraged others to follow suit. By decade's end, the ten-hour day had become the norm in many carriage shops and would remain so for years. Not that such progressive measures applied everywhere; rural shops tended to retain older work patterns longer than their urban counterparts. Apprenticed in rural midwestern shops in the 1850s, both James Sprague and Charles Eckhart regularly put in fourteen-hour days at a time when ten had become widespread. Conversely, some city shops had settled on a twelve-hour day as early as the 1790s. Working hours remained a potential source of conflict in some firms well into the second half of the nineteenth century.[55]

While long hours naturally dragged on at times, craft workers found respite in a variety of ways. Some came from the work itself: the wide range of vehicles being built and the challenge of repairing and refurbishing all brought welcome variety, as did the episodic diversions common to preindustrial labor. Stratton recalled with amusement the shop's blacksmith, who periodically opened the door to boom out the lines to popular songs. "O Betty Arden, she is my darling, &c," evidently amused villagers across the river as much as it did his fellow craftsmen. Craft shop customs served a similar function. Many rituals revolved around alcohol, whether morning and afternoon "drams," additional beverages supplied by new apprentices and journeymen, or the celebratory "frolics" held for departing jours. If a master's temperance convictions precluded these, other diversions presented themselves—like practical jokes. Some had instructional overtones, as when journeyman Stratton left woodwork clamped in a bench vise only to find it next morning sporting a thick coat of axle grease. The same might apply to a hand saw left protruding from a partially sawn plank. Not only did such antics amuse onlookers, they reminded the victim not to strain bench screws unduly or leave sharp tools lying carelessly about. Others merely provoked laughter, as with sending an apprentice out for mythical "strap oil," or secretly nailing down the lid of a departing jour's tool chest. Finally, while certainly not to everyone's liking, being put to farm chores or other outside work at least had its rewards in distraction and variety.[56]

Money remained the preferred reward. Independent master carriage builders received theirs from vehicle sales, either in the cash-up-front that everyone favored, or in payments, which most had periodically to accept. Prices realized for horse-drawn vehicles varied by time, place, quality, demand, and many other factors. These, combined with the bewildering complexity of early American finance and monetary systems, make broad-based price comparisons exceedingly difficult. In the Colonial period nearly any wheeled vehicle was valuable, though persistent demand and more abundant supply made farm carts and wagons reasonably affordable. Carriages, however, remained a luxury item well into the next century. Entry-level two-wheeled chaises and gigs, while hardly inexpensive, cost far less than four-wheeled carriages. Heavier vehicles like coaches occupied the top of the scale, commanding a premium from clients wealthy enough to afford them. Brewster's gratification at the sale of a fine carriage in the 1820s flowed from what he regarded as providential circumstances, but $550 in cash also contributed to his mood. Shop proprietors could not bank all or even most of this money, to be sure. Besides paying for food, fuel, rent and raw materials, they had also to supply wages to the men who worked alongside them.[57]

Shop owners in all trades varied in paying by the piece, hour, day, or month. Journeymen and, increasingly, apprentices might receive regular daily or monthly wages, while tramping jours and unskilled laborers added during busy times might expect piece or job work. In many shops, however, piece work reigned among the most skilled hands for its extra potential profit. While these arrangements will be explored later, all were common in the wagon and carriage craft shop. Forms of payment likewise varied. Especially in cash-poor rural locations, payment in scrip, sometimes called "orders," redeemable at local boarding houses or dry goods stores predominated prior to the Civil War. Early in the nineteenth century, rural and urban shops alike often used scrip to avoid procuring scarce hard money in a largely barter economy. Scrip also lent itself to irregular pay periods. Brewster's 1820s adoption of regular cash payments every Saturday was thus doubly unusual. Not only did his men heartily approve, he found it vastly simplified his bookkeeping. This was only one of Brewster's numerous early business innovations, and like the others it eventually became standard practice in the trade at large. Store orders lasted longer in rural locations but died out by the late 1870s.[58]

What did carriage makers pay for wages? Those working alongside or for the master craftsman—and this increasingly included apprentices and day laborers in addition to jours—received a range of pay. Stratton records $1.25 as a good daily wage for a carriage workman in the 1830s, an amount that had increased to $1.50 or more by 1860. This corroborates another source indicating that while highly skilled antebellum body makers could earn as much as $3.00 daily, the average amounted to half that. Government statistics provide more detail. In 1850 daily carriage shop wages in Indianapolis were $2.00 for body makers ($1.75 for lesser skilled woodworkers), $2.00 for blacksmiths, $1.50 to $2.00 for painters, and $2.00 for trimmers. In 1865 a New Haven body maker could average $2.75 per day, a blacksmith $3.00 per day, painters and trimmers $2.75 per day. These data reveal at least a degree of transregional wage similarity and show modest increases during the midcentury decades. While exceptions occurred, such figures compare favorably with other forms of skilled work. Not surprisingly, however, they still proved grounds for contention. The collision of old craft methods with the 1830s cash economy mortally wounded the craft system, especially as apprentices began receiving wages. More and more apprentices in all trades demanded them during the 1820s and 1830s, and more and more craftsmen found themselves forced to pay to attract and retain qualified boys.[59]

A distinct annual demand cycle only added further instability. Regardless of whether they owned a small shop or a large factory, American wagon and carriage makers lived for the first flush of warm weather and the fabled "spring trade." Mild days and hardening roads set minds on travel, so spring and summer saw customers ordering new vehicles and depleting stock built up during the winter. Warm weather also flooded shops with repair work. Though continually urged to bring in repair jobs during the slow winter months, wagon and carriage owners invariably waited until spring to roll out their vehicles, only to find them in need of adjustment or mending. Most craftsmen found themselves compelled to take on temporary help in order to keep pace in spring and summer, even if it only meant hiring unskilled day laborers for the more menial tasks.

Winter brought just the opposite conditions. Ideally, the carriage maker took advantage of this slow season to turn out vehicles for the following spring, yet doing so required sufficient savings to fund expenditures in raw material and wages during these months of light or nonexistent sales. Many carriage builders failed to accrue adequate financial reserves, and many shops had to run on reduced hours, employing fewer hands during the coldest months despite the slowly dying custom of longer winter hours. Sleigh manufacture and repair might take the edge off the chill in northern climates, but much depended on the local popularity of sleighing—not to mention the amount of snow. A light winter could spell disaster for the shop counting on runnered vehicle sales to whisk it through. While the larger shops, and later the factories, eventually accumulated sufficient reserves and extended markets to smooth out the variation, a degree of this seasonal cycle remained a part of the trade throughout the nineteenth century.[60]

Such regular fluctuations were accompanied by more unpredictable variations in regional and national economies. Brewster recalled with a shudder the lean years of the War of 1812, when not a single paneled carriage left New Haven's carriage shops. Carriages were luxury items people willingly did without during hard times. A host of less dramatic but equally harmful events frequently beset small business owners like carriage makers, including market saturation, sharp price wars with local competitors, and simple economic unpredictability. While innovative carriage makers found ways to cope, part of the solution often involved cutting expenses and dispensing with marginal workers. Masters in apprenticeship's heyday could always assign idle hands to nontrade tasks or hire them out to others. With the waning of traditional apprenticeship during the period 1820–50, shop owners increasingly reduced work hours and the number of shop hands. Although both responses surfaced in nineteenth-century craft shops, American carriage makers appear to have preferred the latter. Trying to keep a full complement of hands on reduced hours spread the hardship over the entire shop and risked the loss of particularly prized men. Carriage builders preferred to lay off the more peripheral workers rather than experience a potentially fatal skill hemorrhage. Many shops eventually retained a year-round core of skilled workmen, adding seasonal help when appropriate. This accommodated the tramping jours and unskilled day laborers who, by inclination or necessity, formed part of the American seasonal labor force.[61]

If a steady stream of orders heralded the kind of prosperity sure to put a gleam in any carriage maker's eye, it also mandated additional attention to administrative matters. This often resulted in two additional features, the first of which was an office. Though an unused workbench might have initially accommodated the limited paperwork of a small craft shop, increasing trade meant more paper. "Notes" for work sold on time, contracts for large orders, trade journals and catalogs, sales flyers, and such began accumulating and required a space of their own. Many carriage makers irritated their wives through episodic occupation of the kitchen table, but a place in the shop made more sense. This might mandate little more than a desk in the corner of one of the departments, but a larger firm usually set up a separate office space. Partitioned off from adjoining work areas, it featured a desk, a few chairs, an oil lamp, and perhaps a wall clock or calendar. Usually located on the ground floor with a street entrance, the office became the place to receive customers, answer mail, pay bills, and attend to the relatively minor but needful paperwork that came with operating a small nineteenth-century business.[62]

The same logic often led to a dedicated space for completed work. Of course all but the smallest shops found themselves storing vehicles for future sale, and it was only a matter of time before they admitted potential customers to inspect the wares. Dust and dead flies sold few carriages, so the resulting tidying and improved lighting saw the storage room become the carriage repository, nineteenth-century equivalent of the automotive show room. Serving the dual functions of storage and sales, it soon became an expected feature of the more prosperous and ambitious establishments. Daniel Platt's coastal Connecticut shop featured a nicely finished repository as early as the 1820s. Such on-site locations made obvious sense where the shop itself doubled as a point of sale, but high-end carriages sold best in cities that featured clusters of them in the most favorable locations. A few might be connected to a single large carriage shop, while others sold the work of a number of builders, especially those from out of town. Platt sent most of his work to a New York City repository, and such establishments often placed "country-made work" alongside that of city carriage builders lacking their own show space. Many of the larger city builders eventually erected their own repositories in more affluent parts of town in order to appeal directly to potential customers.[63]

Sales naturally occupied an exalted space in the carriage builder's world, and all of them relied on personal contacts, referrals, and reputation to move merchandise. These cost nothing and, given an advantageous location, favorable economy, and dearth of aggressive competition, might in and of themselves suffice. The product was also a sort of advertisement. James Brewster related how he clinched a particularly lucrative sale merely by placing a newly finished carriage in front of the shop. It caught the eye of a "very gentlemanly person," an Englishman newly arrived from Barbados. Surprised at its local origin, he told Brewster that it compared favorably with English carriages, and he subsequently purchased it from the proud young carriage maker. Every carriage builder hoped that shiny new examples of his work would attract this kind of attention, and many affixed small brass name plates to their carriages. While motivated in part by understandable pride, carriage makers purchasing plates from suppliers like Ezra Stratton had definite commercial ends in mind.[64]

This raises the issue of product quality. The stereotypical assumption about craft-built products has always been their inherent superiority. The trade's mechanization starting in the 1840s and 1850s only reinforced this conviction, which sometimes contradicted the historical reality. What was the historical reality when all wagons and carriages originated in hand methods? As Brewster's Englishman could attest, craft work indeed produced remarkably well-made vehicles. "We made a job then to last," Charles Eckhart sighed, "and they did last—some of them lasted so long you wished you hadn't made them so good."

Other carriage makers had no such regrets; if craft work could mean quality, it proved equally capable of turning out junk. Lack of appropriate design or drafting ability, poorly trained or indifferent craftsmen, excessively hurried work, insufficient or improperly seasoned timber, weak glue, defective iron castings, blacksmith-scorched wood, prematurely dried paint, old varnish, cheap leather, sloppy stitching—any of these or other shortcomings might contribute to quality ranging from rough-but-passable to barely-able-to-roll. Although such results might stem from lack of skill or other benign causes, they could also be very much intentional. In the event that a desire for profit overcame his ethics, a carriage builder might cut as many corners as possible to save expense. "Men of brains are never tempted to look for cheap carriages," Ezra Stratton insisted. "They value their own lives too highly to trust themselves in these cheap got-up traps, suggestive of spoilt timber, glue and putty." He ruefully admitted, however, that the practice was as old as the trade. So while craft skill often equaled craft quality, the influence of other factors, some quite far removed from purely technical considerations, could play an even more decisive role. The reality was that carriages of all degrees of quality were available and caveat emptor remained ever the rule of wise purchase.[65]

Most carriage builders eventually opted for more explicit forms of market cultivation. City directory listings might produce results, as many urban builders attested. Ever alert for a commercial edge, Cyrus Saladee managed to get his name in the directories of many of his places of residence, though more stationary carriage makers also found it expedient to do likewise. Most directories featured advertising, and carriage makers joined other tradesmen in hawking their wares in their pages. In 1845 Jacob Lowman, dean of Cleveland's modest carriage trade, sprang for a full-page advertisement in *Peet's General Business Directory*, complete with all the essential information about his shop and rounded out with a woodcut of a closed coach. Newspaper advertising had been available from the very beginning. Colonial carriage builders like Boston's Adino Paddock used the local paper to boast of the ability to build vehicles "equal in Fashion and Goodness to the latest Models from England." James Brewster peppered his local newspaper with everything from invitations to examine his carriages to want ads for seasoned spokes. June readers of the 1824 *Connecticut Journal* discovered that right in New Haven they could purchase new and used carriages, as well as "all kinds of articles in the Coach-Making line," and that the proprietor promised that "all orders will be punctually attended to from any part of the union, and thankfully received." Advertising reached beyond the builder's personal acquaintance, allowed him to emphasize certain aspects of his work, and presented him with the opportunity to differentiate himself from his competitors. Word of mouth might suffice, but few carriage makers passed on the chance to shore up their reputations with strategically placed advertising (fig. 2.4).[66]

Dealing with competition was only one of many hurdles confronting the carriage maker. For all the technical acumen required to build horse-drawn vehicles, it took skills of an entirely different order to sell them. Business success depended on knowledge that did not necessarily come with even the finest technical education. Having spent three years as a journeyman and built a shop in his hometown, Stratton entered the business with a youthful naïveté he recalled ruefully some thirty years later. Discovering that "boss coach-makers had their troubles as well as other mechanics," he found many of them squarely in the realm of business. Immediately on closing the shop at 6 P.M., Stratton made visits to press clients who had overdue bills. Other matters consumed more time, whether incessant price haggling or redundant complaints over trivial matters. Competitors selling carriages at or even below cost cut deeply into his market share and reduced his profits to an alarming degree. Clearly there was more to running a successful carriage shop than skill with hand tools. The reality confronting Stratton as a new carriage maker was that his training had failed to prepare him fully for the rigors of commerce. In 1860 he wrote that he had "dreamed that once in business, I would soon make a fortune, and yet I find that after thirty years' toil the fortune is still beyond *my* grasp."[67]

Fortune did not elude James Brewster. If we can deduce anything about his success, it hinged in great part on his ability to negotiate his way through the increasingly complex world of American finance. Given what we know about craft apprenticeship and his own youthful experiences, it seems likely he received no more instruction in business than Stratton had. Brewster appears to have had a more innate entrepreneurial sense, a penetrating business shrewdness that served him exceedingly well his entire life. Yet success did not come to him instantly, and like Stratton he worked long hours at bench and desk to overcome frustrations. The strain and hardship made him "depressed in spirits" and hopeful of his eventual escape from continual toil.

Though his account of his early years dwells less on troubles than Stratton's does, it does not indicate their absence from his working life. In his memoirs, Brewster tends to skip over difficulties in his eagerness to draw moral lessons from his experiences. Stratton remained content to dwell on the difficulties, but his account is all the more enlightening for that. It would be unfair to compare these two traditionally trained craftsmen based solely on their respective entrepreneurial skills. Stratton's ultimate allegiance to his adopted trade was tenuous at best, and his later defection to journalism reveals a deeper devotion rooted in his childhood. Brewster displayed not only a more abiding loyalty to the craft but also a burning desire to couple it with maximum commercial success. Good business sense had always been a prerequisite of the successful master craftsman, and Brewster possessed it in abundance. What separated these two men were their respective means of overcoming the commercial hurdles that had, in various forms, always confronted master craftsmen.[68]

Both men struggled mightily to become their own masters, and both achieved that goal. Beginning as awkward, timid youths, they endured the tedium of workplace chores and the ridicule of older apprentices, survived homesickness, bowed to the obsequies of archaic custom, fumbled with unfamiliar tools and strange materials—and thereby climbed hand over hand up the time-worn rungs of their craft. Hardened muscles, maturing minds, and increased technical facility marked the end of their novitiate, and they celebrated by taking to the road as tramping jours. Work, wages, and saving filled their days as they prepared to surmount the final obstacle and fulfill their destinies according to the venerable script. Established at last as their own masters, they found that technical skill did not imply financial savvy, that commerce had its own apprenticeship, and that despite years of training they still had much to learn, and they did learn, even as they taught. Like Charles Chapman and Daniel Platt before them, they took on apprentices and hired journeymen as the trade passed down the generations in seemingly endless percolation.

As the antebellum years found Brewster and Stratton in midcareer and as newcomers like Sprague and Eckhart embarked on the same journey, omens of change had already appeared on the American scene. Water and steam power coupled with the productive success of textile mills may at first have seemed far removed from the quiet confines of the traditional carriage shop, but they heralded a profound upset in human affairs that would ultimately leave little undisturbed. It all happened so quickly. Within the confines of an average life-span, the drudgery of the draw knife and the world of the traditional trades would seem sufficiently remote for old men like Colonel James Sprague to recall them with something approaching nostalgia. We know this transforming power as industrialization, and it wrought changes of greater magnitude than any craftsman had ever known.

[3]

From Shop to Factory

The Industrialization of a Trade

> With the ripping saw, we can split more stuff in half an hour than could otherwise be done in a day. With the band-saw, we can saw out more intricate sweeps in an hour than can be done in twelve by hand. The planer takes off half an inch of material while we are gauging our jack-plane. The spiral vertical dresser will dress two axle-beds while we are dressing the edge of the drawing-knife, and at the same time will do it with precision.
>
> J. L. H. Mosier, 1875

> Had the body maker of a quarter of a century ago been told that almost every piece of timber entering into the construction of a body would in time be dressed to shape and fitted to its place by machinery he would have laughed.
>
> *The Hub*, October 1897

The Cook family's carriage-making factory must have been an impressive sight to passersby at Grove and State Streets in New Haven in 1860. A huge three-story building ran along both avenues, adjoined by a slightly more ornate four-story wing and a few smaller structures in and around the open-ended courtyard. An elevator tower peeped above the flat roof, and a soaring brick chimney announced the presence of the large powerhouse at the center of the property. The casual observer might have assumed it merely another of the regionally numerous textile mills, but the firm's enormous sign announced that the works belonged to G. & D. Cook & Company and that the establishment was a carriage factory. Instead of turning out bolts of colorful cotton cloth, this imposing plant produced horse-drawn vehicles of astonishing variety and in quantities that no ordinary carriage shop could ever hope to equal. According to the firm's illustrated catalog of 1860, the Cooks produced an average of ten carriages a day, or one every hour, production figures that would make the traditional wagon and carriage builder shake his head in amazement—or his fist in rage. George Cook, David Cook, and Hannibal Kimball were just three of an increasing number of men establishing carriage factories, and theirs would soon be dwarfed by many others. Industrialization had taken the reins of horse-drawn vehicle manufacturing, and G. & D. Cook & Company's factory was only one sign that the trade was in the throes of profound change.[1]

The term "industrialization" derives from " Industrial Revolution," which originally described the dynamic changes taking place in eighteenth-century Great Britain. One of the great watersheds of human history, the Industrial Revolution marked the divide between small-scale, traditional craft production and large-scale, rationalized factory production. Steam engine and textile factory became its apotheosis, progenitors of the British industrial city with its sooty skies, massed factory workers, and "dark Satanic mills." However, mill-dominated industrial cities were only one manifestation; reliance on hand tools, skilled craftsmen, and small shops continued alongside complex machinery, machine tenders, and clamorous factories. The same mix of old and new characterized America's industrial transformation as the revolution crossed the Atlantic late in the eighteenth century. Not that Colonial America had been entirely devoid of industry; understood in terms of its intensity and necessary skills, large-scale natural resource processing such as charcoal making and iron smelting certainly qualified as industrial endeavors. Still, sustained industrialization in the form of rationalized factory production came only after independence. America had its first true industrial experience with textiles, as New England mirrored Old in erecting spinning and weaving mills along its abundant waterways. Also as in England, industrialization embraced more than cloth production. The form, if not the entire concept, of the factory spread to fields as disparate as the manufacture of firearms, clocks, and sewing machines during the 1830s, 1840s, and 1850s.[2]

Even so, in 1860 more Americans labored in homes and small shops than in factories. For all its connotations of rapid, unprecedented change, the Industrial Revolution actually proceeded in fits and starts, the rate varying widely from trade to trade. In its most fundamental sense, industrialization consisted of the process by which human society went from small-scale, low-output, simple manufacturing technologies, to large-scale, high-output, sophisticated production techniques. It included such long-term forces as the growth of a market economy and establishment of the cash wage, as well as significant market expansion caused by the nation's dramatically multiplying population. Local conditions also contributed to its uneven pace and sometimes bewildering complexity. Some trades such as cloth and shoe making became industrialized early in the nineteenth century, while others, like carpentry and blacksmithing, retained their traditional forms for decades. A region's degree of urbanization and wealth also affected its transition from craft to industry.

In its application to a far larger and more ethnically and geographically diverse nation than England, the fabric of industry stretched over a much broader area in America, and tears became evident in many places. Older technologies and ways of working existed alongside radically new production methods even as some crafts disappeared with astonishing speed. The complexities of the process acting on a nation as broad and diverse as the United States make classification difficult, but at industry's essential core are the deeply entwined agencies of the factory system and mechanization. Whether examining the rise of shoe factories, the development of barrel making machinery, the production of cheap wooden clocks or expensive steel plows, factories and machinery formed the key agents of change. Wagon and carriage making proved no exception.[3]

Andrew Ure, the great Scots chemist and industrial apologist, once penned what still serves as a useful working definition of the factory system. He characterized it as "the combined operation of many orders of work people . . . tending with assiduous skill a system of productive machines continuously impelled by a central power." Ure wrote this in the 1830s, when the English textile factory had shown the world the undeniable promise of this new means of production. Historians have since argued about the exact origins of the term "factory," in part due to its ambiguous historical use. In the nineteenth century people haphazardly applied it to nearly any large productive establishment, including those we might now characterize as mills or shops. Historians have argued, with varying degrees of success, that the name might legitimately apply to medieval monastic manufacturing and other large-scale forms of production. Though the historical and etymological record could conceivably support such a catholic use, Ure's definition best characterizes the edifices springing up in trans-Atlantic manufacturing during the late eighteenth and early nineteenth centuries. In this sense, the archetypical factory originated in the British textile industry during the eighteenth century and migrated to New England in the early years of the nineteenth. Large numbers of workers and banks of centrally powered machinery in specially constructed buildings proved a versatile combination with great potential in a range of trades, including wagon and carriage making.[4]

Most accounts of the factory system's origins in the carriage trade begin with a young New Englander named Jacob R. Huntington. A Massachusetts carriage maker who amassed a fortune during the trade's rapid 1860s industrialization and expansion, Huntington became widely lauded as carriage building's factory system pioneer. Though most accounts overstate his influence, he did have undeniable success as a popularizer, making his experiences useful for outlining industrialized horse-drawn vehicle production. Huntington was born in the Massachusetts mill town of Amesbury in 1830 and returned there in 1844 after a brief Maine sojourn. Amesbury, West Amesbury (later Merrimac), Salisbury, and the other towns of the Merrimac river valley were harnessing their plentiful waterpower for industrial purposes. Besides having mills for grinding grain, sawing lumber, and crushing linseed, the area also boasted textile mills. Huntington found employment as a wool spinner in one of these during the late 1840s.[5]

A restless and ambitious youth, Huntington soon moved on to other opportunities and 1853 found him employed as a painter in C. H. Palmer's West Amesbury carriage shop. He evidently started with little more than informal, on-the-job training, for no account mentions a formal craft apprenticeship. In a move reflecting both his entrepreneurial streak and an increasingly common trade practice, Huntington subcontracted for Palmer. Agreeing to accept a job for $100, he completed it in six weeks, earning enough to soon venture out on his own. Many years later *The Carriage Monthly* characterized him as having conceived "a far-seeing and comprehensive scheme for supplying the public needs in the matter of vehicular transportation, and at reduced cost to the user," but this was an exaggeration. Whatever his contributions to developing low-cost carriages, Huntington's primary goal consisted of making money. In the early 1850s he allegedly declared to friends his intention of becoming Amesbury's richest man in ten years' time; even if apocryphal, the anecdote supports his well-known tendencies. If he lacked traditional craft training, Jacob Huntington abounded in business savvy and a sagacious reading of a rapidly changing trade. These rewarded him with great riches and the lasting envy of many would-be imitators. Ambition and perhaps a little avarice, more than any conviction about the social benefits of cheap buggies, account for his bold strides in factory carriage production in the 1850s.[6]

Renting a room in a West Amesbury tannery, Huntington commenced work on his first complete vehicle, rolling a plain, open carriage out the doors eight weeks later. He sold it to a Salisbury man for a little over thirty dollars, an inauspicious start. Thirty dollars for eight weeks' work hardly equaled decent pay, and Huntington may have even failed to cover his costs. No matter, for at a time when a light carriage cost anywhere from two to three hundred dollars, a thirty-dollar carriage attracted notice. During the next several months he turned out a baker's dozen and invested the returns in his business. Though he continued to perform much of the work himself, he hired a handful of carriage trimmers, blacksmiths, and painters to enable him to increase production to over three hundred vehicles during his second full year of business. Despite the venture's incipient success, Huntington abruptly sold out to employee James Hume and moved to Cincinnati in 1858. Undaunted by the existence of several established carriage factories, he formed a partnership with former Amesbury employee Elbridge S. Feltch, set up shop on Elm Street, and began manufacturing carriages. This too proved a short-lived experiment. Advised by his physician that his daughter's poor health could only be repaired by moving back east, Huntington returned to Amesbury in 1859.[7]

Finding Hume prospering at his old location, he started over in space rented from a local machine shop. Huntington commenced with about two dozen workmen and in six months turned out some two hundred carriages. He continued to plow profits back into the enterprise and managed to survive the disastrous loss of southern vehicle sales caused by the Civil War. Recognizing the need for alternative markets, he began shipping carriages to Australia, achieving stability during the lean war years and setting the stage for aggressive postwar expansion. Huntington prospered enough to purchase a competitor's factory in 1869 and survive the fire that burned him out just a few months later. Despite drastic underinsurance, he mustered the funds to rebuild on a much larger scale, erecting a four-story factory in just sixty days in 1869. That factory made his final fortune, and he retired in 1875 a very wealthy man. He remained a well-known citizen and generous philanthropist, spending the next quarter-century looking down over the scene of his triumph from his mansion on a local prominence known as Carriage Hill.[8]

Though exaggerated over the years, Huntington's success story was widely known in the trade during his own lifetime. Even if he did not single-handedly invent large-quantity carriage production, his substantial factory, immense profits, and fine mansion represented the possibilities of industrialized carriage manufacture. How then did a relatively inexperienced young man rent a room, build a carriage, and amass a fortune? The much higher prices his contemporaries realized for their own work reflected simple economics. The sum they commanded for each vehicle was necessary to fund expensive and time-consuming craft work. Lacking a traditional craft background, Huntington embarked on his career unencumbered by preconceived notions of the proper way to build a horse-drawn vehicle. In a letter written to *The Carriage Monthly* a few years before his 1908 death, he recalled that his "system of duplication" allowed him to manufacture annually as many as a thousand vehicles for a significant profit. He alluded to economies of scale: large-quantity production of uniform goods at attractive prices, or, to put it another way, selling more for less to achieve increased sales and higher profits. The old entrepreneur failed to elaborate, but the details were no secret to his readers. Huntington had merely applied the principles of the factory system, and this, along with allied advances in machinery and ready-made parts, formed the technical core of his manufacturing success.[9]

The factory system's organizational core consisted of careful arrangement of work. The carriage trade had always divided labor into four basic branches: woodworking, metalworking, painting, and trimming; what changed now was the nature of work in each. Whereas the carriage shop featured apprentices and journeymen capable of performing all work in their respective branches, the carriage factory utilized workers specializing in a single or closely related set of tasks. What had once been a series of complex operations completed by a single tradesman became a set of distinct steps each performed by a single worker. If the craftsman flourished as master of many skills, the factory hand succeeded by doing one thing exclusively. Specialization begat speed, enhanced accuracy, and boosted output to levels once flatly declared impossible.

The old name for this was "systemization" or "systematization." In 1866 Ezra Stratton wrote a piece entitled "Systemized Carriage-Making" in which he noted the tremendous productive gains made by those American carriage factories adopting an elaborate division of labor. Historians refer to it as rationalization, defined by one as "the introduction of predictability and machine-like order that eliminates all questions of how the work is to be done, by whom, and when." The underlying notion derives from mechanistic Enlightenment philosophy, whose far-reaching principles had profound impact on many areas of human life. Rationalization entered the eighteenth-century world of industry via French military engineering and English manufacturing. Yielding advances as varied as improved draftsmanship, interchangeable parts, and a growing emphasis on routine and order, rationalization became central to the industrial revolution. A few late-eighteenth-century American inventors and mechanics like Oliver Evans employed the concept, but rationalization achieved widespread application only as American industrialization blossomed during the following century.[10]

New Haven's G. & D. Cook & Company emphasized its systematic division of work "in such a manner that each man had but a single part to perform." The firm attributed its daily production rate of ten complete vehicles largely to dividing work among twenty-four departments. In the factories of Cook, Huntington, and the others, one woodworker prepared the materials, another sawed out parts, and still others assembled components. In some cases the total number of hands remained the same, but it often increased in order to expand output. Regardless, the principle remained the same: assignment of individual workers to single tasks performed repeatedly at speed. Productive gains were possible with hand tools and methods, but the most dramatic advances linked subdivided tasks and machinery. Cook & Company's organization derived in part from its heavy commitment to mechanization and its potential to restrict the tradesman's range of activities even as it boosted production.[11]

Task specialization flowed from and contributed to the craft system's decay. The diminished skill quotient necessary to perform simplified tasks, combined with cash wages paid to even the rankest beginner, accelerated erosion of the hallowed craft structure that had long governed the trades. An apprentice weary of compulsory drudgery could avail himself of remunerative work at the local carriage factory; a journeyman unable to amass sufficient capital to start his own shop might instead ply his skills as a department foreman; the master craftsman disillusioned with the reality of small business ownership could find relief as superintendent of a carriage manufacturing plant. Factory owners before and after Jacob Huntington drew many of their skilled hands from traditional trade shops in the United States and abroad, even as they trained their own workers outside the craft system's confines. They also hired less-skilled workers for wages that many trained craftsmen disdained. Both skill levels fit into the carriage factory's division of labor, which by the 1860s was hailed by some as an ideal arrangement.[12]

With task specialization came greater uniformity in a vehicle's constituent parts. Huntington's use of the phrase "system of duplication" indicated his awareness of uniformity's productive advantages. Proprietors of the first carriage factories learned the benefits of a simplified product made in great quantity—quantities obtainable only by making the same item the same way every time. Carriage factories did not entirely eschew variety, nor did Huntington and his cohorts manage to produce vehicles with completely interchangeable parts.

In the wider context of American manufacturing, interchangeability began as a goal entirely divorced from economic and production advantages. Originally a military desideratum, federal armories found that interchangeable parts for small arms offered distinct tactical advantages by simplifying repair. True interchangeability actually proved very expensive for most nineteenth-century manufacturers, but in wagon and carriage factories greater uniformity helped extend the division of labor: uniform parts meant ease of assembly and increased production. Uniformity, like the craft system's decay, was both a cause and an effect of task specialization in the factory system. Advances in uniformity came partially from the accessory industry's ready-made parts. Making components for some unknown vehicle necessitated a degree of standardization, and vehicle manufacturers enhanced their own production speed through the ability to "bolt on" an outside-produced item of consistent dimensions. Eventual availability of nearly all vehicle components proved a boon to manufacturers eager to cut costs in a particular department or eliminate a production bottleneck created by complex component fabrication. Ready-made parts ensured greater uniformity, eliminated machinery and personnel necessary to make them in-house, and allowed workers to concentrate on assembly. Early factory owners like Jacob Huntington grasped these advantages and acted accordingly.[13]

Two factors appear to have influenced Huntington the most. One was his textile mill experience. It seems likely that his early impressions of the factory system remained with him long after leaving the Salisbury Mills for the carriage trade, where he applied some of the same principles to horse-drawn vehicle manufacture. The second factor was his at least passing acquaintance with those antebellum New England industries responsible for what came to be called the American System of Manufactures. Firearms manufacturers and clockmakers adopted a simplified and standardized product, uniform parts, and production broken down into discrete steps performed by specialized factory workers. This "system of duplication" affected far more than just guns and clocks, and if Huntington and his contemporaries failed to embrace the entire system, they nevertheless applied many of its principles. Commenting in 1860 on the methods of a large New Haven carriage firm near his home, pioneer clockmaker Chauncey Jerome said that "much of the work [was] being done by machinery, and systematized in much the same manner as clock-making." Clearly the concept permeated other trades.[14]

Jacob Huntington's lengthy, high-profile retirement and frequent retelling of his early triumphs naturally flavored local perception of his role as "father of the carriage industry," a notion furthered by trade journal retrospectives written by men too young to recall the events in question. As Jerome's remarks suggest, however, Huntington shared his pioneer status with numerous predecessors, contemporaries, and successors who remain little more than names in old city directories.

Historians will probably never know for certain who first applied factory principles to wagon and carriage building, but the extant evidence points to the industrializing Northeast, particularly New York, New Jersey, and southern New England. New Haven's James Brewster and Albany's James Goold both experimented with some of the organizational aspects of factory production in the 1830s and 1840s, though they did so largely without machinery. In the 1850s carriage makers such New Haven's George T. Newhall and G. & D. Cook & Company and Amesbury's S. R. Bailey continued the trend, but in conjunction with increasingly intensive mechanization. Before the end of the 1860s most eastern carriage-producing cities had at least one sizeable carriage factory, and places like Merrimac, Amesbury, and New Haven boasted several. Huntington and others spread the methods to the Midwest, where they contributed to that region's nationwide leadership in the trade. In Cincinnati Thomas Coghill and Charles Lohman were just two of many locals who extended Huntington's ideas; in 1872, Lowe Emerson and J. W. Fisher began turning out buggies on an unprecedented scale using refined factory principles introduced a generation earlier. Despite the rapid establishment of carriage factories during the period 1850–70, Cincinnati retained large numbers of small carriage firms. Ten years later, however, small shops employed less than 15 percent of the city's wagon and carriage workers; the overwhelming majority labored in factories employing over a hundred workers each.[15]

Mechanical Ingenuity

If any concept rivaled the factory system for its contributions to industrialized wagon and carriage production, it would have to be mechanization. Revival of the trade after the Civil War put "the machinery question" on the front pages of trade journals in the 1870s, but machinery had been making inroads for several decades. Proprietors of the larger firms in the 1840s and 1850s experimented with the limited machinery of the day, and by the 1860s more wagon and carriage factories were pursuing mechanization as a logical extension of rationalized factory production. The concept of mechanization required motive power, if only that provided by a muscular arm or leg. Indeed, simple hand-powered machinery played a vital role in the mechanization of many trades, a development still largely unexplored by historians. Hand-cranked, treadle, or velocipede-equipped machinery made valuable contributions to shops lacking central power sources, and in some cases they remained even after steam propelled more substantial equipment. Animal power proved even more capable; horses could be harnessed to a capstan-like device called a whim, sweep, or gin, geared to transform plodding animal motion into useable shaft power. Simple versions existed in antiquity, and more sophisticated examples drove American agricultural machinery during the nineteenth century. Manufacturers made use of live horse power even after the advent of affordable steam engines; Mt. Pleasant, Pennsylvania's Gruber Wagon Works powered its machinery by horse for nearly fifteen years before purchasing a steam engine in 1884. While not nearly so widespread, other animals sometimes served as well. During the 1880s *The Hub* mentioned a clever blacksmith with four dogs trained to take turns propelling an eight-foot wheel operating his bellows. This innovation saved him "a considerable sum yearly in wages," though the journal said nothing about the expense of dog food.[16]

The Industrial Revolution initially ran on the power of falling water, and America's abundant rivers and streams ensured water power's preeminence until the 1860s, long after Great Britain had embraced the steam engine. Simple undershot, overshot, and breast waterwheels had powered saw- and gristmills since Colonial days: in 1790 there were more than seven thousand water-powered mills in the United States. Fifty years later water-powered saw- and gristmills alone numbered more than fifty thousand. The early textile factories of the period 1790–1840 employed exceptionally large waterwheels to drive long rows of carding, spinning, and weaving machinery, a practice gradually emulated by many other manufacturers. Waterpower had distinct economic advantages, requiring both a smaller initial investment and lower operating costs than a boiler and engine. Disadvantages included complications caused by low water and flooding, as well as immobility and damage from winter ice, but its locational requirements were its greatest drawback. Manufacturers in arid regions or in places where the best water rights were already taken had little recourse other than to resort to what eventually made the waterwheel obsolete: steam.[17]

Originating in eighteenth-century England to power mine drainage pumps and imported to America for the same purpose, the steam engine progressed from a fuel-hungry, slow-moving giant to a compact and powerful source of industrial power during the first half of the nineteenth century. Incremental advances by English mechanical engineers, supplemented by more limited contributions of American inventors like Oliver Evans, resulted in smaller, more powerful, and more efficient engines with obvious appeal for industry. Steam power promised all-season reliability, flexible capacity, and geographical freedom. The last of these virtues had the greatest American impact as industry migrated from New England to the Midwest. Advances in quality, number, and power led to stationary steam engines' increasing industrial use during the 1830s and 1840s, but not until the 1860s did steam exceed waterpower in American manufacturing generally. It did so despite greater initial cost, more complicated maintenance, and the added expenses of fuel and attendance. Any manufacturer opting for steam had to be sure its productive advantages would support these greater outlays.[18]

Such economic calculations fueled debate in the carriage trade as steam power became popular during the late 1860s, finally edging out waterpower during the following decade. As expanding markets and increased regional competition pressed horse-drawn vehicle manufacturers, mechanization appealed to the larger firms desperate to maintain market shares. Steam seemed ideal for this application; beyond its locational flexibility, more efficient boilers and engines dampened cost objections even as they also reduced power's shrinking percentage of total manufacturing cost. Vast increases in outlay for machinery, buildings, and raw materials made a steam engine's purchase price seem less formidable. Continued design improvements and reduced prices led to their use in even smaller enterprises; late-century shops were able to take advantage of compact, integrated engine-and-boiler units that became widely available during the late 1880s and 1890s (see Table 3.1).[19]

Human, animal, water, and steam power all gave motion to the dramatically increasing amount of machinery found in American wagon and carriage factories after midcentury. The main forms of woodworking machines used in England and the United States originated in late eighteenth-century England. The earliest American machine builders were blacksmiths, millwrights, and carpenters, as well as inventor-entrepreneurs like Thomas Blanchard. While machinery continued to come from these sources and from users themselves, an increasing percentage came from dedicated machinery manufacturers. By the middle of the century numerous firms supplied other industries with wood- and metalworking machinery, many of them located in the old New En-

TABLE 3.1. Power Used in Wagon and Carriage Manufacture, 1870–1900

	Steam Engines	Water Wheels	Electric Motors	Gas/Gasoline Engines
1870	279	363	—	—
1880	282	68	—	—
1890	1,058	165[a]	—	—
1900	1,226	130	145	351

Sources: Adapted from U.S. Bureau of the Census, "General Statistics of Manufactures," *Census Reports: The Ninth Census of the United States, 1870* (Washington, D.C. 1872), 3:424; "Power Used in Manufactures," *Census Reports: The Tenth Census of the United States, 1880* (Washington, D.C., 1883), 2:12; "Power Used in Manufactures," *Census Reports: The Eleventh Census of the United States, 1890* (Washington, D.C., 1895), 6:760–61; James K. Dawes, "Carriages and Wagons," in U.S. Bureau of the Census, *Census Reports: The Twelfth Census of the United States, 1900* (Washington, D.C., 1902) 10:312
[a]Excludes turbines and "other water wheels."

gland industrial cities and the industrial Midwest. Ohio boasted some of the largest woodworking machinery makers, including J. A. Fay & Company, the Egan Company, and Lane & Bodley of Cincinnati, as well as the Defiance Machine Works, Salem's Silver Manufacturing Company, and many more. By the late 1860s these firms supplied much of the machinery used by wagon and carriage manufacturers (fig. 3.1).[20]

A horse-drawn vehicle's fundamentally wooden construction made woodworking the primary focus of mechanization, and carriage woodworkers often had machinery even when the rest of the establishment remained largely unmechanized. Wagon and carriage woodwork consumed huge quantities of lumber, and vehicle builders stockpiled it far in advance to permit open air seasoning for anywhere from one to three years, depending on its thickness. Though purchase of seasoned lumber eventually became an option, most firms found it more economical to dry and process their own. Ripping wide planks into narrower boards had been a job for the hand saw, but as powered saws became available, wood shops quickly took advantage of their tremendous utility. Shops unable to afford their own machine frequently had their heavy sawing sent out to another shop. Woodshop men used table saws to rip wide hardwood boards into narrow body framing stock, and design innovations such as tilting tables for bevel cuts, "T" squares for cross-cutting, and even dedicated crosscutting power saws all eventually came into production. Band saws turned out curved work for body frames and shafts, while accessory rollers made them useful for reducing thick planks into thinner boards in a process known as "re-sawing."[21]

Planers were almost as important. These machines smoothed rough planks into dimensional lumber using one of two different designs. Poughkeepsie, New York's William Woodworth patented his planer in 1828, and it was widely copied almost immediately. Machines of this type sufficed for general applications where consistent dimensions mattered less than smooth surfaces. The other principal design, the Daniels, excelled at producing accurately dimensioned stock even from warped and twisted planks. Carriage makers also had the option of choosing additional variants based on power and size: small "pony" planers capable of handling short or thin stock were very useful, as were large models capable of the roughest work. Like saws, planers saved enormous amounts of time and effort. Traditionally woodworkers spent long hours preparing stock, and power planers relieved shop hands from many tedious hours of "shoving the jack plane" down every single board. Shapers employed swiftly revolving cutters to form moldings and curved work in the manner of a modern router. Though more specialized and less common than draw knives and spoke shaves, shapers produced curved work that was impossible to create except by those traditional and much slower hand tools, and they excelled also in edge work for joinery (rabbets and dados) or decoration (beads and moldings) traditionally prepared with specialized hand planes.[22]

Mortising and tenoning were easily mechanized. Hand powered, lever-operated mortisers appeared in the United States in the 1830s. J. A. Fay, founder of one of the major nineteenth-century woodworking machine manufacturers, began making mortising and tenoning machines in Keene, New Hampshire, during this period. Designed originally for window and door fabrication, their ease of use soon won them converts in the other woodworking trades. Such simple wood-framed machines were inexpensive yet remarkably efficient, and they remained popular in small establishments even after large firms adopted faster and more powerful reciprocating mortisers. Sanders gained in popularity for both shaping and finishing applications. Each of the three main types (disk, drum, and belt) had its own advantages, yet all shared the same flexibility as the other general-purpose woodworking machinery. More sophisticated sanders constructed like planers came on the market by the 1880s and found their way into the carriage trade. Likewise, woodturning lathes increased in size, power, and sophistication, such that this old and respected device achieved new levels of use in wagon and carriage factories.[23]

While woodworking boasted the most intensive mechanization, blacksmithing also benefited from assorted machines. Some were ancillary to actual metalworking, such as mechanically supplied air to forges. The simplest improvement substituted a geared blower for the traditional wood-and-leather bellows, and large firms with steam power sometimes harnessed it to a centrally located blower supplying air to each forge via piping. Drilling presented another opportunity for mechanization. Smiths generally preferred hot punching for its expediency, but this technique could not meet all requirements. An early alternative consisted of the beam drill, which employed a brace and bit weighted by a heavy overhead timber, but this eventually gave way to the far more efficient post drill, which became standard shop equipment from the 1860s. Designed for vertical mounting on column or wall, post drills were geared vertical-drilling machines powered by hand and given momentum by a heavy horizontal or vertical flywheel. The more expensive models boasted multiple-speed gearing and self-feed devices. Incremental improvements during the latter part of the century resulted in rapid speed changes, quick-adjustable tables, automatic oilers, and a wide drive pulley to accept lineshaft belting. In conjunction with good drill bits, such devices could easily handle nearly any metal drilling job likely to be encountered in wagon and carriage work. Punch presses worked even more quickly, and both hand- and belt-driven models found their way into the larger establishments. Some blacksmithing departments also installed metal lathes, and one authority on the subject declared that "no better auxiliary ever entered the carriage smithing-room." His enthusiasm derived from the lathe's advantages for exacting work such as drilling out and fitting a kingbolt and socket. The pivot point of a vehicle's front axle assembly, this unit had to be fabricated to close tolerances, and the lathe eliminated hours of tedious forging and filing.[24]

Successive well-placed hammer blows shaped hot metal on the anvil, and blacksmiths spent more time swinging hammers than at any other task. Not surprisingly, then, mechanical hammers began to appear on the market to ease the toil, improve accuracy, and greatly speed the work. The earliest, usually foot powered, mimicked hand methods and consisted of a heavy hammer mounted on a mechanically raised helve. Though some shops utilized these designs for decades, improved power hammers crowded them out of the larger firms. These too had venerable pedigrees, with James Nasmyth's 1839 steam hammer being the most famous example. Derivations followed in great profusion, though in much smaller sizes. In ceasing to mimic hand methods, power hammers moved away from handles, levers, and springs to ram-powered designs that permitted enormous power without sacrificing control or accuracy. Besides reducing some of the bone-jarring labor, power hammers also operated more swiftly. Additionally, with rudimentary dies a power hammer became a drop forge capable of forming small metal parts at a single blow. Where a smith and helper might fashion a dozen axle clips an hour, a drop forge did the same in a few minutes.[25]

Grinding and sanding underwent similar transformations, as blacksmiths and machinists experimented with new ways to work cold metal. Heavy sandstone grinding wheels served throughout the shop for tool sharpening, but newer, high-speed emery wheels could also smooth up castings and grind down burrs left by drilling. They also became an increasingly common alternative to filing. Velocipede-driven grinders could attain a very high rate of revolution, and manufacturers improved the foot treadle by the addition of clutches, dual wheels, tool rests, and more. Horizontal emery grinders reliably flattened fifth wheel bearing surfaces and also excelled at cleaning sprue off ready-made castings. Where an apprentice might spend an entire day stripping rust and scale from six or eight tires with a wire brush or emery cloth, a horizontal emery grinder reduced the time to less than an hour, earning the apprentice's undying devotion.[26]

Painting and trimming offered fewer opportunities for mechanization. The single biggest mechanical innovation in the trimming shop was the sewing machine. Household models, initially the only ones available, sufficed for light work but could not accommodate the awkward shapes and heavy materials of dashes and folding tops. Sewing machine manufacturers exploited the potential market for specialized commercial machines, however, and trimmers soon had the benefit of heavy, crane-necked models designed especially for carriage work. The first dash stitcher came on the market about 1856, the product of New York's Nichols, Leavitt & Company. Operated by hand crank, it required two people in use, but the Singer Manufacturing Company followed in 1865 with an improved model needing only a single operator. Sewing machines powered from line shafting increased stitch uniformity and left the operator with, as one workman recalled, "legs enough to wend his way home" besides (fig. 3.2). Though a much later development, tufting machines saw use in wholesale factories able to take full advantage of a highly specialized machine capable of installing dozens of seat cushion buttons at once. Painting lent itself least to mechanization, though byproducts such as centrally heated drying rooms certainly helped. The availability of ready-made varnish as early as the antebellum period eliminated the giant cauldron and fire dreaded by paint shop apprentices. Finally, ready-mixed paint pioneered by Cleveland's Sherwin-Williams Company in 1880 rendered pigment grinding and paint mixing unnecessary.[27]

The buildings housing these evolving processes took on what became a familiar industrialized shape. Though the very term "factory," merely an abbreviation for "manufactory," originally described a function rather than a building, this distinction disappeared as the word became synonymous with the structure in the early nineteenth century. The American factory building took its earliest design cues from the late-eighteenth-century English textile mill, which was shaped by the twin requirements of centralized power and its transmission throughout the structure. The necessity of housing a waterwheel sufficiently powerful to propel the maze of lineshafting, pulleys, and leather belting leading to the machines dictated an architecture that became the model for nineteenth-century factory design. The textile mill usually consisted of a large, rectangular masonry building of two or more stories. While cheaper frame construction predominated in the early years, masonry dampened vibration and offered superior fire resistance. Multiple stories with minimally interrupted floor plans allowed long runs of overhead lineshafting, which transmitted power to rows of carding, spinning, and weaving machinery. Running power to successive upper stories via gears and shafting (common in England) or wide belting (standard American practice) permitted duplicate floor plans limited only by the available horsepower and the foundation's load-bearing capacity.[28]

Despite some very real disadvantages, mill architecture ably served America's early factories, and the model spread beyond textiles into other industries. When James Goold expanded his shop's capacity in the mid-1830s, he did so by constructing a new four-story brick building that mirrored many of the textile mill's features, if none of the machinery. Likewise G. & D. Cook's 1860s plant reflected what had already become classic American factory architecture. Not all carriage manufacturers opted for masonry construction. Frame buildings were cheaper, and carriage-making centers like Cincinnati and Amesbury relied primarily on frame construction for decades. Eventually, however, the very largest wagon and carriage factories achieved a size and profitability that made more substantial building materials cost-effective, especially for their greater fire resistance. Studebaker Brothers, Ohio Carriage Company, Banner Buggy, and dozens of others gradually replaced their frame factories with brick and stone.[29]

Wagon and carriage factories had a central power source, usually steam, located in the basement or in a separate engine house. A coal-fueled boiler, supplied with water from a well or city connection, generated the steam that drove the engine, which in turn transmitted its power via lineshafting to the floors above. The basic building contours were dictated by this power system, another reason factories like Cook's resembled textile mills. Interior arrangements varied, but a general pattern emerged, just as it had with shops. Blacksmithing frequently took place in the basement, often sharing space with the engine and boiler, while woodworking occupied the ground floor or the floor above. Lineshafting itself consumed a significant percentage of an engine's available horsepower, and a closely situated woodworking department reaped the benefit of extra power. Trimming and painting continued to occupy upper stories, and though a freight elevator might have replaced the old exterior ramp, outside drying platforms often remained at even the larger factories. G. & D. Cook & Company had a sizeable catwalk bridging two wings of the plant. The proximity of the smokestack and its promise of gritty downdrafts apparently failed to deter the owners, who were likely desperate for additional drying space.[30]

Most carriage factories had sizeable lumber storage facilities, the best of them covered sheds where huge quantities of lumber cured for a year or two. Other outbuildings might include a coal shed, iron warehouse, or even a shipping platform abutting a rail spur. Provided the steam engine did not occupy the main building, a separate powerhouse and exhaust stack might round out the additional structures, which often formed an "L," "U," "H," or hollow square. The office normally occupied a first-floor space with its own street entrance, though some of the very largest firms had an entirely separate office building. Furnished with roll-top desks, ledgers, file cabinets, and the other necessary equipment, this area sometimes included a drafting room with tables and large chalkboards. A space to receive salesmen, customers, and curious visitors formed part of the ideal arrangement, and the usual assortment of coat closets, washrooms, and storage areas rounded out the features. Many firms maintained an on-site repository even when having another in the city's commercial district, and these frequently occupied one or more stories of the main or office building. The ideal repository boasted neatly varnished floors and plenty of windows and skylights to accent gleaming iron and glossy varnish.[31]

Industrialized Carriage Making

As in the craft shop, the process of building a vehicle in a carriage factory began with a plan. Wagon and carriage shops had long relied on informal sketches, which skilled craftsmen used as a general model, filling in the details as they worked. Prior to the 1850s few American carriage builders relied on anything more sophisticated, and such methods persisted for years. As late as the 1890s *The Blacksmith and Wheelwright*'s editor insisted that practical instruction by experienced workmen was superior to "abstruse theory and complicated rules," but long before this time improved drafting had become intrinsic to the design process in many firms. As an 1890 manual on the subject put it, "the day of the 'cut and try' rule has passed, and the man who hopes to become a good workman must have some information regarding drafting." As we have already seen, serious change in began in the 1850s with the American debut of the French Rule, so called for its popularization by three Parisian carriage makers. Originally trained as a cooper, Brice Thomas found wagonmaking more to his liking, and after a period of military service punctuated with geometrical and mathematical study, he returned to the trade and Paris around 1830. By this time numerous carriage makers there were beginning to apply geometric principles to carriage drafting, and Thomas became keenly interested in their efforts. Albert Dupont came to Paris in 1848 and found work in a prestigious carriage shop; he spent evenings with Thomas studying body making, and their advanced knowledge of the subject soon attracted the attention of others, including a carriage maker named Henry Zablot. During the 1850s the three devised a system that completely transformed carriage draftsmanship.[32]

Informal instruction gave way to more organized efforts in 1858, when Thomas established a carriage-making technical school in Paris. He subsequently left this venture to found *Le Guide du Carrossier*, which became one of the leading French carriage trade journals. During the late 1860s he eagerly used his new medium to consolidate and publish all he and his colleagues had learned, and the serial articles became enormously influential. Dupont took Thomas's place as the school's headmaster, and the institution continued to teach draftsmanship to a series of ambitious young students. During the 1850s the American carriage trade press began carrying information on the French Rule when Cyrus Saladee introduced readers of *The Coach-Maker's Illustrated Monthly Magazine* to the subject in 1855. Recognizing the method's great potential and growing trade interest in the same, Ezra Stratton launched a similar series in *The New York Coach-Maker's Magazine* a few years later.

Publication of what some considered a valuable secret proved just one more sign of the vast changes taking place in the trade. In 1866 a carriage body maker wrote to Stratton thanking him for spreading knowledge of the French Rule, noting that as an apprentice he would have paid a considerable sum to learn such methods. As it was, he later ended up paying a Bridgeport carriage maker ten dollars to learn its details. That people were willing to sell and others were willing to pay for access to the French Rule says much about its worth. The 1871 sale of Stratton's journal to *The Hub* prevented the series' completion, but reader interest prompted editor George Houghton to take up the subject anew. Houghton himself attended Dupont's Paris school, and while *Hub* readers benefited thereby, others followed the editor's example and journeyed to France themselves. Herman Stahmer, long-time Brewster & Company draftsman, took a three-month course of study at the Dupont school at his employer's request, and his new knowledge greatly enriched that firm's work in the 1870s, 1880s, and 1890s. Though the total percentage of Paris-trained American carriage draftsmen remained small, they nevertheless disseminated improved drafting as they worked in a variety of leading U.S. carriage firms. By the late nineteenth century a simplified American version of the French Rule had become widespread, and some claimed the Americans surpassed even the French in carriage design.[33]

As Ezra Stratton fondly noted, the French Rule had nothing uniquely French about it, it being merely a system ("rule") of scale drawing, a means of making detailed and accurate drafts to a predetermined scale. By regulating and providing dimensions for the many compound curves and rendering the entire drawing to uniform proportions, carriage makers could obtain a vastly more detailed and exact representation of the final product. These drawings typically included detailed side, front, back, and bottom views aimed at accurate portrayal, which in turn greatly reduced guesswork in construction—precisely the impetus for their development in the first place. Eliminating the inaccuracies of the older side elevations permitted separate, simultaneous component fabrication with the assurance that assembly would proceed smoothly. In those firms able to take advantage of this potential, it represented enormous gains in production efficiency (fig. 3.3).[34]

Where the craft shop relied on the master craftsman's sometimes limited drawing ability, large carriage factories found themselves seeking a trained draftsman's detailed plans in order to coordinate an increasingly complex production process. Drawn steadily into the administrative side of their enterprises, factory owners gradually turned to tradesmen specifically trained in draftsmanship, in some cases even outside contractors. German-born Adolphus Muller, newly landed in 1860s New York City, began his distinguished career as a freelance carriage designer shortly after his arrival. Advertising in *The Coach-Makers' International Journal*, he soon found his services in gratifying demand. Whether flowing from the pens of skilled body makers or freelancers, these detailed drawings served as construction guides, for the whole point of the French Rule was to impose order on the production process. As Isaac Ware explained to his readers in the 1870s, "this manner of laying out work gives the builder an opportunity to provide the material for the wheels, to order the springs and axles, in fact everything pertaining to it, while the body is being built." In other words, more exact plans permitted simultaneous parts fabrication and, consequently, more efficient production. Such efficiency could help even the smallest shop, but multiplied many times over it became vital to a smoothly functioning wagon or carriage factory (fig. 3.4).[35]

Having settled the style, size, and related details of the vehicle on paper, the draftsman turned the job over to the workmen, who proceeded to render the design in wood, iron, paint, and trimmings. As in craft production, a horse-drawn vehicle began in the woodshop, but mechanization had transformed this work area. Where the craft shop's universal reliance on hand saws and hand planes made the bench the primary work site, mechanized carriage factories substituted rows of free-standing woodworking machines. Vise-equipped workbenches remained, but as stations for the finer shaping and finishing operations and component assembly. While carriage woodworkers generally continued to own their own tools, they gradually discarded the more elemental implements whose functions had been usurped by machinery. Woodwork began with stock preparation, which involved pulling suitable lumber from the firm's drying sheds and finish planing, edging, and cutting it to length. If table, band, and crosscut saws and power planers and edgers now dominated this process, many of the time-honored measuring and marking tools remained: the two- and four-fold boxwood rules, wooden straightedges and rulers, the steel framing squares and the rosewood-stocked try squares all transferred measurements from plans to freshly dimensioned lumber in preparation for the work ahead. Awl scratches, pencil marks, and marking gauge lines provided concrete guides for the ripping, crosscutting, and band sawing that brought the various parts to life. Patterns continued to serve as layout guides for the more complex shapes, and full-scale blackboard elevations remained useful in creating patterns.[36]

While the typical small shop carriage woodworker could efficiently use most, if not all, of the department's machinery, the larger factories often assigned a single operator to each machine, having the stock rather than the worker move through the department. Though machinery lent itself well to such practices, the degree of application varied widely. A small firm with only a table and band saw might simply make them available to all shop hands, while large rationalized factories tended towards dedicated operators. Many firms divided woodworking into gear and body making, with wheel making sometimes included as a third division. Gear makers fabricated axle beds, body blocks, reaches, sidebars, wooden brake components, and all other parts appropriate to the vehicle under construction. Body makers turned out the body framing (sills, cross members, pillars, belt rail) as well as body panels, seat risers, and any wooden rails or moldings, while wheel makers fashioned any hubs, spokes, and felloes or bent rims not purchased ready-made. In all three instances fabrication consisted of shaping and joint preparation. Most horse-drawn vehicle architecture relied heavily on the intrinsic strength of the mortise-and-tenon joint, but where the craft worker had once used mallet, chisel, and backsaw, the factory hand employed powerful, precise mortising and tenoning machines. Woodworkers also drilled fastener holes, rough sanded, and finished up head blocks, spring bars, and other carved parts prior to assembly.[37]

In the craft shop assembly kept pace with parts production, as woodworkers fabricated each successive item and fit it to the vehicle under construction. Rationalized carriage factories could fashion large numbers of parts to predetermined specifications, making stock accumulation feasible when desired. Some woodworkers were assigned to particular machines, others to assembly. There were no absolute rules, however, and carriage manufacturers experimented with all manner of arrangements. Whatever the case, assembly required hammers, screwdrivers, drills, glue pots, and brushes. Vehicle assembly consisted of seating body pillars into sills, fitting panels into place, gluing up axle beds, and all the other operations necessary to construction. Assembly proceeded quickly, its pace often checked only by the setting of glue and loosening of clamps before gear and body moved to the next step.[38]

Ironing fortified the vehicle's basic wooden structure. Blacksmiths received gear and body, usually separately, and began applying the necessary metal reinforcements. Smiths forged iron axles for those vehicles requiring them, fabricated and installed axle clips, stays, braces, and reinforcing plates using bolts, rivets, and screws. If the wheels were an in-house product, the blacksmiths or their helpers ran iron or steel strapping through gear-driven tire benders, forge-welding them into hoops before hot or cold shrinking them to the wheels. They followed a similar pattern in applying hub reinforcing bands, tire bolts, felloe plates, and related wheel hardware. Bodies required structural items such as corner braces as well as accessories including steps, rub irons, seat and foot rails, folding top hardware, and the brackets, loops, braces, and springs necessary to hang the body on the gear. Application of wheel, gear, and body irons required the presence of the finished wooden parts before blacksmiths joined all three components to form a complete carriage.[39]

Carriage blacksmiths working in large factories faced task subdivision. If the firm produced its own wheels, some smiths inevitably engaged in full-time wheel work while others specialized in gear ironing, body ironing, or both. Like any carriage factory department, the smithy offered ample work for skilled and unskilled hands alike, and blacksmiths continued to use helpers for a variety of tasks. Each man still had his forge, anvil, and standard assortment of hand tools, but as in woodworking he also had access to a growing corpus of machinery. Power hammers and small drop forges provided significant assistance in shaping hot metal. Vise-equipped benches attested to continued hand-filing and fitting, though crank-, pedal-, or power-driven emery wheels had diminished their use. Post drills and lever-operated punches sped the work along, while the various nuts, bolts, and related fasteners proceeded out of box, keg, or crate courtesy of the hardware supplier.[40]

The combined efforts of the woodworking and blacksmithing departments built the vehicle; it remained for the painting and trimming divisions to render it fit for sale and use. The advent of patented putties and wood filling had moderately streamlined surface preparation, but both dry sanding and wet rubbing were still necessary to prepare the wood to receive the first coat of paint. Just as in the past, many coats of paint and varnish were needed for a durable and appealing finish. Cheaper work had fewer of them, but even a minimal paint job called for primer, paint, and varnish and the necessary drying time between each. The majority of the space in Walborn & Riker's enormous painting department was consumed by bodies and parts in various stages of completion—as much as two months' worth of the firm's annual output. During drying they remained vulnerable to even the most commonplace mishaps. As one despairing trade journal asked, "which one of the sister branches in the coach shop is liable to be tormented by so small and insignificant a creature as a single house-fly?" Deeper colors, finer finishes, striping, monograms, and the other features of higher quality work required even more time and effort. If factory production of horse-drawn vehicles had a bottleneck, it was painting. Major throughput breakthroughs in the form of spray finish application and oven curing would not occur until long after the automobile had replaced the horse-drawn vehicle.[41]

Mechanization may have had little impact in the finish department, but the division of labor often did. Painters were less likely to follow a job from rough lead through final varnish coats and instead concentrated on a single or pair of closely related tasks. Less-skilled hands applied wood filling and rough lead while experienced craftsmen applied paint and varnish. Some factories divided their painters between wheel, gear, and body, the latter conceded to be the most critical. Painters applied color coats, shifted the work to drying rooms, lofts, or platforms, and brought in dry work for rubbing and brushing on of the next coat of paint. Meanwhile those with the best eyes and steadiest hands applied the many feet of striping and related ornamentation. Even when assisted by ready-made decalcomania from the accessory industry, stripers still relied primarily on striping pens and the kind of coordination that only years of experience bequeathed. Finally the varnishers applied the first of several varnish coats; work came to them from the painters, the drying rooms, or sometimes from the trim shop in the case of jobs having leather or other trim requiring a protective varnish coat. Then varnishers proceeded through the application, drying, and careful intermediate rub downs so necessary for the trademark high gloss.[42]

Trimmers generally began work once paint and varnish had cured, and like their brush-wielding coworkers performed their tasks largely without mechanical assistance. Patterns, knife, and scissors still dominated cutting, as trimmers measured and cut out tops, head liners, cushions, and the rest from sides of leather, bolts of cloth, and rolls of coach lace. Cutting took place on broad tables, while workbenches supported seat cushions being stuffed and stitched. Trimmers moved between tables and benches to trestle-mounted seats, bodies, as well as freestanding carriages. In many firms some of the cutting and stitching took place even while the vehicle was being painted; for example, trimmers could mount up the top bows on a specially prepared trestle and complete the folding top while the painters finished the body. Provided they had been painted in advance of the rest of the carriage, seats might receive their cushions and trim while body and gear dried in the loft. Accurate drafts and careful measurements helped, and some shop hands were sufficiently skilled to cut, punch, and otherwise prepare tops and side curtains without direct access to the vehicle. Still, many trimmers preferred working on a complete and painted vehicle or at least an unhung, painted body; having the vehicle on hand permitted minor adjustments, which could make the difference between a tight, neat job and a slack, sagging one.[43]

As with woodworkers and blacksmiths, painters and trimmers often pushed work back and forth. While the painters generally completed most of their tasks prior to sending the vehicle along, a few had to await addition of cloth or leather; dash rails and shaft tips had first to be covered and stitched before receiving a protective varnish coat. The exact sequence depended on how the firm structured work flow. Particularly when stitching items into place on the body, or working in awkward spaces in closed carriages, hand work remained central, even as heavy-duty sewing machines sped assembly of cushions, dashes, and other items capable of being lifted to the machines. Like painting, trimming was shaped less by innovation than by matters of scale; where a typical carriage shop had a single work table or bench, the carriage factory had a half-dozen. If the well-equipped shop boasted a sewing machine, the factory had eight or ten. More benches, more sewing machines, more trimmers, more curled hair, cloth, leather, and miles of thread signified the industrial age in the trimming department.[44]

The way in which a given wagon or carriage moved through each division says a great deal about the nature of the American horse-drawn vehicle factory. Regular flow of material and product gives the factory its characteristic flavor, and the process of rationalization mandated this kind of efficiency. However, knowledge of such considerations often misleads; some historians have erroneously equated carriage factories with Fordist mass production, an understandable conclusion given the often impressive production figures of some late-nineteenth-century carriage factories. Manufacturing statistics published by such early pioneers as G. & D. Cook leave the impression that only an assembly line could produce such quantities, particularly when phrased as a specific number of vehicles per minute, hour, or day. Yet assembly took place without an assembly line, whether static or moving, and finished vehicles seldom left the work rooms on precisely metered schedules. The nature of work in even a heavily mechanized and rationalized carriage factory retained some of the departmental interplay and irregular workflow characteristic of the old craft shop. "Nowhere in the factory was the absence of the assembly-line principle more evident than in the paint shop," Ben Riker recalled of his father's pony wagon firm. "It was almost a block long on Springfield Street . . . and in the course of their many paintings the various parts shuttled back and forth across it half a dozen times or more." While work in the woodworking and blacksmithing departments lacked the uneven work rhythms bequeathed by long drying sessions, it too embodied some of this same nonlinear quality.[45]

Even in the absence of in-line workflow, however, wagon and carriage factories did achieve throughput gains distinguishing them from their craft shop predecessors. Machine fabrication of parts in advance, purchase of ready-made parts, improved materials handling, and incremental efficiencies in the division of labor all bore abundant witness to advancing rationalization. The centralized control and planning potential inherent in drafting permitted radical work sequence alteration: parts fabrication could take place independently of assembly yet with full confidence of a perfect fit. This control held great potential even when not fully realized in practice. Repeatable accuracy in wood and ironwork was the result of not only sophisticated wood- and metalworking machinery and related application of jigs and fixtures but also rudimentary gauging and inspection routines. Highly rationalized carriage producers like the Ohio Carriage Manufacturing Company emphasized rigorous raw material and finished goods inspection as a means of monitoring quality and preventing production bottlenecks. If a factory-built carriage was nothing more than the sum of its parts, those parts had to meet consistently high standards in a smooth-running factory.[46]

Many of them did run smoothly, producing thousands of wagons and carriages every year. By virtue of a manufacturing ethos capable of quantity production of uniform goods, highly rationalized late-nineteenth-century wagon and carriage factories qualified as mass producers. While absence of assembly lines and related refinements restricted them in retrospect to the non-Fordist variety, the largest of these plants produced far too many vehicles to be anything but mass producers. By the mid-1890s firms like Cincinnati's Alliance Carriage Company averaged over ten thousand units annually, a figure met or exceeded by many other competitors. Even in the face of such impressive output, flexibility remained. Long reliant on the sales-generating whims of fashion, the industry remained wisely focused on customer tastes. Even the giant Studebaker advertised its ability to alter established products according to customer dictate, whether something as simple as paint color or as complex as a change in body dimensions. Many of these customer-derived options actually fell within the bounds of company-prescribed choices, as with the selection of a body, seat, or suspension style from a codified option list. Yet many factories engaged in actual custom work, if only occasionally, and for an additional price. However, most customers were perfectly satisfied with a standard product, which met their needs as it did those of thousands of fellow purchasers.[47]

"Where Do They All Go?"

Where a finished vehicle went from the factory depended on its intended market. Carriages slated for rail transport to a repository or mail order customer or by ship to a foreign port required disassembly and crating. Early rail shipment had once involved sheet-wrapped vehicles on flat cars, the well-remembered "ghost trains." Crating proved more secure and cost-effective, permitting many more vehicles per railcar and increased protection from theft, vandalism, and the elements. Crating entailed removal of wheels and poles or shafts, folding or removal of the top, and wrapping of axle nuts, top props, and related small parts. In many cases the crate was essentially built around the carriage, leaving the axle ends projecting to act as convenient lifting points. The specific dimensions of the completed container varied with the requirements of the particular railroad. The Carriage Builders National Association devoted a great deal of time to standardizing crate size and negotiating favorable rates. Some firms could package a vehicle in more than one way depending on its destination. The Ohio Carriage Company recommended its thirty-four-inch crate for long distances, with a taller version suitable for regional delivery. Many carriage factories had their own shipping departments, but some contracted with an outside entity. Ezra Stratton closed out his last few years as an active tradesman by crating Brewster & Company vehicles in his Elizabeth Street shop. Those slated for display in a firm's adjacent or cross-town repository required far less elaborate arrangements: employees simply rolled them there.[48]

This brings us to the question invariably asked by passersby observing the orderly rows of new wagons and carriages fresh from the factory, whether Cook & Company's in 1860 or Ohio Carriage's in 1910: "where do they all go?" The answer was, truthfully, "everywhere." Capitalism required markets, and American carriages met ready sales in and beyond its own borders. While the large late-nineteenth-century carriage factories cultivated sometimes extensive foreign markets, they were hardly the first to do so. Carriages from antebellum New England shops had once been found as far afield as Mexico, the Caribbean, and portions of Central and South America. The same features that attracted buyers in New Hampshire, Alabama, and the Oklahoma Territory resonated with customers in Saskatchewan, Cape Town, and New South Wales. America had no monopoly on rugged terrain, bad roads, rocks, dirt, and mud, and the tractive limits of the horse remained the same the world over. Lightweight, durable, factory-produced wagons and carriages appealed to people anywhere conditions matched their design ethic. The attractive pricing that encouraged American purchasers prompted foreign buyers to reach for their own wallets.[49]

Getting word to such far-flung potential customers led to new sales and marketing departures. Manufacturers continued to rely on the tried-and-true sales patterns of their craft predecessors, whether direct sales to individuals or intermediary transactions through repositories and retailers. A few important distinctions accompanied the industrialization of the trade as it applied to marketing and sales, however. One was more a general orientation than a specific technique and it became known as the wholesale vehicle trade: factory production of large numbers of inexpensive vehicles for quantity sales to retailers. In a sense the term was a misnomer in that the trade used it to refer to a production ethos as much as to a sales method. While a significant portion of factory-produced vehicles flowed through wholesale channels, carriage factories engaged in a variety of other tactics ranging from aggressive direct marketing through proprietary repositories and mail order, to licensed dealerships granted to implement vendors and small town merchants. In the final years of the industry, the trade press made much of Jacob Huntington's alleged role as father of the wholesale trade, but large-scale wholesale production and sales stemmed not from one man but from the productive triumph of industrialization. Huntington hardly monopolized this achievement.[50]

What formed the backbone of the wholesale vehicle market? It consisted of a variety of arrangements, from licensed dealerships to contracts with the big midwestern mail order houses. Carriage repositories had always been ready sources of vehicles "made for the trade"; the development of wholesale carriage making only accelerated this trend. Even proprietary repositories frequently purchased outside work to meet unexpected demands, particularly for some style or type not normally produced by the parent company. For example, the three largest late-nineteenth-century pony wagon manufacturers—Walborn & Riker (St. Paris, Ohio), the Colfax Manufacturing Company (South Bend, Indiana), and J. A. Lancaster & Company (Merrimac, Massachusetts)—all sold goods to proprietary repositories. Purchasers as diverse as Studebaker and Columbus Buggy found it expedient to buy such specialized work and remarket it under their own labels, a common practice from the very beginning of the American carriage trade. While many manufacturers produced at least some goods designed for resale, true wholesalers devoted nearly all their output to that end. Engaging in little or no direct sales themselves, these firms mass-produced wagons and buggies for retail sales all over the United States and abroad. Busy city repositories sought out wholesalers capable of producing the fancier carriages attractive to wealthy urban clients, while small town dealers stocked the wagons and buggies more in keeping with their rural markets. Other wholesale customers included hardware, feed, and general merchandise stores and agricultural implement dealers, all of whom frequently offered a selection of horse-drawn vehicles alongside their regular lines of goods.[51]

A large percentage of the wholesale market reached the consumer via mail order. General merchandise retail giants such as Sears, Roebuck & Company and Montgomery Ward had transformed Americans' buying habits by the 1880s. The mail order revolution began in the last quarter of the nineteenth century as catalog firms courted the vast rural market for manufactured goods ranging from shirt buttons, overalls, and pocket knives to furniture, parlor organs, and reapers. Impressively weighty catalogs brimming with detailed illustrations and carnival-barker prose seduced an entire generation with promises of quality goods at prices that usually undersold local stores. Wisely sensing the growing interest in attractively priced buggies and other light carriages, not to mention the perennial call for carts and farm wagons, mail order houses canvassed the carriage industry for factories able to produce large lots of low- to medium-priced vehicles. They found many takers. The 1897 Sears "wish book" informed customers that the firm bought its vehicles from "representative factories" in Columbus and Cincinnati in Ohio; Jackson, Pontiac, and Charlotte in Michigan; Racine, Wisconsin; Indianapolis, Indiana; and Abingdon, Illinois. Montgomery Ward's fall/winter 1894–95 catalog was less specific, but it did mention vehicles made in Indiana and Illinois and several buggies and carts built by Studebaker, and the retailer eventually carried a selection of Walborn & Riker pony vehicles. The latter, like the majority of vehicles sold this way, bore no name save the seller's. Not that the manufacturers really cared; low bidders able to meet large orders and smaller firms supplying specialized work benefited enormously from trade with the mail order giants.[52]

The mail order retail boom contributed to another sales and marketing innovation in the American carriage industry, the direct sales revival. Craft-produced wagons and carriages had most often passed to the consumer via face-to-face transaction. Whether generated by local advertisement or word of mouth, direct sales formed the backbone of a typical shop's revenues. However, direct contact between producer and consumer could never generate demand sufficient to absorb the output of a factory turning out several thousand vehicles annually. Direct appeal to potential customers nationwide through the U.S. Postal Service, however, promised results of an entirely different magnitude. Taking a page from the big retailers, many large carriage manufacturers developed their own attractive catalogs designed to seduce the wariest farmer or jaded urban consumer into purchasing a new buggy or carriage through the mail. A few of the more prosperous urban carriage makers of the 1850s and 1860s had turned out limited numbers of illustrated product catalogs; while these were likely neither the very first of their kind nor geared principally to mail order, they heralded a trend. Advances in illustration, printing, and mass mailing encouraged others to follow suit, and the genre blossomed from the 1870s. Lavishly illustrated vehicle catalogs thudded into post office boxes and mail slots in the 1880s and 1890s as more large vehicle producers joined the mail order revolution.[53]

It soon became apparent that general merchandise mail order houses held no monopoly on extravagant sales pitches. Cincinnati's Alliance Carriage Company sent out their "Successful Salesman," a free, vest-pocket-sized booklet extolling the company and its products. First published in 1888, the seventh edition went on its "Annual Triumphal Tour" in 1895. In eighty pages of illustrated copy it included a virtual factory tour ("a lucid representation of a number of the principal departments"), shameless self-promotion ("The Alliance Carriage Co. saves Consumers half a million yearly"), maintenance tips ("keep your tires tight"), and pointed reference to the advantages of direct purchase ("save all profits and get better goods"). Engraved illustrations showed bustling office scenes and a stolid board meeting, a capacious stock room, and frenetic clusters of blacksmiths, painters, trimmers, packagers, inspectors, and salesmen. Readers learned about the firm's Famous Combination Spiral Spring, discovered the secrets of superior axles, and were provided a copy of the guarantee pasted on the seat of every buggy. Should the reader still harbor doubts, the booklet closed with twenty-four closely typed columns of customer names and addresses, followed by three pages of the florid testimonials so characteristic of Victorian advertising. The sole purpose of this barrage was to persuade readers to send for Alliance's illustrated catalog ("200 pages, and free as water") from which they could order with complete confidence and satisfaction guaranteed—at least for two years from the date of purchase.[54]

Such naked pleas paled in comparison with the logical rigor and appeals to reason marshaled by a turn-of-the-century competitor. If general merchandise mail order houses had their Richard Warren Sears, the carriage trade had its H. C. Phelps, and both prospered accordingly. Phelps was president of the Ohio Carriage Manufacturing Company, a Columbus maker of buggies and other light carriages. The firm's sprawling South High Street factory produced thousands of its "Split Hickory" brand vehicles annually, nearly all of which found owners via the U.S. mail. "There is no one between you and me but Uncle Sam," as Phelps preferred to put it, "and his postage stamps are the cheapest thing in the world." Like his direct-marketing competitors he excoriated the middleman, whether general mail order firm or traveling salesman. "This catalogue and our advertisements . . . ," he boasted in 1910, "are the only salesmen that we employ."

Phelps himself, who appeared throughout company literature beaming intently at the reader through steel-rimmed spectacles, was obviously the firm's primary marketing asset. Nothing short of a master of the medium, his catalog prose took the reader by the arm rather than the throat: "Isn't from $26.50 to $40.00 of your own money in your own pocket worth more to you than in some retailer's pocket?" Should one harbor a secret desire for a more expensive vehicle from a dealer, the canny pitchman retaliated with another of his devastating queries: "Would the buggy be any better? Would it wear any longer? Would it look any nicer because you paid the retailer $30.25 more than you would pay us?" Phelps and the reader both knew the answer. As if to assuage any remaining doubts, he reminded skeptics of his twenty years in the business, his firm's peerless reputation, and that "back of every vehicle and harness is the big factory you see on the opposite page." Thousands of customers let Phelps persuade them to become "Split Hickory" owners, and the 1913 catalog implied that 167,000 purchasers since 1901 could hardly be wrong.[55]

Even as the giant wholesale carriage factories embraced mail order, a significant portion of the industry continued to sell vehicles through other channels. Many continued to rely on the tried-and-true third-party repositories that had existed since the early nineteenth century. Wagon and carriage makers continued to sell a modest volume of work through those handling horse-drawn vehicles as a sideline, whether agricultural implement dealers, hardware merchants, livery stable owners, or horse traders. However, another form of repository sold far more vehicles than any number of part-time agents or small town entrepreneurs. Whether located in the same building, across town, or in another city, proprietary repositories formed an important part of many firms' marketing efforts. Many medium and large vehicle manufacturers boasted proprietary repositories in major American and foreign cities. In 1883 the Studebaker Brothers Manufacturing Company had ones in Chicago; St. Joseph and Kansas City, Missouri; Salt Lake City; and San Francisco, and agencies—licensed individuals authorized to take orders for Studebaker products—"throughout the United States and Territories." Basically a form of networked dealerships, such venues perpetuated retail sales by the manufacturers themselves at the same time that the factory continued to ship hundreds of carloads to wholesale purchasers.[56]

The success of the proprietary repository depended greatly on brand recognition, which was itself another outgrowth of the industrialized trade. In a market crowded with similar products from a host of sources, advertising and the creation of name brands became important strategies. Competing firms shamelessly copied each another's claims and distinctives, just as they had with styles and designs; prior to federal trademark laws, such practices generated heated disputes in and out of court. Much of the bitter rivalry between James Brewster's sons revolved around a dispute over the rightful manufacturer of the "celebrated Brewster wagon" and whether such a trade name could even be considered proprietary. Reputation had always sold wagons and carriages, but word-of-mouth encomiums paled beside the sales generated by a successful brand-name advertising campaign. In the last few decades of the nineteenth century, trade names like "Split Hickory," "Columbus," and "Solid Comfort" vehicles, the "Cozy Cab" ("much more handsome than the old style buggy"), "Mikado Road Wagon," and dozens more all vied for consumer patronage. Trademarks, visual devices unique to the product or company, also worked well. Some firms used simple monograms like the Hackney Brothers' "H.B." superimposed over a carriage. Others combined name and symbol, like the banner and wagon wheel used by both Thornhill Wagon and Studebaker. Still others opted for symbolism, whether obvious like Anchor Buggy's predictable choice, or not, as with Fouts & Hunter's life buoy ("Life Savers for Land Travelers"). Like so many other late-nineteenth-century manufacturers, the American carriage trade found great utility in the devices of modern advertising.[57]

The whims of fashion sometimes created their own sales windfalls.

Carriage makers had long known that shifting consumer tastes frequently worked to their advantage. Even desultory attempts to keep pace with the latest Paris or New York carriage styles often boosted sales figures, yet this also had its dangers. Carriage makers always ran the risk of overestimating a trend or failing to anticipate a reversal. Purchase of too many components to build a popular style, combined with a sudden change in consumer taste, frequently saddled builders with parts or even entire vehicles that refused to sell. One Cincinnati carriage maker recalled the misfortune of purchasing large quantities of Brewster springs just before such suspension fell from favor. "We stacked them in the basement where they stayed for years, dead stock. When floods came we didn't bother to move them, just hosed them off after the floods went down, and there they stayed, and rusted." Characterizing dead stock as "the bane of the carriage business," he went on to relate a successful scheme to combine several lots of these long-disused parts into a run of popular vehicles. While much dead stock never came back to life, canniness sometimes paid off handsomely for those able to stimulate consumer demand for an item otherwise slated for the scrapyard or town dump.[58]

With the increasing rationalization of late-nineteenth-century factory production, many carriage manufacturers moved from accommodating fashion to a more aggressive approach. By the turn of the century, some firms were incorporating regular, planned change into their production schemes by promoting the concept of annual models. Taking the traditional seasonal manufacture and sales cycle and combining it with special emphasis on the incremental improvements found in any industrial product, carriage manufacturers tapped into the rich vein of consumer fashion anxiety. According to one of H. C. Phelps's pitches, purchasers of Split Hickory vehicles could be assured of obtaining "strictly 1910, up-to-date goods," not the "shopworn, out of style, and out of date" vehicles carried by many retail dealers. Bargain hunters might seek out "new old stock" as a matter of course, but keeping pace with style required a different strategy. "New" sold, and "newest" sold even more, maxims not lost on Phelps and his contemporaries. For all the acclaim historians have heaped on General Motors for its 1930s "discovery" of the annual model change, the fact remains that Alfred P. Sloan Jr. traveled a path blazed by the American carriage industry decades before.[59]

A natural question to ask, then as now, was to what degree industrialization affected the average quality of American horse-drawn vehicles? During the 1870s the term "Cincinnati buggy" described the products typical of large factories in the Queen City and elsewhere. A period newspaper carried the following exchange between a Cincinnati "cheap-buggy manufacturer" and a dissatisfied customer:

> "That was a nice buggy you sold me the other day. It broke down the first time I rode in it."
>
> "Rode in it!," exclaimed the buggy maker. "Why, great heavens, man! You don't mean to say you rode in that buggy?"
>
> "Of course I did!"
>
> "Well, no wonder it broke down. . . . It's a most excellent selling buggy, but it isn't a riding buggy at all."

Wholesale work was indeed built to sell, and quickly, the better to make room for more. Low prices frequently took precedence over quality, a truism well known to detractors of early Cincinnati factory work. Many years later one commentator recalled that Cincinnati carriage maker Charles Lohman's aim consisted chiefly of making "something that would look like a buggy and sell." He cited as proof Lohman's axle fabrication method: "he would take a bar of 1-inch square iron, forge the ends to represent spindles, cut threads on [the] ends of the spindles and form cheap nuts for them." Such expediency hardly credited its creator, but it did achieve a commercial end by allowing Lohman to undersell his more traditional and quality-conscious rivals. Whatever his detractors said, the public bought all he and his factory-producing brethren could make. This was not without precedent, for antebellum factory pioneers like Jacob Huntington and the Cooks had done much the same thing. Some averred that much of the pre–Civil War "cheap work" had been even more inferior because, as one Newark tradesman recalled in 1866, "most of it went South, and was never heard from afterwards. Now it is home trade, and we have to be more particular."[60]

Some of these quality issues stemmed from the initially considerable gap between public demand for low-cost carriages and the limited ability of the trade to produce them without reducing quality to unacceptable levels. Continued personal income growth, an expanding capitalist economy, and industrialization spreading rapidly westward all contributed to post–Civil War demand for low-priced vehicles. The allure of breaking into such untapped markets tempted carriage makers to oblige with goods often incapable of meeting even minimal owner expectations. A coat or two of just-dry paint and varnish, cheap cloth, and split cowhide frequently masked improperly seasoned timber, flimsy hardware, split panels, and other signs of drastic economy and hasty workmanship. Yet they still sold, so much so that early wholesale centers like Cincinnati became permeated with an almost gold-rush atmosphere during the 1860s and 1870s.

The market seemed limitless, and the potential for quick profits persuaded unprecedented numbers of entrepreneurs to become buggy manufacturers. Many of these industrial pioneers had been brought up in the trade and possessed the accompanying training, but others had only superficial craft knowledge. Instead they abounded in business sense, at least modest capital, and willingness to substitute experimentation, innovation, and expediency for more time-honored tools and techniques. Frequently these men remained in the business for a relatively limited time, as intense competition, price wars, and backlash from poorly constructed work forced some into bankruptcy and others into buyouts. In the late 1860s and early 1870s, cities like Cincinnati and Amesbury boasted dozens of firms whose life spans amounted to less than five years—sometimes much less. The wise proprietors rode the wave and got out: Jacob Huntington sensed the limits of the trend and retired in 1875 a wealthy man. Yet he too had felt the heat, losing two factories to fire and a lot of money besides. It still comes as a surprise that he accrued his final fortune just six years before his exit from the trade.[61]

Factory owners were not the only ones taking advantage of the situation. Unemployed craftsmen, discontented journeymen, and half-trained apprentices flocked to the factories where they found work suited to their sometimes truncated abilities. Skilled craftsmen became department foremen or works superintendents, while others found jobs in the various departments. Some of the more ambitious also plied their trade on the side: "After Emerson, Fisher & Co. started," one account recalled, "some of their body makers, after working all day, would go around to Auel's factory and make a little beer and cigar money by turning out piano-box bodies for 'shyster buggies.'" Such were the temptations of capitalism for the industrial worker. Many of the earliest carriage factories seeking the low-cost market indeed had little alternative to cutting corners, and their "shyster buggies" injured the reputation of early factory work. Time and manufacturing improvements showed that price and quality did not necessarily have to conflict, however. Rapid mechanization from the 1850s and 1860s and the increasing availability of ready-made parts enabled manufacturers to make lower-priced vehicles at acceptable quality levels. Shoddy work still existed, as firms of all sizes continued to produce inferior products that sold. The issue of product quality ultimately had less to do with the presence or absence of machinery than it did with the quality of materials, methods, and labor. Factory production no more automatically equated poor quality work than shop production guaranteed the highest standards of craftsmanship. Machinery might enable a carriage maker to turn out larger quantities of flimsy, defective vehicles, but the reason behind inferior workmanship had little to do with the choice between jack plane or Woodworth planer.[62]

Perception of a factory-built vehicle's quality depended greatly on customer expectations, and purchasers frequently held new vehicles to unrealistic standards. Many common forms of owner abuse and neglect, ranging from failure to reset loose tires and inadequate axle lubrication, to hard driving and overloading, became unjust cause for complaint about product quality. Matters of vehicle type and appropriate use might also play a part. As early as the 1860s Ezra Stratton noted farmers' tendency to complain of newer, lighter buggies' lack of carrying capacity. Disdaining such "quill wheeled" vehicles in comparison with the heavier work of the 1830s and 1840s, they failed to grasp the light buggy's specialized nature. The proliferation of purpose-built vehicles was an important outcome of industrialized carriage production; customers subjecting them to use outside their design parameters wrongly attributed lack of performance to poor-quality construction. There were exceptions even here, for some relatively inexpensive carriages gave excellent service even under trying conditions. One New Jersey man wrote to the manufacturer to boast of his mass-produced buggy's rugged endurance, noting that in ten years' hard use he sometimes conveyed as many as five adults at a time—"not small ones, either." Few buggies were designed to carry this kind of load, and his endorsement could easily have turned to condemnation had the vehicle collapsed. Still, the anecdote attests to the sometimes remarkable endurance of well-made factory work.[63]

In time both carriage manufacturers and sellers recognized the wisdom of providing a more explicit reference to product quality in the form of grading. Some of the larger carriage factories adopted this practice after the 1870s, and it eventually spread throughout the trade. The most important factors in a firm's grading were the quality of the raw materials and ready-made parts, the manner of construction, the degree of finish, and the amount and type of accessories. The grading system used by Walborn & Riker in the early 1880s typified the genre, Grade 1 being its own product built to be "first class in every particular, and warranted equal in style, finish and durability to any work made in this country." Grade 2 consisted of outside work made to its specifications and finished in its own factory according to customer directives, and

Grade 3 was outside work the firm finished itself and "sold as cheap as this class of work can be bought anywhere." Sears, Roebuck & Company devoted a special prefatory segment to explaining its AA, A, B, and Special B grades of work: AA was the best available work made from the finest materials; A was "strictly high grade" and allegedly better than that carried by many others; B consisted of work matching the standard grade carried by most wholesale houses, repositories, and retail dealers. Special B was "made more for show than for service," and, despite its inferior materials and painting, appeared in the catalog "for just what it is to meet a certain class of competition." All but the Special B grade work, which had no warranty of any kind, carried a two-year guarantee.[64]

The matter of guarantees by this time had been subject to intense scrutiny by the carriage industry itself. The CBNA tackled the thorny question at its inaugural convention in 1872, and the following year it adopted a standard guarantee for voluntary use by the trade. Printed on a slip intended to accompany each vehicle sold, it stipulated that the carriage in question would be guaranteed for one year from the date of purchase, provided the vehicle did not see commercial use. The seller agreed to make good on material and workmanship defects while the purchaser agreed that paint damage from manure ammonia and excessive exposure to the elements, as well as tire wear, fell outside the agreement. The association officially adopted the guarantee in November of 1873 and urged carriage makers to do likewise. Adoption was purely voluntary, and some firms elected to use their own written or verbal versions, or eschewed guarantees entirely, while others welcomed it as protection from unjust claims. The document remained controversial, and consumers, dealers, and manufacturers all periodically expressed dissatisfaction. The CBNA made various changes to the official guarantee, itself the subject of animated debate at the conventions of 1886, 1905, and 1906. By the latter date many carriage makers felt the amended agreement exposed them to unjustified dealer and customer claims. After 1907 the CBNA recommended discarding the official warranty, though the document continued in sporadic use for the remainder of the carriage era.[65]

Grading and guarantees were designed to protect the interest of manufacturer, dealer, and consumer alike by making an explicit connection between price and quality. What were those prices that so attracted consumers? Despite the great deal of attention given to the low-priced factory produced buggy of the 1850s and 1860s, "low cost" was relative. Wholesale pioneers like Jacob Huntington managed to sell stripped-down buggies for under $50 in the late 1850s, about half the going price of an average-grade no-top buggy, which then sold for $110 to $150. While such factory work was indeed a considerable savings over costlier shop-built products, it still seemed dear to most consumers just a few decades later. Industrialization contributed to a dramatic decrease in the price of a typical light carriage over the last third of the century. During the 1870s a savvy shopper could purchase a new folding top buggy for around $90, though better materials, workmanship, and options still normally took the price up above $100. During the 1880s retailers carrying the products of the Columbus Buggy Company sold no-top models for $100, folding tops adding $50 to the bill. These prices brought the purchaser a typical, middle-of-the-road buggy: more expensive than flimsy "shyster buggies" but a considerable savings over a shop-made custom carriage.[66]

During the same decade some Cincinnati buggy manufacturers selling no-top buggies had managed to bring the price down as low as $34.50, cash, direct from the manufacturer. Many buggy factories sold low-cost work "in the white" to carriage makers who finished and sold them at modest profit. Such "three for a hundred" jobs sold well even if the quality paled in comparison to more expensive vehicles. Even better work declined in price during the aggressive marketing climate of the late nineteenth century. The 1894–95 Montgomery Ward & Company catalog included a page of top buggies ranging from $84.50 for the Special-grade Heavy Concord Piano Box Buggy down to $44.50 for a Standard-grade Elliptic Spring Top Buggy, but customers looking for a rock-bottom price on a no-frills ride could purchase a "common grade," no-top road wagon for as little as $28. Never slow to retort, the 1912 Sears catalog hawked "THE WONDER OF THE BUGGY WORLD," a leather quarter-top buggy for $34.95, not including the cost of shipping. This cost presumably was not objectionable given the firm's promise that "the freight will amount to next to nothing as compared with what you will save in price." Like any good mail order house, however, they could go lower. Another page announced that "$22.35 IS OUR LOWEST PRICE for a reliable road wagon," and that this "barely covers the cost to build with but our small percentage of profit added." The catalog informed readers that a $16 or $17 buggy might be possible by using "cull wheels, cheap materials, and cheap finish," but the product would be entirely too unreliable. The same catalog did, however, offer a Portland cutter for $16.95, shafts included. One wonders what Charles Kimball would have thought.[67]

While it would be correct to state that the overall trend in vehicle prices was downward, a number of variables continually impacted the equation between price and quality, making it anything but a strictly linear decline. Some cheap vehicles were just that and fell apart accordingly, while others provided good value for the money. Some of the most elite custom carriage makers like Brewster & Company actually received more money for their work as time went on. In the realm of medium- and low-priced carriages it became easier for the consumer to find an acceptable quality vehicle at an acceptable price, where once the choice had been between an expensive shop-made carriage or a pair of new shoes. If anything, industrialization increased the number of choices throughout the entire process of making, selling, and purchasing a horse-drawn vehicle. All parties concerned had to cope with a wide range of variables in attempting to balance quality and cost.

This brings us back in the end to the very essence of the transformative process we know as industrialization. If one outcome of industrialized wagon and carriage manufacture consisted of this vastly more complicated commercial climate, it was because the very process of industrialization was just that, and one fraught with a good deal of irony, ambiguity, and unanticipated consequences. Industrialization and the attendant concepts of the factory system and mechanization were less destinations than they were stations along the way, processes that continued to change even as they dominated American manufacturing and business. One of the many ironies of this transformation was the preservation of older ways of working. Yet the small shop's survival owed as much to another development only touched on in the above account. The advent of ready-made parts had a profound impact not only on shops but also on factories and thus the overall contours of the entire wagon and carriage industry. This entitles it to detailed consideration in a chapter of its own.

[4]

The Coming of Parts

Specialty Manufacturing in the Wagon and Carriage Industry

Finished materials for wheels and wagons can be found in almost every town, at such prices that it would be foolish to make hubs, spokes, and felloes at the shop.

Practical Carriage Building, 1892

There are no extensive wholesale manufacturers of carriages situated at Cleveland, but there are many that do a good business in a fine grade of work for the retail trade. The city, however, can boast of some of the largest carriage hardware concerns in the United States.

The Hub, October 1895

IN 1887 a sales representative for Cleveland's Eberhard Manufacturing Company took an order for over a million castings from a Cincinnati carriage manufacturer. Though implying that the eight-year-old firm had handled larger orders, he or someone else thought it worth broadcasting to the trade. The transaction is significant for what it tells us about the wagon and carriage industry during the last quarter of the nineteenth century. Besides indicating that somebody in Cincinnati was making enormous numbers of buggies, the order demonstrates that supplying wagon and carriage parts had become a sizeable and lucrative business of its own (Table 4.1). Firms like Eberhard were in many ways the industry's crowning achievement, representing unparalleled size, productivity, and rapid growth. More significantly, however, what had come to be called the "accessory trade" irrevocably changed the way horse-drawn vehicles were made. Wagon and carriage manufacturers no longer had to fabricate every part of a given vehicle. Enormous plants presented the vehicle builder with an astonishing variety of ready-made parts, from bow sockets to carriage bodies. They were the key to considerable productive gains for the carriage factory and spelled the salvation of the small shop struggling to compete with much larger firms. The advent of "parts" did as much as the factory system, mechanization, or any other single factor in determining the nature of industrialization in American wagon and carriage manufacturing.[1]

In the eighteenth and early nineteenth centuries, necessity compelled wagon and carriage makers to fabricate nearly every part of a vehicle in their own shops. In time, however, some of the more complex items became available from outside sources. Wheel making, while not beyond the ken of the skilled vehicle builder, required exacting talent and a considerable amount of time. Elliptic springs called for high-quality steel and metallurgical knowledge beyond that of some carriage shop blacksmiths. Good shops with skilled workmen could certainly make these parts but only at a limited rate. Larger establishments, especially city firms tapped into a more concentrated demand for horse-drawn vehicles, often sought ways to expedite production, providing yet another impetus for specialty manufacturing. It was not at all uncommon for an entire company to be organized around a single product: the Royer Wheel Company, Union Spoke Works, American Patent Leather Company, Fitch Gear Company, Tomlinson Spring Company, Mount Carmel Axle Works, Baltimore Buggy Top Company, McKinnon Dash Company, the Rubber Cushioned Axle Company, Millersburg Fifth Wheel Company, the Bow Socket Manufacturing Company, the Cleveland Top Company, Standard Varnish Works, the Bridgeport Patent Leather Manufacturing Company—to name just a few. Often, however, these enterprises expanded their lines to include other components, some related by function, others by raw material or production process. Well before the end of the nineteenth century, the accessory trade had become an intrinsic part of the carriage trade.

Some items used in wagon and carriage construction had nearly always come from outside the shop. One old carriage blacksmith recalled a common reaction when Philadelphian Thomas Skelly began turning out machine-made bolts in the 1850s: "Well! well! get a 2 in. × ¼ in. bolt for five cents, the iron cost me half that amount." Even early vehicle builders often purchased nails, screws, tacks, and similar fasteners from outside sources on the basis of similar logic. Cloth, leather, thread, and the other trimming materials obviously came into the vehicle shop from elsewhere, as did oils and pigments for varnish and paint. New York City had a paint and varnish works at least as early as 1754, and Philadelphia followed before the end of the century. With nails, cloth, paint, and similar items coming from outside manufacturers, even the most traditional craft shop was not so self-sufficient as one might think.[2]

Other vehicle accessories were available by the late eighteenth cen-

TABLE 4.1. Comparison of Wagon and Carriage and Accessory Industries, 1900

	Wagons and Carriages	Wagon and Carriage Materials
Number of firms	7,632	588
Capital invested ($)	118,187,838	19,085,775
Number employed	62,540	15,387
Total wages ($)	29,814,911	5,987,267
Cost of materials ($)	56,676,073	13,048,608
Value of product ($)	121,537,276	25,027,173

Source: Adapted from James K. Dawes, "Carriages and Wagons," in U.S. Bureau of the Census, *Census Reports: The Twelfth Census of the United States, 1900* (Washington, D.C., 1902) 10:306.

tury. One such example is coach lace, a narrow, intricately woven silk or wool fabric used to trim carriage interiors. Its American production began in late-eighteenth-century Philadelphia, a city soon famous for textiles of all kinds. German immigrant Frederick Hoeckly handwove fringe and lace there in the 1790s, and in 1815 his eventual son-in-law, William H. Horstmann, established the firm that would one day be acknowledged as the nation's oldest coach lace manufacturer. Horstmann learned the trade in his native Germany and, on coming to Philadelphia, combined old world craftsmanship with American ingenuity. He imported German braiding and plaiting machinery in 1824, a silk-weaving Jacquard loom the following year, and by the early 1830s equipped his looms with power. Philadelphia's coach lace business developed alongside that of Newark, New Jersey, and by the 1830s both Massachusetts and Connecticut firms had joined them. A Lancaster, Massachusetts, company developed formidable power looms that made it the top manufacturer during the 1850s, to be bought out in 1857 by Horstmann's sons.[3]

Simple iron vehicle hardware had been on the market nearly as long as coach lace. By 1810, Pennsylvania had at least one iron tire manufacturer, upstate New York another by 1820, and many ironworks and rolling mills produced rolls of strap iron for vehicle wheels and sled runners. In fact, the wheels themselves constituted another vehicle part revealing specialty manufacture's surprisingly early origins. Practitioners of a craft as ancient as carpentry and joinery, wheelwrights specialized in wheel construction. They also built vehicles, generally the heavier and plainer varieties used on farms. Intimate familiarity with woodworking and blacksmithing made wheelwrights especially versatile, and they frequently engaged in nonvehicle woodwork and the odd-jobbing typical of a village blacksmith. Members of this trade constituted the very earliest American horse-drawn vehicle makers: in 1629 Richard Ewstead and Richard Claydon arrived in the Massachusetts Bay Colony equipped with the tools and skills of the traditional English wheelwright. Consequently, even Colonial American horse-drawn vehicle manufacturers often had the option of purchasing their wheels ready-made. By 1820, when wheels formed a sufficiently large specialty to appear in the U.S. census, market shifts and an evolving vehicle trade intensified the wheelwright and carriage maker's symbiotic relationship. As increased competition forced antebellum horse-drawn vehicle manufacturers to seek productive gains, one solution was to outsource some of the more complex and difficult-to-produce parts. Wheels were the obvious choice.[4]

This increased demand ultimately turned the wheelwright shop into the wheel factory. Special-purpose woodworking machinery designed for wheel fabrication in the 1830s and 1840s opened the door to mechanization, which, in concert with rationalized factory production, gave birth to large-scale wheel manufacture during the 1840s and 1850s. This period saw the establishment of E. E. Bradley (ca. 1845) and the New Haven Wheel Company (1845) in New Haven, Connecticut, and S. N. Brown & Company (1847) in Dayton, Ohio. Boston had at least one wheel factory by 1850, and others included Bridgeport, Connecticut's Wheel and Wood Bending Company (1853); Phineas Jones & Company (1855) of Newark, New Jersey; and Zwick & Greenwald Wheel Company (1859) in Dayton, Ohio. Delaware was home to the Wilmington Wheel Company (1860) and many more followed after the Civil War. Though retaining a presence in the traditional southern New England carriage centers, large wheel-making concerns increasingly located in Ohio, Michigan, and Indiana, where they could tap that region's plentiful hickory supply.[5]

Wheel design evolved along with these companies, and machine-produced, proprietary variants gave birth to the American "patent wheel." Though the U.S. Patent Office recorded dozens of varieties, two proved especially popular. An enterprising Tennessee carriage maker named James D. Sarven began experimenting with lightweight wheel construction during the 1850s. The traditional wheel hub was of sufficient size to permit staggered spoke mortises for strength, but it seemed increasingly out of proportion to lighter vehicles. Sarven developed a much smaller design featuring unstaggered spokes abutting to form a solid wooden ring around the tiny hub. A pair of through-bolted iron flanges countered lateral stress and compressed the unit into an exceedingly strong, compact, and much lighter whole. Granted a patent in 1857, the carriage maker doubtless anticipated great acclaim, but cashing in on his brainchild proved an exercise in frustration.[6]

Because of inherent craft conservatism, it took some twenty years for what would become the single most popular patent wheel to achieve truly widespread use. Frustrated by early rebuffs, Sarven sold the manufacturing rights to two firms that produced his wheel during the 1860s and 1870s. By 1872 a total of three highly mechanized companies were succeeding: St. Louis's Woodbury-Sarven Wheel Company, Cincinnati's Royer Wheel Company, and New Haven's New Haven Wheel Company. The increasingly attractive prices and discovery of the design's real advantages finally began convincing carriage makers, and other entrepreneurs began to manufacture similar wheels. One of Sarven's biggest competitors was the 1867 patented Warner Wheel. Owner of a modest wheel factory, A. Warner had also wrestled with the limits of the traditional wooden hub; he looked to improved metal fabrication for the answer, employing a reduced-size wooden hub inside a cast iron, one-piece housing incorporating cast-metal spoke mortises. Its increased durability began attracting sales until Sarven initiated a patent infringement suit in the early 1870s. Though the designs actually differed materially, the court ruled in Sarven's favor in 1872, helping ensure his market dominance. Still, the Warner Wheel eventually became popular, and both designs were widely manufactured after their respective patents expired (fig. 4.1).[7]

Entrepreneurial activity in wheel manufacture testified to the growing popularity of the factory-made wheel among wagon and carriage makers. Phineas Jones's Newark factory supplied the expanding trade in his own state, as well as in metropolitan New York. In the late 1850s and early 1860s this firm sold wheels to such prestigious carriage makers as Brewster & Company and J. B. Brewster & Company. While many rural and urban vehicle builders continued to manufacture their own wheels and the traditional wheelwright soldiered on, mass-produced wheels claimed a sharply increasing market share from the 1840s. The term "wheelwright" remained in American use to denote small-scale wheel makers and repairers, while in England it continued to refer to the traditional wheel- and wagonmaking shop. However, craftsmen on both sides of the Atlantic increasingly had to compete with durable and popular American factory-made wheels.[8]

Mass-produced wheels did not please everyone. Many vehicle builders disdained "factory wheels" and continued to fabricate their own. Even the most fiercely independent carriage maker, however, could be tempted to save time and labor by purchasing wheel components like hubs or hub blanks, presawn felloes, and amazingly consistent machine-made spokes. These frequently formed the basis of what many carriage makers nevertheless marketed as hand-made wheels. Wisely exploiting this chink in craft conservatism, wheel makers sold mass quantities of the same parts entering into construction of their own products. Specialized wheel-making machinery developed in the 1840s and 1850s made bulk fabrication possible at low prices. By 1860 this segment of the accessory trade had grown dramatically, encompassing more firms in more states than the manufacture of any other single vehicle part. In terms of geographic distribution, the same general trends appeared: eastern stalwarts like Massachusetts, New York, and Pennsylvania continued to host many firms, but by the 1870s midwestern states like Ohio were moving to the forefront. In that year the Buckeye state led the nation with nearly sixty wheel making firms, a portent of things to come.[9]

Whether shop-made, factory-made, or something in between, these wheels traditionally revolved on wooden axles. Because wood lacked hard wearing qualities, axle ends normally sported iron strips called clouts or axle skeins. An improved arrangement consisted of a hollow iron cone, called a thimble-skein, which completely covered the vulnerable wooden axle ends. The carriage blacksmith could fabricate these from wrought bar or sheet iron, or they could be purchased ready made. Available in a variety of styles and sizes from dozens of manufacturers, thimble-skeins helped ensure the longevity of the wooden wagon axle long after they had been replaced by all-metal versions in light vehicles. Improvements in friction reduction and metal fabrication, combined with attempts to shed weight and bulk, gave birth to the iron and later the steel axle. Some of the earliest and most successful versions originated in late-eighteenth-century England, including the mechanically sophisticated Collinge, as well as the mail axle that was used on heavy mail coaches. Axles eventually came in dozens of designs incorporating friction-saving devices that included oil grooves, lubricated washers, grease ports, and more—all an expansive demonstration of American mechanical ingenuity.[10]

The 1830s and 1840s brought a handful of full-time axle makers into the trade. One of the first was W. H. Saunders of Hastings-on-Hudson, New York, who started an axle works in 1832. Because he could easily ship his goods down the Hudson, Saunders enjoyed an enthusiastic reception by New York City carriage makers appreciative of his consistent product quality. He appears to have been one of the earliest American importers of English mail axles, though he eventually produced his own version. Saunders soon shared the field with a northern competitor, the Concord Axle Company. Begun in 1839 in the New Hampshire village of Fisherville, the firm eventually moved to Penacook on the outskirts of Concord, where it took on its familiar name and became a major American axle producer. J. S. and S. J. Mowry, of Greenville, Connecticut, entered the axle business in 1847, followed by the New Haven Spring Company in 1853. Like many such concerns, the latter company also produced wagon and carriage springs. David Dalzell began his working life as a carriage trimmer in Albany and Hudson, New York, before starting his own firm in 1845. Situated in the Massachusetts Berkshires, in the town of South Egremont, Dalzell acted on an employee's suggestion that a metal lathe would permit them to fabricate their own axles. Demand for these soon outstripped that for carriages, and so Dalzell refitted his shop for full-time axle production. By the 1860s his firm had achieved a national reputation by virtue of quality and earlier endorsements by noted customers like James Goold, Jason Clapp, and James Brewster.[11]

The field became increasingly crowded from the 1860s, with more firms manufacturing an even greater assortment of axles and related goods. Some produced the usual range of common carriage axles, from plain grades up through complex and expensive mail axles; others pioneered entirely new ways to reduce wheel friction. Radically improved bearings were a late development in this vein, the most important stemming from an established inventor and carriage maker. German-born Henry Timken came to America as a child and became a St. Louis wagonmaker in the 1850s. After a brief California gold-seeking sojourn and Civil War military service, he returned to carriage manufacture and patented a buggy spring that was widely used during the 1880s. A restless attempt to retire to the west coast ended in 1891, when he returned to St. Louis to establish a large wholesale carriage factory and address the problem of friction.

Horse-drawn vehicle manufacturers had long sought to decrease draft through improved axles, and experiments with various types of steels, hardened bearing surfaces, oil ports, and other features had yielded incremental improvements and an avalanche of competing patents during the preceding decades. By the early 1890s, however, the axle industry had approached the limits of conventional bearings. Ball and roller bearings, achieving increased application in various industries by this time, appeared to hold promise, yet the former's limited surface area and the latter's inability to withstand lateral thrust limited their use to light vehicles. Well aware of these developments, Timken and his sons undertook experiments during the period 1895–98, culminating in U.S. patent 606,635. Combining the superior working area of the roller bearing with a taper to accommodate side pressure, Timken's tapered roller bearing made inroads into the heavy wagon industry after the 1899 formation of the Timken Roller Bearing Axle Company. Initially located in St. Louis, the firm soon outgrew its facility and moved to Canton, Ohio, in 1901. Although originally developed for heavy wagons, Timken's invention later found widespread use in the automobile industry.[12]

At the end of the eighteenth century, carriage suspension consisted primarily of leather thoroughbraces or steel whip, S-, and C-springs, which absorbed at least some of the shock generated by horse motion and rough roads. Limited early American metal production meant that steel vehicle springs had to be imported from England and France, and their high cost limited their application to only the finest carriages. The overwhelming majority of vehicles jolted along without suspension of any kind. Domestic spring manufacture using imported iron and steel originated during the early nineteenth century. Philadelphians William and Harvey Rowland began producing carriage springs around 1806, and as American steelmaking gradually increased its capacity, domestic spring production accelerated accordingly. Located initially in eastern urban centers these early spring makers courted carriage builders supplying America's socioeconomic elite. Philadelphia had at least one spring works by the early 1840s, and in Connecticut, Bridgeport and New Haven had their own during that and the following decade. Some of these firms had already pioneered axle production and employed their metal fabrication capability to turn out springs. Others started with springs and found in them alone a veritable fortune.[13]

By the 1850s American carriage manufacturers were already experimenting with design changes that would soon revolutionize horse-drawn vehicles. The desire to decrease weight without sacrificing durability found material outlet in improved suspension. Few other subassemblies contributed more to a light and flexible horse-drawn vehicle, and no truly workable system was possible without springs. Springs and related hardware became integral parts of a vehicle's undercarriage and, when built around patented combinations of springs, sidebars, and other components, became known as "patent gears." Although the Civil War dampened such innovation, the process resumed during the late 1860s. The late 1860s and the 1870s witnessed the creation of the majority of the patent gears that became staples of the American horse-drawn vehicle industry. New designs passing through the patent office after this time tended to be refinements of concepts first developed right after the war.[14]

A full list of these spring and gear patents would run to many pages, and many of them flunked the test of practical use before passing into oblivion. The 1870s trade press featured lavish advertisements for new suspension designs promising devastating advantages over lesser-equipped competitors. Cyrus Saladee, never at a loss for superlatives, claimed his 1877 triple-spring system possessed nothing less than "perfection, simplicity, lightness, strength, durability, ease, elegance of style, and safety." Indeed, as one wag commented many years later, there was virtually nothing else left to claim. Hubris aside, many of the best designs did embody sufficient advantages to ensure their widespread adoption. Perhaps the most successful of all was the sidebar gear. Eschewing the formerly common direct body-to-spring linkage, sidebar gearing sported two transverse wooden bars supporting the body and terminating on both ends in steel springs. Mounted at right angles to the bars, these springs in turn bolted to the gear. This resulted in a lower-mounted body and reduced sensation of horse motion; it also shielded the horse from excessive spring play. Sidebar gears made for a more stable and smooth ride without adding weight or rigidity, all in a package much lower and trimmer than its predecessors. Ideal for vehicles receiving regular use on rough roads, the sidebar gear quickly became the number one form of buggy and other light carriage suspension.[15]

While many variations rolled out of specialty factories over the years, the best known were the Brewster Side Bar and the Timken Patent Cross Spring. Brewster's patent really inaugurated the rage for sidebar rigs. Devised in the early 1870s by the oldest son of the famed carriage maker, this design eliminated a problem associated with earlier sidebar suspension: the inadequacy of connecting the body directly to the bar. Brewster added a second set of half-springs below the body and attached to the sidebars, the ends of the latter running into an additional set of half springs in the usual manner. Close mounting of the extra spring set retained the overall low profile while greatly strengthening the body-to-gear attachment. Brewster patented the arrangement in

1872, and its popularity set off an "improved gear fever" lasting through the end of the decade. Henry Timken's Cross Spring, patented in 1877, achieved similar success. Instead of sandwiching the crossbars between two sets of half springs, Timken put all springs above the bars. Placing the body on a pair of brackets each astride a pair of crossed springs, one end of these bolted to the bracket while the other attached to the sidebars with the latter riding on the gear. This "x" profile springing proved more flexible and perhaps handled variable loads better than Brewster's system. In any case, the Timken Patent Side Spring Gear formed another important contribution to American light carriage design.[16]

Springs, shackles, sidebars, and the related hardware used in patent gears could all be purchased ready-made by the 1870s. Several southern New England and mid-Atlantic states had spring manufacturers before the Civil War, and, by the end of the 1860s, Pennsylvania had at least ten, while Connecticut and New York had seven each. New Jersey, Massachusetts, and several others joined an increasing number of midwestern competitors. Some manufactured springs alone, whether for the wagon and carriage industry, street railways, the railroads, or both. Others took it upon themselves to produce or purchase the myriad iron, steel, and wooden parts required to build gears, often under license by various patentees. In the early 1880s the Cleveland Spring Company informed *The Hub*'s readers that it had recently become sole manufacturers of the "New Total Eclipse Springs and Irons," one of Saladee's answers to the sidebar gear. Producing this one assemblage required purchase or fabrication of over a dozen different parts, yet the company offered for sale a variety of other springs and gears (including its "American Chief: King of the Tribe") requiring many more, all while ensuring steady work for its expanding plant.[17]

Gear manufacture frequently led to the production of other composite wood-and-metal components like bodies, shafts and poles, or top bows. As early as 1850 Boston's A. M. Wood Company began turning out bodies and gears, followed by Dann Brothers and Company eight years later. The two firms soon had competition; Philadelphia had a handful of individual buggy- and carriage-body makers during the 1860s, as did other cities. Most of these ventures were founded by former body makers from conventional carriage shops, and many failed after a few short years, due in part to the entry of larger firms into what had become an increasingly crowded and volatile market. At least a few began making wooden carriage parts not because they were steeped in traditional carriage making but because they were woodworking firms with the necessary machinery. New Haven had a folding chair manufacturer that began making axle beds, head blocks, and sidebars for horse-drawn vehicles in the 1860s, and it surely was not the only one to branch out in this way.[18]

Impressive numbers of firms began building carriage bodies in the 1870s and 1880s, some producing semifinished wooden stock for use in carriage gears and bodies, while others turned out more complete products. These often included the necessary iron hardware and reinforcements but usually came unpainted. Some, like the Pioneer Pole and Shaft Company (1865), of Piqua, Ohio, or the Excelsior Seat Company (1878), in Columbus, concentrated on only a few items. Others turned out complete gears and bodies. Like many similar firms, New Haven's H. G. Shepard & Sons (1880) was located in one of the old carriage-making centers but faced newer midwestern competition. Detroit had the C. R. Wilson Body Company (1870) and Flint the W. F. Stewart Company (1881), while St. Louis was home to the Powitzky & Collins Carriage Woodwork Company (1888). Even little towns claimed their share: Struthers and Bellefontaine in Ohio had both a gear wood and a wagon body maker, and there were many more whose names and lists of products have long since been lost to time.[19]

Firms scrambled to meet the demand for still other horse-drawn vehicle parts. Dashboards were designed to screen the driver from road debris, mud, and the effluvia of his motive power. Several firms made them in quantity, though the best known was probably Buffalo, New York's McKinnon Dash Company, established in 1876. It by no means monopolized the trade, as many smaller firms such as the Cleveland Dash Company, the Bennett Manufacturing Company, H. Scherer & Company, and others vied for a slice of the profits. Another Ohio firm, W. C. Reynolds of Columbus, began buying out its smaller competitors during the 1880s to become one of the nation's largest dashboard manufacturers. Lamps were another item of production. Once primarily imported, carriage lamps began to come out of American firms during the 1830s. New York, Bridgeport, and Newark all had short-lived carriage lamp shops, but a stable American contribution to this specialty came with the formation of two key firms. New Haven's Cornwell & Cowles (1838) and Newark's Robert & Eagles (ca. 1842) succeeded by making durable copies of the most popular European designs. By the 1850s, the products of these and several others competed successfully with imported lamps, and when cheap buggies arrived in the 1870s, American lamp makers produced inexpensive lighting for a market segment largely untapped by Europeans. By the 1880s and 1890s, a cluster of firms ranging from successors of the New Haven and Newark companies to newer enterprises like the Richmond Manufacturing Company (ca. 1870) and the Newark Coach Lamp Company (1891) met the bulk of orders coming from the American trade.[20]

Doing One Thing and Well

No product escaped the accessory industry's attention, perhaps because of the potentially terrific profits lurking in even the most mundane item. Take the lowly bow socket: a product of inventive minds, sheet metal, and proprietary capitalism, the stamped-steel bow socket reaped its inventors a considerable fortune. It also divided two brothers and created one of the longest running patent battles in the industry. Isaac Newton and John A. Topliff were born on a Connecticut farm in the early nineteenth century, and each made his way to the Midwest before the Civil War. Sometime soon thereafter, John established a carriage shop in partnership with George H. Ely in Elyria, Ohio, a small manufacturing town southwest of Cleveland. Isaac opened his own shop in Michigan, working there for ten years before returning to Elyria in 1869 to help his brother produce carriage hardware. Topliff & Ely specialized in hollow sheet steel fabrication, manufacturing bow sockets for buggy sidebars, their patented load-equalizing crossbar (a suspension component), and a hollow metal shaft socket patented by a Michigan inventor. Isaac demonstrated the latter to *The Hub*'s editor during a New York visit in December 1875. Their most popular and lucrative product was the tubular steel bow socket, the item that garnered official acclaim and notable patronage at the 1876 Centennial Exposition. By 1877, the little company shipped several hundred sets a day, with demand skyrocketing.[21]

Carriage makers had traditionally fashioned folding top bows from hardwood. Side bows consisted of three or four wooden uprights sharing a common pivot on the carriage body's belt rail and flaring to individual connections with the top bows, curved wooden ribs supporting the top fabric. This resulted in a durable but heavy folding mechanism. The Topliffs used newly available sheet steel to form rigid and much lighter tubes to take the place of the wooden uprights. What the brothers had done, though they might not have conceived of it this way, was utilize recent metal fabrication techniques to advance lightweight vehicle construction. A great deal of American carriage architecture's lightness derived from substituting metal for wood: steel elliptical springs in place of the heavy wooden pole and perch, slender wrought-iron fifth wheels in lieu of massive hardwood, steel elliptic seat springs instead of cantilevered wooden supports. Though the lightweight American wheel owed much to hickory, it also relied on iron bands, flanges, and bolts to relieve the tiny hub of excess wooden mass. The Topliffs' steel bow sockets contributed to this ongoing process and accounts for their spectacular success.[22]

This same success proved a powerful source of fraternal strife. In the fall of 1879 Isaac left the firm to establish his own rival bow socket company; his two-story brick factory in Cleveland soon ran to capacity. The following year Topliff & Ely sued I. N. Topliff for alleged infringement on John Topliff's 1875 patented bow socket design, initiating a series of lawsuits that would drag on for fourteen years. Meanwhile, though Isaac's plant already completed about 150 sets of bow sockets daily, he expanded it to include a rail spur and added a night shift, employing sixty hands by the early 1880s. By this time, he offered his customers a stylish leather-covered variant, and, despite diversifying into other carriage hardware, this remained his primary product. So it must have been particularly gratifying for Isaac when the U.S. Circuit Court ruled in his favor in 1884.

While Topliff & Ely appealed, Isaac issued a twenty-page catalog and stepped up his advertising campaign. The publicity certainly helped: he sold over 100,000 sets of sockets during the first six months of 1886, and the following January shipped 25,000, his greatest monthly shipment thus far. Though all classes of carriage builders patronized his firm, huge orders from the big Cincinnati buggy manufacturers accounted for several carloads monthly. Also, the law continued to support his version of events: in May 1887, the U.S. Supreme Court upheld the lower court's ruling. Isaac celebrated by inserting a notice in *The Hub*, featuring an illustration of a Roman gladiator, cribbed from a popular 1870s chromolithograph, with his foot on the throat of his prostrate foe. Aside from his helmet and shield, the victorious warrior brandished a bow socket, while his opponent clutched the shattered remains of what was obviously an inferior article.[23]

One might have assumed this closed the matter, but when John stepped off the Detroit boat in Cleveland in late 1887, a sheriff handed him a summons notifying him that Isaac was instituting his own patent infringement suit. At some point, Isaac had granted Topliff & Ely the right to manufacture his tubular bow socket in return for royalties, an agreement evidently abandoned by the Elyria firm. Still, production at both plants continued all the while; Isaac's success enabled him to acquire a home on Euclid Avenue, Cleveland's famed "Millionaire's Row," and by the 1890s had the luxury of turning plant operations over to his son-in-law in order to pursue other business ventures, leisure, and travel. Well before his death in 1904, Isaac Newton Topliff had used the appeal of a seemingly mundane product to attain a level of entrepreneurial and industrial achievement little short of astonishing.[24]

Whether a result of their legal misfortunes or their inability to successfully compete with the I. N. Topliff juggernaut, John Topliff and George H. Ely apparently lost control of their company during the late 1880s, though the firm continued to prosper. Meanwhile the court battle raged on, with the law once again ruling in Isaac's favor. Topliff & Ely again appealed, and the case wound its way through the courts during the 1890s. Finally, in August 1894, *The Hub* announced "the end of a gigantic litigation," in which the Ohio Supreme Court upheld the decision awarding Isaac approximately $100,000 in royalties. This brought to an overdue close one of the longest running patent suits in the history of the industry. Yet it must have seemed an anticlimax; just a year earlier a small trade press notice announced formation of the Carriage Specialty Company. Headquartered in Cleveland and comprising the I. N. Topliff Manufacturing Company, the Topliff & Ely Company, one other Ohio firm, and three more in upstate New York, the consortium posited itself as a "joint selling office" handling the goods of all six. In other words, intense competition, escalating price wars, and plain old market saturation made for strange bedfellows. It ultimately made little difference, as the arrangement appears to have failed during the depression of 1893 and its constituent firms petered out one by one after the turn of the century.[25]

The Topliffs' experience illuminates one end of the spectrum of possibilities in the accessory industry: building a virtual empire around a very specialized part. Other equally industrious individuals explored the other extreme, quantity production of general-purpose goods not only for carriage making but for other trades and industries as well. By the second half of the nineteenth century, paint and varnish manufacture had entered the age of big business and become a vital part of the accessory trade. Perhaps the oldest firm catering specifically to American carriage makers was C. Schrack and Company, established in Philadelphia in 1815. New York City's early varnish producers included John W. Masury and Son (1835), but its best known firm actually began elsewhere. Simson, Valentine & Company, established in Boston in 1832, produced varnish for carriage makers, railroad- and streetcar builders, cabinetmakers, and anyone else requiring a clear, glossy finish. By the 1860s, the firm, now known as Valentine & Company, had become a major producer and further enhanced its visibility by moving to New York City in 1870.[26]

It had a great deal of competition, ranging from Bridgeport's Parrott Varnish Company and Newark's Murphy Varnish Company to Felton, Rau, and Sibley of Philadelphia and Chicago's Heath and Milligan Manufacturing Company. At the beginning of the 1880s, Newark had more varnish makers than any other city, while New York and Philadelphia led in the number of paint firms, followed by St. Louis, Chicago, and Cleveland. The largest individual enterprises hailed from the industrial Midwest and were a byproduct of the infant oil industry. Western Pennsylvania had been the only significant crude oil source since the 1850s, and John D. Rockefeller's brilliant, ruthless business acumen made Cleveland the nation's refinery. The Forest City Varnish, Oil, and Naptha Company, established in 1862, produced varnish and japans, as well as solvents like naptha and gasoline. It shared the field with the American Paint & Oil Company; Billings, Taylor & Company; the Cleveland Varnish Company; the Western Reserve Varnish Company; and the Forbes Varnish Company. Carriage makers, along with many other domestic and industrial users, eagerly patronized these firms.[27]

Two members of the Cleveland paint and varnish trade shared the distinction of becoming the largest of their kind anywhere. Francis H. Glidden entered the business in the early 1870s, and though his company would one day be famous for paint, its primary product was varnish. Varnish sales to industries such as carriage making and agricultural implement manufacture accumulated profits much more rapidly than did targeting the domestic trade. During the 1870s Glidden joined forces with two successive partners before settling down as the Glidden & Joy Varnish Company in 1883. By this time, the firm's east side plant turned out a range of varnish and related products that sold in the United States and abroad. In the spring of 1888, *The Hub* relayed the news that the Glidden & Joy Varnish Company had been awarded a sizeable varnish contract by a French railroad concern; steady orders of this magnitude garnered the firm as much positive publicity as profits. By the time Glidden Varnish occupied its new seventeen-acre plant in early 1908, it claimed to operate the largest varnish works in the world.[28]

The other famed company began with Vermont native Henry A. Sherwin. The ambitious eighteen-year-old moved to Cleveland in 1860 and invested a modest sum in Truman Dunham & Company, a downtown retailer of paint-making supplies and window glass. This arrangement lasted until 1870 when Dunham and George O. Griswold took over the linseed oil works; Sherwin joined forces with Alanson T. Osborn and Edward P. Williams to form Sherwin, Williams & Company, which eventually became the Sherwin-Williams Company. Osborn soon retired, but Williams remained. A Cleveland native, college graduate, and Civil War veteran, he brought capital and experience to the venture. Initially located downtown, the firm soon moved to Standard Oil's old cooperage building in the riverside industrial area, there to commence production of ground paint colors and oils. The firm achieved lasting fame in 1880 through an innovation that would reshape the way paint was made and used. Sherwin-Williams's improved pigment grinding permitted nearly indefinite particle suspension, and this, along with fine tuning of the ingredients' proportions, netted a major breakthrough: the first reliable ready-mixed paint, packed in cans and ready for use. It would set the standard for the entire industry.[29]

Like other paint users, wagon and carriage builders patronized Sherwin-Williams for its increasingly broad color selection, as well as its convenient product. Of the many steps necessary to complete a vehicle, painting required the most time, and cost-conscious vehicle builders embraced anything promising speed. This, perhaps more than any other factor, encouraged Sherwin-Williams to divest itself of its retail store in 1882 to concentrate solely on manufacturing. The company vigorously promoted ready-mixed paint, emphasizing both its time-saving virtues and the high-quality finish ("Saves Time Without Any Risk") obtainable with a minimum of effort. The year 1884 brought incorporation and expansion, with sales agencies established in Chicago, New York, and other major cities. By late in the century, Sherwin-Williams had become one of the world's largest paint manufacturers. The wagon and carriage trade press tracked its successes, reported its comeback from disastrous fires, touted its new products, and testified to the importance of another member of the wagon and carriage accessory industry.[30]

Though its largest manufacturers would eventually become synonymous with another industry, rubber tires were an important, if late, feature of the accessory industry. Centuries before anyone thought of applying the substance to a wheeled vehicle, Central and South American Indians discovered that the milky sap of certain tropical trees coagulated into an elastic substance they called "caoutchouc." Primarily a curiosity, it saw limited use in sixteenth-century indigenous cultures as balls for various sports, as well as in small amounts of crude waterproof apparel during the following century. Samples brought back to Europe aroused curiosity and earned it the names "rubber" and "India rubber" for its ability to erase pencil marks. Discovery of solvents that permitted it to be spread on cloth gave it some success as a fabric waterproofer during the early nineteenth century, but the product still melted when warm and cracked when cold. It remained to be seen whether there could be any means of permanently fixing the material when American hardware man and inventor Charles Goodyear became entranced with rubber during the 1830s. A series of experiments and accidental contact with a hot stove in 1839 revealed that heat in conjunction with sulfur and white lead permanently stabilized raw rubber. The process, called vulcanization, made working and fixing rubber into various products practical at last, and Goodyear had the satisfaction seeing his vulcanized rubber goods win a medal at the Crystal Palace exhibition in 1851.[31]

By this time, the hapless inventor had become enmeshed in a series of horrifically expensive patent battles that would occupy him for the remainder of his life. Meanwhile vulcanized rubber saw use for everything from rubber boots and raincoats to waterproof sheeting and life rafts. Goodyear had also discovered that extended vulcanization yielded a hard rubber that could be cast into items as diverse as watch fobs, paperweights, and book covers. By the 1850s and 1860s, both hard and soft vulcanized rubber products started to appear in the American wagon and carriage industry in the form of whip sockets, "anti-rattlers" (rubber cushioned shaft couplers), prop blocks and bow separators (cushioning folded top bows), brake blocks, and rubber-coated cloth used for lap robes, storm curtains, and tops. In the 1860s, Philadelphia's Flesche & Perpente sold rubber carriage and railroad car trimmings, while New York's Odorless Rubber Company handled carriage roofing material and other goods guaranteed not to reek of sulfur ("A Trial is Respectfully Solicited, Which Will Convince the Most Prejudiced"). Rubber step pads protected the japanned finish of iron carriage steps, rubber carpeting insulated and protected wooden carriage floors, and rubber-coated cloth shielded passengers from rain and snow.[32]

As early as the 1830s the French had considered applying rubber to carriage wheel rims, and other inventors had also been intrigued by the idea of providing a quieter, smoother ride that would also protect the vehicle from road shock. Though Englishman Robert W. Thompson took out an 1845 patent for a pneumatic carriage tire, he had greater success with a simpler, solid design. Improved rubber production and vulcanization helped Thompson apply solid rubber tires to steam traction engines in the 1860s. Their expense and lack of durability limited their use, and it took several 1870s patents for solid rubber bicycle tires to suggest a workable application to carriage wheels. By the 1880s, rubber tires had become increasingly common on London's hansom cabs; despite objections from those fearing that their quietness endangered pedestrians, this same quality endeared them to customers, muting concerns over high cost and limited service life. In 1866 Brewster & Company's Channing M. Britton observed solid rubber carriage tires in Berlin, and by the 1880s that firm offered them as an option (an expensive one at approximately $150 per set) on its carriages. During the 1880s, New York City's Brewster & Company and Healy & Company and Chicago's C. P. Kimball & Company were the only significant American carriage makers to turn out rubber-tired carriages, all of them using imported English tires.[33]

The chief mechanical difficulty with the rubber carriage tire lay in reliable attachment to the wheel. Though not the only ones to succeed, two Americans came up with solutions that proved especially popular. In 1895–96 Arthur W. Grant patented a rubber tire incorporating a pair of steel wires that, in conjunction with a flared steel rim, firmly clinched tire to wheel. In 1901 James A. Swinehart patented another solid rubber tire that used a pair of exterior side wires to clamp the rubber to the metal wheel rim. Both products and their many imitators proved superior to English designs and contributed greatly to the gradual American popularity of the rubber carriage tire. Some versions incorporated hollows for additional cushioning, and the late 1880s efforts of an Irish veterinary surgeon named John Boyd Dunlop revived Thompson's idea of a hollow rubber tire containing compressed air. The pneumatic tire gave the best ride of all and became standard on safety bicycles by the early 1890s.

Durability remained a factor, and in the carriage trade racing sulkies and other light vehicles could be equipped with pneumatic tires while heavy carriages received cushioned or solid designs. One of the earliest American companies to supply them began in the late 1860s. Dr. Benjamin Franklin Goodrich, a former Civil War surgeon who quit medicine to dabble in real estate, purchased a small New York rubber firm operating under a license from Charles Goodyear. Manufacturing and business difficulties followed, and in an effort to distance himself from his eastern competitors, he moved to Akron, Ohio, in 1870. Producing primarily fire hose, belting, and miscellaneous smaller items, the firm began prospering and, after a few additional structural changes, incorporated in 1881 as the B. F. Goodrich Company. It began producing solid rubber tires for the high-wheeled bicycles of the 1880s and pneumatic tires for safety bicycles during the following decade.[34]

A record-breaking 1892 run by racing trotter Nancy Hanks pulling a pneumatic-tire sulky publicized the concept, and B. F. Goodrich soon added pneumatic carriage tires to its line. By now the firm shared the field with many others eager to cash in on a rubber product of proven popularity. Eastern concerns like the Hartford Rubber Works Company (1880) in Connecticut, New Brunswick's India Rubber Company (1896) and Milltown's International Automobile and Vehicle Tire Company (1899) in New Jersey, and Batavia's Sweet Tire and Rubber Company (1893) in New York, vied for place with midwestern firms ranging from the Milwaukee Rubber Works Company (1903) and Indiana's Kokomo Rubber Company (1895) to Chicago's Calumet Tire Rubber Company (1897) and the Victor Rubber Company (1898) in Snyderville, Ohio. Many rubber firms were already located in Akron, a city that as early as the 1890s showed signs of leadership in rubber tire production. Besides Goodrich, Akron was home to the Diamond Rubber Company (1890), the Consolidated Rubber Tire Company (1894), and in 1898 two local brothers organized the Goodyear Tire and Rubber Company. They were soon joined by the Firestone Tire and Rubber Company (1900), the Stein Double Cushion Tire Company (1902), and several others. While many horse-drawn vehicles would continue to roll on steel tires, the quiet, smooth ride of a rubber-tired carriage appealed to more and more consumers. By 1900 solid rubber tires had become common on higher-grade carriages, and by 1902 the rubber tire manufacturers collectively achieved annual sales of more than $5 million.[35]

Malleable Iron and the Vehicle Hardware Sector

Paint and rubber were two products especially indebted to the wagon and carriage industry for its patronage, and both were heavily Ohio-based. Yet if any one segment of that state's industry could claim closest ties to the carriage industry, that distinction fell to the hardware manufacturers in and around Cleveland. In its most common use, "hardware," sometimes called "findings," referred to the myriad small metal items required in fitting out homes and commercial structures. Ranging from nails, screws, and other fasteners to hinges, door knobs, and shelf brackets, these general-purpose goods weighed down the shelves of local hardware stores nationwide. The term also encompassed items as disparate as cutlery, hand tools, light machinery, and the many parts necessary for their fabrication and repair. Hardware manufacturers sold not only to distributors and merchants but also to industry. Considerable profit flowed from wholesale hardware supply, and cutthroat competition reigned. Cleveland owed its hardware preeminence to a vibrant iron and steel industry. In existence since the 1840s, the city's iron and steel sector entered its golden age with the 1850s discovery of the vast and fabulously rich iron ore deposits of the upper Great Lakes. Immediately exploited by ambitious entrepreneurs and firms, this ore traveled down the waterway and into the smelters of lakeshore industrial cities like Cleveland, where it emerged as a highly useful raw material for numerous products.[36]

Screw, bolt, and nut making had become an important Cleveland industry by the 1870s. Many small fastener plants were located in the flats, the industrial district on the banks of the Cuyahoga River. Some, like Bradley, Lewis & Company, turned out carriage and tire bolts; others, like the Colwell & Collins Norway Bolt Company, produced a more general range of fasteners from high-quality imported iron. Some of the largest originated in New England but had moved to the Midwest to take advantage of the iron ore boom. Lamson, Sessions & Company started in Connecticut in 1866 and came to Cleveland three years later. The firm produced a variety of fasteners, including the ubiquitous carriage bolt, and outlived many of its competitors. Likewise, the Upson Nut Company, also a Connecticut firm; established in 1854, it transferred its main works to Cleveland in 1870. Others followed that decade, including the Union Steel Screw Company, the Kirk-Latty Manufacturing Company, the Cleveland Bolt Manufacturing Company, and National Screw & Tack. These companies produced general-purpose goods, but Cleveland had other hardware manufacturers wed specifically to the carriage industry. Two in particular would exploit an important metallurgical discovery to become the twin giants of the carriage accessory industry.[37]

Wagon and carriage hardware was a catchall term encompassing most metal vehicle hardware. Whether designed to protect wood from wear (iron tires, shaft tips, rub irons), as a structural reinforcement (corner braces, stays, hub bands), or as a self-contained subcomponent (brake levers, fifth wheels, folding steps), vehicle hardware emerged from iron stock through efforts of the carriage blacksmith. Through most of the nineteenth century, the standard raw material consisted of wrought iron, a metal that worked well on the anvil and had adequate strength for most horse-drawn vehicle applications. The earliest wrought iron came from bloomeries where ironmakers heated iron ore in a charcoal fire and hammered ("wrought") the hot, spongy mass to expel impurities and create a slab of tough, fibrous iron. Though capable of producing perfectly useable metal, bloom smelting had limited output; while bloomeries continued to supply iron, the fledgling American metal trade soon experimented with more efficient methods. Before the end of the Colonial period, some ironworks had turned to blast furnaces, whose much greater heat completely melted large quantities of ore. Foundry men ran this off into sand molds in the shape of a final product, like a frying pan, or an intermediate bulk unit, called a pig. In both cases this cast iron featured a high carbon content, which made it extremely hard and heat-resistant, ideal for firebacks, cooking utensils, and stoves, but excessive brittleness prevented its use in any application (such as wagon and carriage hardware) requiring impact resistance. Blast furnaces could turn out large quantities of cast iron, but the product had to undergo an additional process to create the blacksmith's preferred raw material.[38]

Iron makers resorted to two methods to convert cast iron into wrought iron. The earliest of them, fining, consisted of reheating cast iron pigs and employing a water-powered trip hammer to work the metal to expel excess carbon and impurities. This "heat and beat" process continued until the mass reached the fibrous, flexible state characteristic of wrought iron. Fineries (also called forges) were adjuncts to many American blast furnaces as early as the Colonial period and permitted iron makers to supply iron useful to blacksmiths, hardware manufacturers, nail makers, and others. Puddling, the second method, consisted of reheating iron pigs to a liquid state and stirring the liquid to burn off excess carbon. Allowed to cool into a solidified but still hot mass, ironworkers then ran it through heavy rollers to expel an additional measure of carbon to create wrought iron slabs. Developed in England in the 1780s, puddling quickly became the standard English wrought iron production method. Owing to a fuel difference (puddling furnaces used mineral coal instead of charcoal), this improvement did not become common American practice until coal became a viable alternative to charcoal during the first half of the nineteenth century. By then, exploitation of Ohio and Pennsylvania coalfields and charcoal's increasing cost made puddling the preferred wrought iron production technique.[39]

Regardless of its specific origins, wrought iron formed the blacksmith's primary working material. Although the skilled vehicle blacksmith could form items like tires, seat stays, and hub bands easily and rather quickly, complex parts like hollow shaft tips, steps, and fifth wheels required much more time and skill to fabricate. The smith might spend as much as two hours fashioning a set of shaft tips and couplings (the primary metal hardware applied to a pair of single-horse shafts), and more complex items took even longer. With the vehicle market's expansion and the drive to increase production, the limits of traditional handwork constituted an increasingly vexing bottleneck. Iron makers' search for a better way to produce ductile iron goods and carriage makers' desire for efficient fabrication met in a product that combined the ease of casting with the quality of toughness. The malleable iron casting was thus a doubly important industrial development.[40]

The process's precise origins are difficult to pinpoint in part because the exact chemical transformation remained a mystery for so long. Though the concept predated him, eighteenth-century French physicist Rene de Reaumur penned the first widely disseminated description of how ordinary iron castings could be rendered tough and shatter-resistant by annealing. Doing so required packing castings with hematite, slag, or other carbon-depriving substances into sealed containers and baking them at high temperature for several days. Annealing robbed iron of its brittleness and gave such castings wrought-iron-like impact resistance. Pervasive trade secrecy and limited scientific knowledge prevented this European discovery from being widely known until a New Jersey man named Seth Boyden introduced the malleable iron casting to America during the 1820s. An iron maker with an experimental flair, Boyden attempted to copy Reaumur's annealing process, which he eventually succeeded in doing. Though neither he nor his immediate successors fully understood the metallurgical particulars—the American process actually involved a different chemical reaction—they were quick to exploit the results.[41]

One of the earliest malleable iron casting applications consisted of harness buckles and related tack, which could now be easily produced in previously unheard-of quantities. Malleable iron manufacturers soon adapted the process to larger items used by railroads, agricultural implement makers, and the horse-drawn vehicle industry. The wagon or carriage maker's advantage in purchasing malleable iron castings was the elimination of time-consuming hand forging, as well as ensuring a much greater degree of uniformity. Because nearly all the shaping took place before the item reached the blacksmith, all that remained were relatively minor operations prior to fastening the part to the vehicle. Just as the table saw and planer drastically reduced the woodworker's stock-preparation investment, so the malleable iron casting reduced the blacksmith's labor in shaping iron; like the woodworker, he became increasingly devoted to assembly. Such efficiency contributed to savings eventually passed on to the consumer. In some industries, such as agricultural implement manufacture, the malleable iron casting did as much as any other technical innovation to make its goods widely available at moderate prices.[42]

Boyden operated his own malleable iron casting works in Newark from 1826 to the mid-1830s, and that city went on to have as many as eight malleable foundries, some operated by Boyden's brothers. Elizabethport had two malleable works by the 1840s, and the process also made its way into New England, which featured a substantial hardware trade of its own. Connecticut hosted several important firms in Plantsville, New Haven, and Bridgeport. J. H. Whittmore and B. B. Tuttle started a malleable iron castings works in Union City during the 1850s, just one of the plants that became part of the giant Eastern Malleable Iron Company. This firm, as well as Bridgeport's Malleable Iron Company, provided many of the men who went on to establish malleable iron castings works in the Midwest during the second half of the century.

The Midwest's growing strength in iron and steel drew the malleable trade there, and Cleveland became home to the first malleable iron works west of the Alleghenies. Formed by a group of industrialists and investors in 1868, the Cleveland Malleable Iron Company was located on the city's east side, where convenient rail access provided transportation and a local Hungarian community provided labor. The plant consisted of a cluster of brick buildings containing the foundry, finishing shop, warehouse, and packing department, all of which underwent seemingly continuous expansion. The company soon became the undisputed leader in malleable iron castings. Heavy demand potential in the rapidly expanding horse-drawn vehicle industry led president Alfred A. Pope in 1879 to form the Eberhard Manufacturing Company to produce malleable iron hardware for the wagon, carriage, and saddlery trades. Born into a New England industrial family, Pope spent part of his boyhood in Ohio, where he worked in his family's woolen manufacturing business. He entered the malleable iron industry around 1869 when he became the Cleveland Malleable Iron Company's bookkeeper.[43]

Pope advanced quickly and by the 1870s became president of the concern. His leadership proved vital to both malleable iron and wagon and carriage makers, while making Cleveland the capital of the American malleable iron industry. Rather than trying to shoehorn an enormous product line into a single company, Pope opted to establish a subsidiary dedicated to supplying the wagon and carriage industry. Both firms shared not only a president but also the Woodland Avenue plant, and they promoted their goods in extensive, market-specific product catalogs. Eberhard began issuing its chunky publications around 1880, and they soon found their way into carriage shops and factories all over the United States. These catalogs reveal the sheer variety of products possible to make by malleable iron casting. An 1880s Eberhard saddlery catalog lists dozens of styles and sizes of trace buckles, tug loops, D-rings, assorted snaffles and bits, and virtually any other metal part used in saddlery and harness manufacture. Vehicle hardware accounted for a far weightier catalog. Here one could purchase relatively simple goods such as rub irons, clips, and reinforcements of all kinds, as well as complex components such as steps, brake systems, and fifth wheels. Later catalogs even grouped the hardware necessary to make a particular vehicle type as a further incentive to order Eberhard products (fig. 4.2).[44]

By the time one of its salesmen accepted the huge order mentioned at the opening of this chapter, the Eberhard Manufacturing Company occupied a thirteen-acre facility with its own railroad spur, foundry, machine shop, packing department, and warehouses totaling twenty-seven buildings. Over six hundred employees made the standard catalog lines of castings and hardware, as well as vast quantities of the same type of goods produced to customer specifications. Though Eberhard emphasized its willingness to accommodate even small quantities of one-off designs, the firm clearly preferred its custom work to revolve around the needs of the larger wagon and carriage factories. Time and money spent working up custom designs and molds met richest reward in runs totaling in the tens of thousands, though the firm continued to accept smaller orders. Despite periodic fire losses, continued high demand enabled continued expansion so that by 1907 the carriage hardware division alone employed more than 1,400 people. Shipping its goods all across the country, Eberhard also established markets in South America, Europe, and Australia.[45]

Further evidence of the malleable iron casting's popularity came in the form of new plants in Toledo, Indianapolis, and Chicago, facilities that eventually merged into a single enormous firm. The National Malleable Castings Company retained Pope as treasurer and his nephew Henry F. Pope as assistant treasurer. Henry Pope had joined Cleveland Malleable Iron back in 1869 as an office boy, and he had been climbing the managerial ladder ever since. Eberhard remained under the same name, expanding all the while. In addition to separate saddlery and wagon and carriage hardware lines, the companies added an "auto specialty" series sometime after the turn of the century. By this time Cleveland had been actively producing automobiles for several years, and though it would eventually give way to Detroit, it retained leadership in auto parts production due in great part to versatile firms like Eberhard. The malleable iron casting's wide applicability ensured this segment of the accessory industry life beyond the horse-drawn vehicle. It also owed something to an experimental bent, as improved steelmaking enabled mild cast steel gradually to displace malleable iron. This transformation was reflected in the 1923 name change of Pope's flagship conglomerate to the National Malleable and Steel Castings Company.[46]

The other Cleveland firm quick to seize opportunities presented by new metals technology was the Cleveland Hardware Company. It too began with the 1870s efforts of a group of hardware businessmen and investors. Sometime during 1876–78 Samuel E. Brown, Leander McBride, and Clarence and Lucius Curtiss established Brown & Curtiss, an ironworks on Cleveland's east side. Like many founders of the largest accessory firms, these men shared backgrounds in business and finance rather than horses and carriages. McBride came to Cleveland just after the Civil War and, like Rockefeller, amassed a fortune in the lucrative dry goods trade. Samuel Brown appears to have been the other prime mover, though the Curtiss brothers brought with them diverse business experience and useful investment capital. The company produced a variety of iron goods, but most of it consisted of wagon hardware. Some forty employees turned out the usual range of braces, brackets, and other fittings, as well as a patent brake system and an iron shear for blacksmiths.[47]

In 1881 the firm incorporated as the Cleveland Hardware Company, a name that would become familiar to thousands of wagon and carriage builders before the end of the century. The Curtiss brothers apparently dropped out of the endeavor, but McBride remained as president and Brown as treasurer and general manager. Relocating to an expanded facility in 1882 enabled the company to field a complete wagon hardware line vigorously promoted through catalogs. Cleveland Hardware utilized conventional wrought iron, but the firm broke new ground by pioneering the use of steel in wagon and carriage hardware. Much tougher than wrought iron, steel had great potential advantages for its long-wearing qualities, but high production costs had traditionally restricted it to cutlery, edge tools, and other specialized applications. The American advent of the Bessemer process in the 1860s made affordable steel a reality, and in the following decades manufacturers explored many new uses for the material. One was for wagon and carriage hardware, and here Cleveland Hardware seized the lead, offering the trade its first taste of steel vehicle hardware in 1886. Initially consisting of a few of the more wear-prone carriage parts such as toe rails and rub irons, the list grew quickly as the firm improved its fabrication methods and production capacity.[48]

Not all wagon and carriage hardware could be formed from the rolled steel blanks that Cleveland Hardware used as a raw material, and wrought iron hardware remained in the line. Still, the firm expanded steel's application and more new items swelled the pages of its mail order catalogs during the 1880s and 1890s. One important distinction consisted of its role as one of the few companies manufacturing a complete line of sleigh hardware. Its wagon line ranged from the usual body and gear reinforcements (box rods, corner braces, coupling plates, tongue plates), wear protection devices (pole caps, rub irons), and the general-purpose hardware required in wagon building (staples, strapping, nuts and bolts) to simple tools like the ubiquitous wagon wrench (five different sizes) and even a complete patent wagon braking system ("the only Rolled Steel Brake made"). Carriage hardware accounted for even greater variety, including body accessories like toe rails and dash feet, gear components such as body loops (dozens of styles) and patent suspension goods like stay braces (both Timken and Brewster versions), and a truly daunting array of rolled steel steps: plain steps, embossed steps, rubber padded steps, shaft steps, body steps, steps for specific vehicles, and steps with company logos. Heavily illustrated catalogs paraded stock items (available in incremental sizes), sample buggy steps bearing famous company marques ("Not for sale generally"), and indications that special needs were most welcome: "We can furnish these Rub-Irons in any length desired when quantity ordered is large enough to justify." Finally, Cleveland Hardware continued to market its metal shear and turned surplus rolled steel into tire stock, in which it did a brisk trade.[49]

Such prodigious output required numerous revisions to its plant. Prior to its rolled steel debut the firm had moved from the old Brown & Curtiss buildings to a new location further south, but this too proved inadequate. In 1887 Cleveland Hardware erected a splendid new plant on East 45th, which would remain its main facility for more than forty years. The railroad connection in all three locations is a testament to the vital importance of transportation to these large accessory firms, not only for the influx of raw materials but also to facilitate daily product shipment. The new factory and rolling mill proved a boon to production and allowed the company to employ well over two hundred people by 1890. Rebuilding after a disastrous fire, Cleveland Hardware erected an updated plant. By 1894 this complex included an extensive machine shop, forge shop, combination rolling mill and warehouse, and a separate warehouse and office building, complemented by a half dozen other structures including storage sheds, japanning rooms, and oil tanks, topped off with a scrap iron yard and serviced by no less than three rail spurs. The buildings housed the banks of heavy machinery required to produce vehicle hardware: the heavy engine and rollers turning out rolled steel; the usual range of milling, slotting, stamping, drilling, and other metalworking machines for parts fabrication, along with the power hammers and dies to forge others; and finishing facilities replete with grinders and a japanning department. This was heavy industry indeed.[50]

By the turn of the century, Cleveland Hardware had attained a preeminent position in the vehicle hardware business. It continued to improve its market standing in part by acquiring competitors; in 1900 it purchased the Coss Company, of the Topliff's failed 1890s joint venture. Shutting down Coss's Mansfield plant and moving the machinery to Cleveland, Cleveland Hardware now added bow sockets to its product line. By now the firm employed close to a thousand workers, and its East 45th location had become sufficiently crowded to warrant a second plant on East 79th Street in 1906. Plant no. 2, as it became known, revolved around the huge brick forge building housing the company's new automobile parts division. In order to satisfy greater strength requirements, the firm began exploring large-scale drop forging, a process of increasing importance in many kinds of hardware; by 1915, Cleveland Hardware had the largest drop forge in the country. While drop-forged auto parts were the state of the art, previous success with rolled steel had shown that malleable iron casting was not the only process for making metal wagon and carriage parts.[51]

Leander McBride's advancing age and other business and philanthropic endeavors led him to relegate day-to-day operations to junior officers who piloted the company through the explosive growth of the 1890s and early 1900s. When McBride formally relinquished the presidency in 1909, he turned the reins over to a man who would oversee Cleveland Hardware's transition from the carriage age to the automobile age. Charles E. Adams entered the company in the 1880s largely on the basis of his work ethic and good-natured enthusiasm, qualities that served him well as he worked his way through stints as secretary, treasurer, vice president, and president. Adams shared McBride's dual passions of industrial business and philanthropy, a combination that made Cleveland dually famous. Beginning in the early years of the twentieth century, Adams embarked on a series of ambitious industrial reform efforts, making Cleveland Hardware into a showcase of industrial benevolence. He promoted his company's interests in a variety of ways; active in the CBNA, he left his stamp on that organization even while steering his firm through the transition to auto parts production. Like Eberhard, Cleveland Hardware embraced the automobile as the key to corporate survival.[52]

The Specialty Manufacturing Ethos

Both Eberhard and Cleveland Hardware exemplified the might of the nineteenth-century wagon and carriage accessory industry, particularly in the Midwest. When the U.S. census first began separate tabulation of statistics for the accessory industry in 1880, Ohio and New York easily led all others in number of firms, with just over sixty each, followed by Pennsylvania (47) and Indiana (44). Yet Ohio's 2,073 workers produced goods valued at over $2 million and Indiana's 1,805 managed about the same. By contrast, New York and Pennsylvania each had fewer than five hundred workers turning out a little over $500,000 of product annually. Clearly, massive plants like Eberhard's accounted for the considerable disparity, yet eastern states continued to maintain a presence in the accessory industry. In 1890 Ohio easily retained first place in number of firms, but Pennsylvania, New York, Massachusetts, and Connecticut continued to play important roles. Nationwide, the accessory industry boasted impressive gains during the period 1880–1900; while the total number of firms increased only slightly, everything from the number of employees to the product value increased substantially. Overall, these figures attest to the importance of the accessory trade to the industry it serviced (Table 4.2).[53]

Statistics tell only part of the story, however. Fabrication of hundreds of thousands of vehicle parts imposed demands and presented opportunities encompassing some of the most important developments in nineteenth-century American manufacturing. Specialty manufacturers pioneered quantity production in the carriage industry, both among their fellow parts fabricators and in the ranks of vehicle makers. Early parts manufacturers quickly discovered the high-volume benefits of standardization, and strong internal and external pressures moved them quickly in the direction of goods that went beyond uniformity to true interchangeability. Specialty manufacturers became the real pioneers of interchangeability in the wagon and carriage industry through a more or less logical outcome of the division of labor and mechanization that characterized their production methods. Efforts to increase output led to breaking down the production process into discrete, easily repeated sequences, often in conjunction with specialized machinery. Rationalization created products that were at first more uniform and in many cases interchangeable. Interchangeability, rationalization, and mecha-

TABLE 4.2. Growth of Accessory Industry, 1859–1909

	Number of Firms	Number Employed	Wages ($)	Cost of Materials ($)	Value of Products ($)
1859	8	25	8,436	12,884	29,790
1869	44	453	144,278	214,544	590,878
1879	412	7,502	2,733,004	4,781,095	10,114,352
1889	539	9,996	4,366,233	7,387,904	16,262,293
1899	588	15,387	5,987,267	13,048,608	25,027,173
1904	632	17,160	7,484,450	16,312,683	30,535,873
1909	622	17,388	7,973,771	18,060,866	34,525,635

Source: Adapted from U.S. Bureau of the Census, "Vehicles for Land Transportation," *Census Reports: The Thirteenth Census of the United States, 1910* (Washington, D.C. 1913), 8:473.
Note: Pre-1870 U.S. census data are notoriously incomplete (there were more than eight accessory firms in 1859, for example), but the overall growth trend is still apparent.

nization formed a trio of interdependent factors, as interchangeability became an actual goal rather than just a byproduct of industrialization. Accessory manufacturers wanted their goods to appeal to as many wagon and carriage makers as possible; in order to ensure this, they produced parts fitting the widest possible range of vehicles. Standardized parts appealed to wagon and carriage factories that sought uniform products capable of seamlessly plugging into their own production sequence. Interchangeable parts appealed to small shops for similar reasons.

There were limits to interchangeability, some of them deriving from the nature of the goal itself. Unlike the American military, which conceived of interchangeability as a tactical goal to be achieved regardless of cost, vehicle accessory manufacturers embraced it as a means of efficient production and a key to broad market appeal. On the one hand, producing interchangeable products followed naturally from the attempt to achieve productive gains; on the other hand, these same standardized products appealed to the maximum number of wagon and carriage makers. Sometimes the goals seemed contradictory: while it made economic sense to produce huge quantities of identical goods, the incredible range of sizes and styles even within a single category necessitated numerous production changes, different raw materials, and huge inventories. It all derived from market demands, where fashion reigned supreme. "In no other country in the world," wrote a trade press editor, "is the carriage builder called upon to furnish the number of new [body] styles as in this." Keeping pace mandated flexibility, the only virtue empowering carriage makers to turn out so many different versions. Likewise, an accessory manufacturer also had to be flexible; while many accessory products fit a variety of different vehicles, some had to be designed for a specific style application. Whether making incremental changes to the dimensions of an existing part or designing an entirely new product from scratch, the accessory manufacturer had to contend with fashion's fickle demands. Such flexibility precluded excessively rigid mass production.[54]

As an example, take item 204 from Cleveland Hardware's 1897–98 catalog: its rolled steel road wagon toe rail was a basic accessory installed in dozens of light carriage styles, and this simple cranked-end rod lent itself handily to mass production. The apparent simplicity is deceptive, however, for part number 204 was in reality eight different items: the rail came in lengths from twenty to twenty-seven inches, in one-inch increments, and the company provided custom sizes falling anywhere in between. Even the simplest mass-produced part often came in numerous variants, if only in terms of dimension. Such permutations frequently went beyond mere matters of length and diameter, however. In ordering from the Columbia Spring Company catalog, one had to list no less than ten different specifications, ranging from the obvious number of spring sets desired and their styles and dimensions to more arcane matters of temper (oil or warranted), finish (bright, half bright, or black), and any "additional instructions." Clearly Columbia was more than a custom spring manufacturer; it carried a huge inventory of springs in its standard A, B, and C quality ranges. Still, even these stock items encompassed considerable variety, and the firm could also make any number of additional alterations. That is, it made all its springs with a bright finish—"unless otherwise ordered." Its elliptic springs left the plant without holes—"unless otherwise stated in order." Finally, barring all else, a customer could commission made-to-order springs based on submitted patterns or drawings.[55]

The inherent contradictions between quantity production and shifting product specifications pose a quandary for historians categorizing the accessory industry's manufacturing style. Most firms were more than custom manufacturers, for they produced too many standardized goods. Yet product variations and continued willingness to make items to order took them well beyond classical mass production. Such mixed-format styles derived from a welter of consumer demands and fashion dictates that offered seemingly endless opportunity to produce new items. Thick catalogs full of every conceivable vehicle part equaled money in the bank, even as they mandated significant capital investment, particularly in the form of machinery. While the accessory industry's impressive gains followed from many of the same organizational techniques used in other industries, mechanization remained the single most decisive factor.

Because the accessory industry produced such a wide variety of goods, the range of machinery covered the gamut of nineteenth-century production equipment. In the case of a malleable iron plant, it consisted of heavy industrial machines such as ore hoppers, blast furnaces, ingot cars, molds, cranes, and tumbling bins. Non-cast-metal hardware involved forges, dies, steam hammers, shears, punches, emery wheels, and japanning ovens. Wooden parts required power saws, planers, and shapers, along with steam-bending equipment, belt sanders, and more. Particularly in instances where no commercially produced equipment was available, many specialty manufacturers developed and built their own production machinery; much of the Topliff's bow socket machinery was produced in house, and they were hardly the only ones to do so. A great deal of the specialty manufacturer's equipment came from dedicated machine builders, as both the woodworking machinery and machine tool sectors became substantial industries. Specialty manufacturers especially relied on incisive application of special-purpose machinery. Performing a single task quickly, precisely, and in endless repetition, special-purpose wood- and metalworking machinery poured out the huge quantities and variety of goods so characteristic of the specialty manufacturer.

While nearly all accessory manufacturers used machinery and most relied on at least some special-purpose variants, wheel making provides the best example of accessory industry mechanization. Over the course of the nineteenth century, wheel manufacture drove the development of the largest corpus of special-purpose machinery in the industry. Once the domain of the highly skilled and specialized wheelwright, nineteenth-century wheel manufacture evolved from an exacting handcraft to an operation involving machinery that by its very sophistication reduced the skill necessary to make strong, serviceable wheels. While general-purpose wood- and metalworking equipment abounded in these firms, complex and highly evolved special-purpose machinery fabricated the primary wheel components. The familiar planers and saws prepared the hardwood stock, but special-purpose machines actually made the spokes, hubs, rims, and the rest. This development was well under way by the 1850s, making wheel manufacturers one of the first in the accessory industry to pursue intensive mechanization (fig. 4.3).[56]

Even very early wheelwrights employed simple machines like the lathe, perhaps the oldest such device commonly used in wheel manufacture. Heavy-duty "great wheel" lathes, with their huge hand-propelled flywheels, had turned symmetrical hubs at least since the eighteenth century in America and earlier in Europe. While the great wheel lathe could be turned to other tasks, it, as with most wheel-making machines, was really a device designed for one or two specific operations. As with other classes of machinery, the earliest wheel machinery ran on human propulsion, with later models driven by water or steam. The complexity of most late-nineteenth-century wheel machinery virtually dictated the power source, as nothing else could hope to set into motion such ponderous gear trains, belts, chains, and cutters. Hub turning was not a technically demanding process and mechanized primarily to increase speed, as the great wheel lathe gave way to rapid and powerful water and steam driven versions. Hubs were one of the earliest ready-made wagon and carriage parts placed on the market in significant quantities: on the eve of the Civil War, a Newark, New Jersey, firm boasted an inventory of about forty thousand of them ready for sale.[57]

Mortising the hub required consistent accuracy, as errors here caused spoke misalignment and other problems difficult to correct in a completed wheel. Hand hub mortising thus required great skill but, like other steps in wheel making, became mechanized. The Hayes Hub Boring and Mortising Machine was invented by an Illinois carriage maker in the late 1850s. As the name implied, the machine both bored the hub and chiseled its spoke mortises, and when powered by lineshafting did so in a minute and a half, and accurately. This last especially impressed Cyrus Saladee, who reported on it in *The Coach-Makers' Illustrated Monthly Magazine.* "The mortises are as near alike in every respect," he wrote, "as bullets from the same mould." While Hayes envisioned great sales, we have no way of knowing how well it succeeded. The concept certainly became popular, for similar devices soon came on the market, one from Troy, New York, in 1857. Run by hand or power, the Guard Wheel Machine also bored and mortised hubs. Henry Brewster purchased one, claiming it allowed his workmen to make wheels better than by hand and at reduced cost; wheelwrights would hear more such boasts in the coming years. Eventually the major woodworking machine manufacturers each produced heavy-duty, high-speed automatic hub mortisers based on the same basic principles. Most employed high-speed rotary motion to drive twin mortising chisels and complete an entire hub automatically. Some also bored the central axle hole, but wheel makers could instead purchase one of the many mechanized hub borers available. Salem, Ohio's Silver & Deming company produced one of the first of these in 1868. A round, three-jaw chuck, resembling a metal lathe's, gripped the end of the hub while the workman turned a "T" handle to advance the cutter. Though relatively simple hand-powered devices, such hub borers became phenomenally successful and remained in use in small shops through the end of the industry.[58]

Lathe turning an irregular object had once seemed an impossibility, but in 1819 American mechanic and inventor Thomas Blanchard created an irregular turning lathe for musket stock production and eventually adapted it to other uses. Samuel G. Reed, a wheelwright in North Brookline, Massachusetts, became one of the first to apply it to spoke fabrication when he obtained permission from Blanchard in 1840. During the following decade, Reed pioneered machine-fabricated spokes and other wheel components to become one of the early factory-made-wheel producers. No longer an old-fashioned wheelwright, he employed as many as fifty workers in his factory in the mid-1840s. Not that he entirely eschewed the old ways; he still passed on much of his knowledge to apprentices, one of whom established his own firm in Elizabethport, New Jersey, in 1855. In 1860 this business moved to Newark, where it became Phineas Jones & Company, one of the largest and best-known wheel manufacturers in the country. Blanchard's invention later spread out to vehicle and wheel manufacturers in Cincinnati and St. Louis. Major woodworking tool fabricators eventually brought out improved versions when Blanchard's patent expired in 1862. The Egan Company's spoke lathe incorporated modifications to increase speed and accuracy, among them automatic indexing and a vibration-dampening iron frame.[59]

Spoke manufacture required more than initial shaping; each spoke had to be sized to fit the hub and felloe or rim at each end. Wheelwrights had originally done this with hatchet and draw knife, but a variety of machines gradually supplanted this practice. One early hand-powered spoke tenoning device consisted of a bed-mounted hollow auger turned by a crank. This produced a uniform, round tenon that fit securely into matching holes bored in the wheel rim. Like hand-powered hub borers, this simple item remained a small shop fixture long after faster, more powerful machines assumed the function in wheel factories. Spoke facing and throating became similarly mechanized via equipment employing cutters or abrasive belts and relying on a variety of jigs, stops, or fences to guarantee uniform results. In the larger firms, spoke driving, once performed by hand with a heavy sledge, eventually gave way to mechanical spoke drivers. These were usually a type of trip hammer that mimicked hand methods. However, easy force regulation and spoke guides ensured more uniform results.[60]

One of the greatest breakthroughs occurred with mechanical wood-bending devices. The structure of wood dictates that fracture occurs most easily along the grain, so curved wooden objects are strongest when comprised of a single bent piece. Ox yokes demonstrate this principle, and small amounts of wood bending for similar purposes had taken place for centuries. However, the limitations of excessive stock breakage and other problems prevented its wider application. Successful industrial wood bending began with Thomas Blanchard, who, taking to heart Samuel Bentham's 1793 observation that wood bent best when longitudinally compressed, patented an enormously successful wood-bending machine in 1849. Compressing the wood beforehand ensured that bending expanded the wood's structure only within its original, uncompressed limits. By honoring this maxim, Blanchard's machine successfully avoided the cracks and splits that had defeated earlier attempts. Though best remembered for his irregular turning lathe, Blanchard made far more money on this invention. He did not labor alone; shortly before Blanchard obtained a patent, a Decatur, Illinois, man patented a wood bender to make plow handles for John Deere and still another patented a wood bender for hame fabrication in 1849.[61]

Blanchard found an excellent market for his wood bender among wheel makers. The bentwood wheel rim was not entirely novel; some ancient peoples constructed chariot wheels on this principle, and the same concept appeared from time to time in Europe through the Middle Ages. For unknown reasons the practice died out, and most wooden wheels manufactured until the nineteenth century had rims composed of felloes. Nineteenth-century American wheel makers found that by combining the springy durability of American hickory with new steam-bending techniques, they could construct rims of only two segments. The first recorded sale of the rights to Blanchard's bending machine occurred in 1852 to two New Jersey men who went on to produce vast quantities of bentwood parts for the wagon and carriage industry. Common horse-drawn vehicle bentwood included shafts, top bows, corner pillars, and much more. Around this same time, Cincinnati's Royer, Simonton & Company acquired the rights for the machine; in 1859 this firm's annual production consisted of a million spokes, 160,000 felloes, and huge quantities of other parts. Meanwhile, Cincinnati's John C.

Morris patented a bending machine on a different principle in 1856, and he and Blanchard soon plunged into a lengthy patent suit. Machinery utilizing both methods proliferated, and despite allegations that the Blanchard machine produced more consistent results, most commercially manufactured felloe benders copied the Morris principle.[62]

Factory wheel making involved many other machines, including rim planers and sanders, felloe borers, and dowelers, but the wood-bending principle proved the most revolutionary. Metalworking also played a role; besides application of hub bands, hub boxes, and sometimes more minor items such as tire bolt and felloe plates, much metalworking machinery concerned the iron or steel tire. Wheels had once been tired with curved iron segments nailed to the outer rim surface; these segments, called "strakes," formed a circular iron tire in the same way that felloes composed a rim. Ready availability of strap iron made one-piece iron hoops all but universal during the nineteenth century, and though steel tires came later in the century, the process was largely the same: blacksmiths forge-welded a measured length of strapping into a hoop, expanded it by heating, installed it over the wheel, and contracted it into place by rapid water cooling. Machinery impacted this procedure as well. Where the smith had once bent the strapping by hand or around a worn-out grindstone, geared tire benders increasingly took over this function after the 1860s. Tire heating eventually moved indoors, as wheel makers found they could construct tire furnaces that burned coal or gas. These had the advantage of closer proximity to the work area, produced more consistent and even heat, and some could accommodate several tires simultaneously.[63]

Eventually, however, wheel makers discovered that enormous mechanical pressure could shrink an oversize iron tire and eliminate the heating process entirely. One of the earliest cold setters employed a large hand-cranked screw to compress a band around both tire and wheel. Shops eventually ran belting to these, as well as to successors that used hydraulics. The advent and growing popularity of patent mechanical cold setters touched off a long debate among wagon and carriage men over hot and cold setting's relative merits. Hot-setting advocates claimed cold setters could not create the proper dish, while devotees of the new machines countered that hot setting burned the wheel. In reality, cold setters made sense for firms that could afford them: unlike hot setting, cold-set tires would not burn the wheel, did not have to be quenched, and largely eliminated having to touch up the paint job. The process required no heat, saved time and fuel, and hastened repair work by not requiring tire removal for resetting. Despite contrary opinions, it appears also to have made it easier to create a consistent degree of dish.[64]

The accessory industry's marketing strategy fell into two distinct categories. The first consisted of retail sales to the trade by a middleman, usually a hardware merchant dealing in either general or vehicle hardware. Handling the lines of the various specialty manufacturers most appealing to local carriage makers, such establishments warehoused often considerable quantities of everything from hub bands to top bows. These concerns formed a convenient source of ready-made parts for the small wagon and carriage shop. Most cities with even a moderate number of carriage makers had at least one of these firms, sometimes several. The rise of large mass-production vehicle factories led to the development of the other main category, direct sales. Unable or unwilling to route their incoming parts flow through a middleman, large vehicle factories went directly to the accessory manufacturers, where their buying power won them significant discounts. It proved a mutually beneficial development, for the large vehicle builder streamlined his parts acquisition while the specialty manufacturer took orders of sometimes staggering proportions. Parts retailers remained, for many small vehicle makers, blacksmiths, and repair shops still required the services of a retailer willing to fill small, sporadic orders. Satisfying this demand by no means meant a marginal business, for some of the larger retailers began utilizing mail order catalogs to service customers nationwide.[65]

One of the largest retailers was the George Worthington Company. Sixteen-year-old George Worthington came to Ohio from New York in 1829 and sold picks, shovels, and other tools to laborers on the Ohio & Erie Canal. He opened a hardware store that experienced phenomenal success; by the second half of the century, it had become one of the nation's leading hardware dealers, publishing enormous catalogs, making tremendous sales, and continuing an expansion process that would not cease until well into the twentieth century. By 1887, the firm was taking in millions of dollars in business, employing around ninety men, and selling its products in several surrounding states. In addition to an extensive general hardware line, it carried a particularly large assortment of wagon and carriage hardware. A period guide touting Cleveland manufacturers gives some idea of the firm's tremendous variety: the ground floor contained the offices and sales space; the second floor, harness and saddlery hardware and machinist's tools; the third floor, tinner's tools, horse blankets, and related goods; the fourth contained iron, tin, and brass housekeeping goods; and the basement held the largest selection of iron, nails, and metals in the entire city. Worthington acquired his goods from manufacturers across the country, but many of them, especially the metal hardware items, were made in the greater Cleveland area.[66]

Despite Worthington's commanding position, retailers like Cray Brothers provided considerable competition. A relative latecomer founded in 1896, this Cleveland firm specialized in mail order sales of wagon and carriage materials, blacksmiths supplies, and related goods. Issuing its first catalog in 1897, Cray emphasized its central location and easy rail access to virtually all parts of the country. The 1902 catalog announced, in best huckster fashion, that "we have customers living over a thousand miles away who have found it pays them to do business with us, and why not you?" What followed was 192 pages of anvils, forges, post drills, tire setters, woodworking tools, vehicle springs, axles, vehicle hardware, dashboards, seat cushions, bodies, and wheels. Cray Brothers carried and identified by brand name many products of the machinery, tool, and vehicle specialty manufacturers, including Western Chief post drills and forges, Stanley planes and levels, Disston handsaws, Silver's hub borers, Timken and Mulholland springs, Valentine and Parrott varnishes, and much more. The firm added a line of automotive specialties around 1910 and carried these along with their older staples. Cray even dabbled in manufacturing, running a factory that produced wheels, gears, and tops. After the debut of Ford's Model T, the company wisely sensed change in the wind and specialized in tops for these popular vehicles. The making and selling of horse-drawn vehicle parts clearly had become as large and as remunerative a business as that of making the vehicles themselves.[67]

Summarizing the major developments in the American horse-drawn vehicle industry for a special 1904 historical issue, a *Carriage Monthly* editor wrote that the members of the accessory industry "are in armed league, as it were, against the carriage builder, and have issued an edict that he shall not invade their domain by manufacturing anything that goes into his work." Of course no such "edict" existed, but the accessory industry did have a tremendous impact on the wagon and carriage maker's work. Vehicle builders adopted ready-made parts to overcome production bottlenecks, to increase output, and to remain competitive with others doing the same thing. Beginning with axles, wheels, and springs in the antebellum years and expanding into virtually every other component in the following decades, the range of goods pouring forth from the accessory industry bore witness to the success of the concept. So did the explosive growth of the participating firms. Inability to meet demand proved a pleasant enough problem for I. N. Topliff and hundreds of other accessory makers in that it enabled them to expand their businesses beyond all expectations. Many, if not most, wagon and carriage builders adopted ready-made parts voluntarily, but the rigors of market competition forced even the reluctant to embrace them in order to survive. Not surprisingly, this inaugurated profound changes in the wagon and carriage industry. To the trade journal editor it seemed self-evident that the horse-drawn vehicle maker "must stick to his last, so to speak, and only assemble." Did ready-made parts really turn the craftsman into an assembler? This question cuts to the heart of the experience of work in a craft that had, in the course of a few short decades, transformed itself into a leading American industry. Before exploring this central question, however, we should first take a closer look at two firms whose histories illuminate industrialization's agonies and triumphs alike.

[5]

An Empire of Taste

The Brewster Carriage Dynasty

So far as my experience goes, the whole secret of good work is seasoned timber, and good taste.

James Brewster, 1826

Wherever vehicles are made, wherever vehicles are used, the name of Brewster is familiar. The name Brewster on the nameplate of a vehicle is always associated with quality, and a vehicle bearing it is assumed to be worth more than one without it.

Varnish, August 1891

Readers of popular equestrian publications frequently see advertisements for antique horse-drawn vehicles, the buying, selling, and use of which forms a small but devoted branch of contemporary horse culture. Collectors often specialize in particular marques, and few have greater appeal than Brewster. The very name retains an aura of quiet taste redolent of Fifth Avenue, scarlet-clad coachmen, and old money. The vehicles imply a kind of craftsmanship equally lost to time, a notion only reinforced by generations of eager collectors. This interest has engendered a mythology about the superiority of Brewster carriages, traditional Brewster craftsmanship, and Brewster distinctives stretching back to James Brewster himself. Yet few of these articles of faith have much basis in fact. Like a Herreshoff yacht or a Wyeth canvas, a Brewster carriage might be the product of more than one firm bearing the name; Brewster quality did not invariably exceed that of the competition; all but the earliest Brewster carriages came from factories rather than craft shops. The irony of this most-recognized name in American carriages is how little anyone knows about the people, places, and events behind the builder's plates. The real Brewster story involves four principal family members in three successive generations, twice that many business partners, from three to a dozen different firms—depending on how one counts—a bewildering number of mergers, one dissolution, several bankruptcies, two major court cases, and what old carriage makers recalled as "the most celebrated and picturesque feud that ever was in the carriage trade." This story requires telling, and not just to separate myth from reality. The Brewsters straddled the divide between craft and industry, and their experiences provide an engagingly personal view of a transformation that affected every corner of the American carriage trade.[1]

Genealogically inclined accounts invariably begin with William Brewster, famed Puritan pioneer and Plymouth Colony church elder. The first of the carriage-making Brewsters, however, was James Brewster, born in 1788 in Preston, Connecticut. A small town in the state's southeast corner, Preston is located in New London County, where James's grandfather once owned sizeable quantities of land. His father became a large landowner as well, but mid-1790s financial reversals whittled down his holdings considerably, and he died when James was quite young. The second of eight children, James enjoyed a common school education until age fourteen, when a wealthy family friend offered to pay for private schooling. He admired books, learning, and educated men and yet refused his mentor's offer; though his mother harbored similar ambitions for him, Brewster made his own decision. "I concluded that the surest way to secure a living was to learn a trade," he would write many years later, "and thus escape dependence on others."[2]

Brewster credited this choice to his early and abiding reverence for Benjamin Franklin. While he doubtless took direction from the author of *Poor Richard's Almanack*, his early desire for financial self-reliance owed something to his childhood as well. His father's financial problems, combined with his mother's struggles to provide for eight children, left a deep impression. They amplified pithy maxims about thrift and self-reliance even as they infused an already painfully conscientious youngster with something like performance anxiety. This, as well as his hero's well-known Boston print shop origins, oriented him toward the trades, and in 1804 sixteen-year-old James turned from books to tools and became a trade apprentice. No extant evidence reveals his method for choosing this particular trade or for learning it so far from home, but February 1804 found him far up the Connecticut River in Northampton, Massachusetts, where he apprenticed to Colonel Charles Chapman, carriage maker. Revolutionary war veteran, militia captain, and successful master craftsman, Chapman enjoyed a regional reputation, which may have drawn young Brewster north. We know little about the man's shop and business, and at least some of his posthumous reputation stemmed from his apprentice's eventual fame.[3]

Later accounts of Brewster's life suggest an intensely pious upbringing, but he described his early years as bereft of significant religious inclination. Years later, he recalled his strong temperance convictions in spite of his indifference to organized religion. He made general reference to his mother's conventional piety, and she may have been responsible for this stance, yet his almost palpable terror of alcohol suggests more than conviction by precept. His loathing of alcohol and the hard-drinking early-nineteenth-century craft culture soon collided, but to his surprise his refusal to imbibe won him grudging admiration. "It was a matter of notoriety," he later wrote, "that there was an apprentice in Colonel Chapman's shop who had a natural antipathy to all intoxicating drinks." Brewster's handwritten autobiography concerns such matters to the exclusion of what to him seemed mere mundane facts of existence, and he insisted nothing in his apprenticeship fell outside the "ordinary routine." He spent five years learning to be a traditional carriage maker, one capable of designing and constructing an entire vehicle with the usual hand tools, working twelve-hour days during the summer and as many as sixteen hours a day during cold weather. Brewster learned heavy coach construction, light passenger carriages having yet to become popular, and some stagecoach work also passed through the shop. Chapman's appears to have been a reasonably busy place, with numerous apprentices and journeymen, and Brewster absorbed additional skills and attitudes from them all.[4]

Brewster's unusual maturity endeared him to his master's wife, who served as a surrogate mother to many of the younger apprentices. Brewster credited Mrs. Chapman's faith in him to his temperance views, but Colonel Chapman also developed a fondness for the serious young man, entrusting him with duties and granting him favors withheld from the other apprentices. Around this time, Brewster evinced a strong interest in military matters, an inclination likely informed by his innate love of order and Chapman's Revolutionary War experiences. Chapman was colonel of the local state militia regiment, and Brewster became a sergeant in the same unit. His performance in shop and drill alike so impressed Chapman that in 1808 he invited Brewster to become his business partner. Touched by the gesture but concerned about Chapman's drinking, Brewster politely declined. The following year he "attained his majority" when he reached age twenty-one, craft apprenticeship's customary terminus. At the "entertainment" Chapman held in his honor, he drank to Brewster's health and made a flattering prediction: "James," he said, "if you are as faithful to your own interests as you have been to mine, you will be a rich man." But he could not resist adding another observation: "I fear there is one thing which will cause you much trouble: you are honest yourself, and will be apt to confide too much in the professions of others." Chapman correctly surmised a character trait that would indeed cause Brewster future vexation. As a final token of esteem, he obtained a diploma from the state's Mechanical Association and presented it to the young man he had come to admire so greatly.[5]

Brewster harbored ambitions to travel to New York City and work in a metropolis already highly influential in the trade, but he first answered the call of a Pittsfield, Massachusetts, shop desperately seeking temporary help. He spent six arduous weeks working exceptionally long hours, but the high pay scale netted him $60. As eager to save money as he was to see his mother before reaching New York, he decided to return to Preston on horseback. On the suggestion of a "professed friend" offering seemingly generous terms, he purchased a horse on credit with the assurance that the balance would not be due until he reached home, where he could pay at a discount. So it was a chagrined young horseman who reached Northampton only to find his note demanding immediate payment. Brewster dutifully fulfilled his obligation, wiping out most of his newly earned savings. In a single stroke he had validated his old master's warning and learned his first lesson about the dark side of business. The remainder of the trip made a sad sequel; only a few miles outside of Preston he mired his horse in a bog. He had no choice but to go for help, returning later with some local farmers who helped him pry the neck-deep animal loose with fence rails. Wet, mud splattered, and deeply discouraged, the hapless traveler reached home where he slept off his fatigue, sold his horse, and borrowed $30 from an older brother. Years later he recalled that "with this capital—a balance of $30, against me—I commenced the voyage of life; and since that time I have not received, as a gift, one dollar from any person." It was a Franklinesque beginning.[6]

Brewster climbed aboard the New York stage that September, but once again events seemed ordained to frustrate his efforts. Delayed by a routine stop in New Haven, he wandered up and down the town's shady streets until happening upon a local coach maker's shop. He learned of the proprietor's need of a journeyman and filled the position himself. The Orange Street firm belonged to the town's first carriage maker, John Cook, who had been in business since 1794. New Haven was still a modest city in 1809, but one with important coastal and blue water maritime trade and a growing reputation for light manufacturing. Determined to live frugally, Brewster roomed at a local boarding house and began saving money to open his own shop. Though initially employed by Cook, Brewster also worked in nearby Stratford. He still managed some diversions, however, and besides reading he attended church regularly, eventually ending up at the Yale College Chapel. Here he took an immediate and lasting liking to Dr. Timothy Dwight's preaching. College president since 1795, the minister was at the height of his powers as a champion of conservative Calvinism, and from this point until Dwight's death in 1817 Brewster remained a regular attendee. During his first year in New Haven, the carriage maker also began courting the woman who would become his wife. James Brewster married Mary Hequembourg of Hartford in September 1810, just a year after his arrival. Marriage ended more than his bachelorhood: later that year he established his own carriage shop, having spent only a little more than a year as a journeyman. While the position brought new challenges, it must have seemed preferable to prying horses out of bogs.[7]

Brewster's first shop was a simple-one story building on the corner of Elm and High Streets. Though he performed the majority of the work himself, he quickly acquired apprentices who helped him meet the needs of the many customers who had followed him over from Cook. He also acquired a business partner. Brewster's memoirs refer to him merely as "Mr. B," "a young man of respectable connections, but not well-informed in the business." Clearly Brewster settled for a cooperative capitalist rather than a fellow craftsman, and this likely contributed to his modest 1811 profits. His partner's new wife felt "mechanics" beneath her, however, and so "Mr. B" sold his share to Brewster who soldiered on alone.

Continued business success allowed him to purchase his first home, a former High Street plater's shop that he remodeled into a dwelling. Throughout his career Brewster built many more houses and assisted in New Haven's development, but the immediate future brought harder times. Mounting tensions between Great Britain and her former American colonies dampened business, especially in luxury items; according to Brewster, not a single quality carriage rolled out of New Haven's shops during the period 1812–15. Yet lean times brought an unexpected bonus in the sudden popularity of light carriages. Many who once would have ordered coaches and other heavy work turned to lighter vehicles that were cheaper to purchase and maintain. Brewster plunged into light wagon manufacture with improved designs until, as he recalled years later, "'Brewster's Wagons' were all the rage." "Wagon" connotes agricultural vehicles, but here it referred to lightweight, multiple-purpose passenger carriages such as pleasure wagons. The four-wheeled design made them easier riding than the two-wheeled chaises and riding chairs that had been the coach's primary alternatives, and they made a great deal of sense in the American market. Brewster was only one of numerous carriage makers rushing to fill the need. As he later pointed out, this trend eventually spawned the buggies, phaetons, rockaways, and other light carriages that soon dominated the trade.[8]

Brewster survived the mid-teens in reasonably good shape, but his constant business finance struggles reveal the master craftsman's often unappreciated difficulties. Despite austere economies and increasing sales, he continued to turn only meager profits, a persistent problem that occasionally plunged him into depression. Brewster found acquiring business savvy a frustratingly slow process. Like so many other new tradesmen, the New Haven carriage maker was discovering that the master craftsman's world contained hidden challenges that had nothing to do with spoke shaves, bench planes, and timber. Too often historians' portrayals of the craft shop ignore socioeconomic complexities by assuming that business consisted of little more than a simple supply-and-demand market relationship. Yet in even the most rudimentary enterprises there arose situations requiring problem-solving skills and entrepreneurial vision that craft training failed to bestow. Much had to be acquired through trial and error, and here Brewster's perseverance served him well. His early discovery of good credit's virtues supplies just one example; his favorite anecdote on the subject concerned his efforts to secure agreeable rates on a small business loan. The local bank president granted the money to Brewster instead of a rival applicant primarily on the basis of character. The president resided near both and knew that Brewster, "to my certain knowledge, gets up two hours the earliest every morning!" Such memories likely warmed the young carriage maker rising in those brittle-cold hours before dawn.[9]

Persistence paid off as the 1820s rewarded Brewster with greater personal and professional success. War turmoil had faded by 1817 when his wife gave birth to their first child, James, known as Benjamin to the family. Meanwhile Brewster's expanding business necessitated his 1821 purchase of John Cook's larger quarters. By this time Cook had taken two sons into partnership to form J. Cook & Sons, Coach Makers, and they moved to a new shop nearby. The aging Cook evidently feared sharp rivalry, but Brewster assured him of his own commitment to honest competition. Cook took the young man on his word, and they continued as friendly competitors and brother tradesmen.[10]

Brewster's business had increased sufficiently to require an accountant in the meanwhile, and he hired Solomon Collis to keep his books.

The following years were busy ones, with the birth of several more children, including the last, Henry, born in 1824, adding to his family while foreign markets added to his bottom line. New England's long-standing maritime trade involved many kinds of commodities beyond the familiar staples. Brewster was only one of several East Coast carriage makers to investigate foreign trade when he began selling carriages in Cuba during the 1820s. Inspired by news of the sales success of a peculiarly Latin carriage called a volante, Brewster determined to build a few. They sold so well that he felt sure a much larger demand existed, a conviction shared by several regional competitors. New York City carriage maker Milne Parker specialized in Cuban and Mexican volantes, sometimes having his blacksmith tow them through town as advertisement. Brewster gained customers in Havana, Matanzas, and Santiago through local repositories, but his agent, a "not a very enterprising or responsible person," proved unequal to the task and sales soon trickled off. If Brewster failed to fully realize his Cuban hopes, the additional money and publicity certainly helped. An 1824 *Connecticut Journal* advertisement announced both his U.S. and "Spanish Market" sales, and during this period Brewster also sent work to Charleston, South Carolina. He noted with pride that his fellow New Haven residents increasingly preferred locally made carriages.[11]

What was Brewster's shop like? The scattered clues suggest a scenario familiar to any carriage maker of that era: a master craftsman, apprentices, and jours using hand tools to construct horse-drawn vehicles. Most, if not all, of Brewster's apprentices lived in his home, and he eventually counted jours among his hands as well. He followed the common practice of placing a jour at the head of each of the four main divisions so that apprentices in woodworking, blacksmithing, painting, and trimming had adequate training and supervision. He expected jours to teach by example, and not just about carriage making: "An oath, or an obscene word . . . ," he warned them, "may affect millions." He habitually thought first of his impressionable young apprentices. Brewster himself carried on in the manner of the owner-operator, working alongside his charges. All put in customarily long hours, Brewster having settled on the fourteen-hour day: twelve hours of labor with an hour twice for meals. The exact daily routine depended on the season, with a later start and evening work during the winter, but the six-day week generally totaled around seventy-two working hours.[12]

One of Brewster's "financial expedients" from this period consisted of Saturday cash wage payments, a plan with two novel features. The first was regular distribution on a fixed day, in contrast to the irregular, as-needed paydays more common in the trades. Much of this had often consisted of scrip, so Brewster's early conversion to cash was another part of a larger plan to put his shop on a sounder business footing. This involved his own growing reliance on cash transactions, elimination of running bank and employee accounts, and generally streamlined bookkeeping. Regular cash payments proved exceedingly popular in his shop even as they economized his time. While a common practice only a few decades later, it ranked as an unusual innovation then, one Brewster later recalled with evident satisfaction.[13]

Brewster and his assistants built carriages from the wheels up. Power and machinery were unavailable and working from rough-sawn lumber required the usual range of hand tools, though Brewster did employ some of the limited ready-made parts then available. He routinely placed want ads for lumber and hardwood spokes; an 1820 *Connecticut Journal* notice typified his approach: "WANTED immediately 3,000 first rate carriage spokes. An extra price will be paid for those that are seasoned." Depending on the size and style of the wheels, three thousand spokes translated into anywhere from 187 to 250 wheels, or enough for roughly forty-six to sixty-three carriages. Though he surely accumulated wheel stock for future use, in this instance his desire for preseasoned spokes suggests immediate application. Brewster emphasized matters other than quality raw materials, however. "The common saying, 'it will do,' or 'it will pass,' will not answer: it has ruined its thousands," he liked to say. Yet he was not an impractical perfectionist, and his growing business acumen tempered his ideals with pragmatic caveats. Witness his intuitive grasp of fashion, what he called "the fancy of men," and the necessity of balancing it against quality: "The utmost care is indispensable that the work be made strong, as well as fanciful and neat," he insisted, for both contributed to trade success. "Work, unless made expressly as cheap work, and to be sold as cheap or depreciated work, should be made in the best manner." Even so great a craftsman as James Brewster conceded that low-cost carriages had their legitimate place.[14]

So much of Brewster's attitude toward his work flowed from his world view that it bears some explanation. While his outlook certainly stemmed from more than one source, by the 1820s it increasingly revolved around religion. Having been instilled from childhood with a fierce sense of moral rectitude, Brewster early adopted regular church attendance as a fixed personal habit. His rigid personal ethics found resonance in Timothy Dwight's strongly logical preaching, and he listened with growing attraction to the same words he later revered in the minister's posthumously published *Theology Explained and Defended* (1818–19). Two of Dwight's theological distinctives were to have decisive impact on Brewster. The first was a thoroughgoing emphasis on reason and logic, whether in rational defense of Christianity or to solicit individual responses to the Gospel. This owed much to a body of thought known as Scottish realism. An eighteenth-century British movement with enormous influence on nineteenth-century American theology, Scottish realism stressed common sense and the mind's God-given ability to grasp reality. Though its rationalistic overtones stemmed partly from the Enlightenment, American divines like Dwight ironically appropriated it as a corrective to that movement's freethinking heretical excesses. Dwight's systematic theology depended on a highly rational framework, and it found particular resonance in the young carriage maker in the rear pew.[15]

According to his own account, Brewster still sensed something missing from his life. Influenced by Dwight's expository sermons, he became increasingly conscious of his inability to measure up to God's standards as revealed in the Bible. This was more than a flare-up of Brewster's familiar self-censure; the early 1820s found him in the throes of a religious crisis. It came to a head during the winter of 1824–25, when he turned from divine law's harsh requirements to the free gift of divine grace. This too owed something to the Yale sermons, for the second influential component of Dwight's theology was an emphasis on revival, itself predicated on a personal response to God. Dwight was one of numerous American ministers and laypeople who precipitated what became known as the Second Great Awakening (ca. 1795–1830), and this wave of piety engulfed James Brewster as it had thousands of others. Not that he converted in revivalistic fervor; his turning point came through the slow working of his conscience, informed by Bible reading, and punctuated by Dwight's evangelistic preaching. As Brewster later explained it to his shop hands, his had been a journey from skepticism to belief:

> In my youth, I said, and fain would have it so, that the Bible was not true, for the reason that unless I repented, and availed myself of the mercy proffered in the Gospel, I was lost. My conscience denied me this resting-place. I then would have it, that as Christ died for all, I was Safe; and I believed that, do what I would, I should be saved. The further I examined, the more strongly I became convinced that my hopes were groundless and delusive; for the Bible told me that my hope was on a false foundation. My necessities carried me to a throne of grace, where,

> by a confession of my sins, and constant importunity for mercy, I trust I found forgiveness.

Ever after Brewster dated his real sense of purpose to this time when he turned from his own inadequacies to embrace the truth of John 3:16. Brewster's remained a highly rational world view replete with "natural law" and the perils of human agency, but he now found them bearable through divine grace. "I am not ashamed to own the Lord Jesus Christ," he told his apprentices, and this proved the key to a crisis that remained solved to the end of his days.[16]

Outward signs of Brewster's inward change manifested themselves immediately. In 1825 he and his family became formal members of New Haven's Church of the United Society. Even before this Brewster required his live-in apprentices to attend services as part of their participation in his own family life, but after 1825 his efforts took on new meaning and intensity. So too did his interest in the welfare of those outside his immediate care. The same year marked the commencement of Brewster's charitable involvement, beginning with New Haven's Alms House, the city's facility for helping the community's poor. This institution suffered from inadequate funding and neglect, and Brewster deplored what he found on his visit. He spearheaded a community effort to improve the physical structure and its inhabitants' living conditions. Town meetings, committees, reports, and a great deal of Brewster's own time and money led to building renovations, a school room, and chapel. In addition to paying the new chaplain's salary, the carriage maker taught a Sabbath school class there for seven years.[17]

His renewed emphasis on helping others also found outlet in his own shop. Trained in a system where the master related to his apprentices in loco parentis, Brewster had always taken seriously his surrogate status. Beyond accompanying his apprentices to church, he maintained a curfew, forbade alcohol, and provided educational and entertaining reading material. None of these measures were unusual for the time and place, even though not all masters took them as seriously. Starting in the winter of 1826–27, Brewster embarked on a lecture series specifically for his apprentices' benefit. His motivations derived from two circumstances. The first was late 1820s national politics and the Jacksonian political machine's growing momentum. Characterizing the Jacksonians as "a set of harpies or demagogues who are doing all they can to instigate the poor against the rich," he objected that this led to class warfare and anarchy. He took even more strident exception to a political ethos that denied that poor young men could better themselves. In Brewster's conception the Jacksonian Democratic faith operated on the assumption that in the absence of massive wealth and power redistribution, only wealthy young men could be upwardly mobile. This was nothing short of anathema to a self-made man like Brewster. His own experiences told him poverty spurred ambition and effort, a conviction he repeated so often that his apprentices said "Boss Brewster says it is a blessing to commence the world poor." Where the Jacksonians saw in poverty an impenetrable barrier surmountable only through collective political action, Brewster found in it a uniquely powerful competitive advantage. The two viewpoints could hardly have been more diametrically opposed.[18]

Brewster's actions emanated less from innate political instinct than from the abrasion of an ascendant philosophy running counter to his own convictions. Brewster's friends and apprentices noted his constitutional disinclination to political strife; one suspects he found that it overwhelmed his already highly strung temperament. While he fulfilled citizenship's basic political requirements, he never relished politics for its own sake. The fire behind his passion lay elsewhere, and politics animated him only as much as it impinged on his deeper convictions. These, the second motivation for his lectures, flowed from his recent embrace of evangelical Christianity. Brewster's tendencies prior to his conversion had already assumed many of the forms familiar to the dedicated believer: adherence to principle, attention to duty, and life lived according to immutable precepts. In light of his newly discovered faith, however, he increasingly acted not just on principle, but out of love. "In the relation I sustain to you," he told his apprentices in the fall of 1826, "I should be morally culpable were I to limit my duty merely to teaching you the art or trade of coach-making." He felt responsible for their habits and morals and wanted them to achieve more than mere trade competence. "You have the means," he went on, "by the exercise of diligence and virtue, of obtaining respect and property, and also of making glad the heart of a father—a mother—nay, more: of securing for yourselves a blessed hope of immortality." Duty and devotion urged Brewster to address his assembled apprentices not merely as workers but as fellow human beings with eternal potential.[19]

Each once-weekly talk took place on a given day at 9 P.M., the end of the winter workday. One imagines the apprentices racking tools, sweeping floors, and tossing aside their aprons before gathering around the stove to hear "Boss Brewster's" latest thoughts. Brewster employed spare-time notes supplemented with extemporaneous comments to hold talks ranging from thirty minutes to an hour's duration. The topics ranged from business practice ("Prominent Points Necessary to Success in Business" and "The Acquisition of Property") and working life ("Consideration of the Mechanical Professions—with Comparisons") to character ("The Necessity of Forming Correct Habits"), underlying principles ("The Influence of God's Moral Law," "On the Popular Error, Chance"), and evangelism ("On the Influence of the Bible, and Respect for the Sabbath," "The Value of the Proverbs, with Scriptural Quotations"). While the content varied, certain themes recurred: the importance of good habits, worthy examples, right conduct, and the tremendous potential of personal agency even in a world constituted by God on fixed principles. A title of a later address summed it up best: "Man, under Providence, the Architect of his own Fortune."[20]

Too many historians dismiss such episodes as the meddling of an oppressively paternal zealot, but Brewster's expectations were rather commonplace. In its ideal form, apprenticeship really did aim to inculcate character as well as skill, even if not all masters took this ideal so literally. Brewster had confidence in his approach and—even if only later—so too did many of his apprentices. During the late 1850s the aged carriage maker invited several former apprentices to a reunion where their remarks validated their master's actions. Robert Nelson Mount, member of an artistic Long Island family, remembered Mrs. Brewster's motherly kindness ("something more than I always merited") and how Brewster himself would stay up to meet a late-retiring apprentice, reminding him "in a mild way" of his transgression. E. W. Andrews spoke for many of them when he said that "the wayward tendencies of some of us were such, that, had it not been for the paternal restraints he imposed, the lessons of wisdom he taught, and the irresistible proofs of the tenderest regard for our welfare he gave," none might have succeeded as they had. He took pains to relate how Brewster's legacy extended far beyond mere restraint. Three years into his apprenticeship in the paint shop Andrews began spending his spare time studying, which eventually helped him pass the bar to become a prominent New York City attorney. In retrospect, many of Brewster's former apprentices recognized the value of his guidance during their formative years.[21]

Back when these boys first sat down for after-hours exhortation, Brewster had started to realize some long-desired success. Part of it derived from incursions into the New York City carriage market. Though still a relatively modest-sized city, New York formed an important venue for urban and rural carriage makers alike. Rural shops located within reasonable shipping distance had long sold their goods on city streets, though not without opposition. Abram Quick, one of the city's leading carriage makers during the 1820s, decried increasing importation of "country work." At a meeting convened to discuss the issue, he put it this way: "Gentlemen, we are in a similar position with our Saviour on the cross: we are between two thieves,—Connecticut on the one hand, and New Jersey on the other." According to Ezra Stratton, who had intimate knowledge of the New York carriage trade, such opposition led to the establishment of carriage repositories specifically for selling country work. While this hardly eliminated all friction, it did grant greater legitimacy to "out of town" vehicles. Around 1823 George Burnie opened what was probably the first such establishment, soon followed by several others. Strangely enough, Quick, for all his disdain of the trend, provided James Brewster a foothold when he leased his Broad Street shop and repository to him in 1827. Located near Manhattan Island's southern terminus, Broad ran through the city's commercial district. Brewster planned to use Quick's repository for sales and the shop for repairs.[22]

Quick's reasons for getting out of the trade do not survive, but he cared enough for his apprentices and jours to arrange for their continued employment. Brewster agreed to retain them so long as they refrained from drinking in the shop, a condition to which all readily agreed. He appointed one of them, an energetic young man named John R. Lawrence, to act as supervisor. Lawrence eased his new boss's anxiety over this experiment in remote business management, a service Brewster would not forget.

While he apparently sold some of his own vehicles through his new repository, Brewster also stocked it with John Cook's work. What we do not know is how he handled a related and potentially thorny issue. As we have already seen, repositories routinely affixed their own name plates to the vehicles they sold, regardless of their specific origin. Third-party repositories, those with no factory of their own, normally created no misunderstandings this way, for the firm's entire stock obviously originated elsewhere. However, proprietary repositories, those owned by a carriage maker, frequently placed their "house label" on their own and outsourced work alike, a potentially misleading but widespread and perfectly legal practice. The decision of how much to reveal about a product's origins lay strictly with the retailer, and he had a number of options. He might simply sell outside work with outside name plates or no plates at all—these were by no means universally used. If a proprietary repository purchased work in the white and finished it in house, it might legitimately claim to have had a hand in its manufacture and so justify its sale under the house label. Even in the case of work purchased fully finished, a repository often described it as "built to our specifications" and its own name tag primarily validated guarantee and repair. When the question arose at all, some repositories gave intentionally vague answers while others simply lied. While it seems unlikely that Brewster engaged in deliberate sophistry, the murky issue would eventually feature prominently in his sons' business experience.[23]

Brewster's star was clearly on the rise by the late 1820s, and he deepened his local charitable commitments accordingly. The same year he established his New York repository saw him commencing work on a brick auditorium in New Haven. He intended Franklin Hall for public functions but especially for one of the city's educational organizations. Founded by a handful of local craft apprentices in 1826, New Haven's Apprentices' Literary Association (which became the Young Mechanics' Institute in 1828) was the local manifestation of the popular educational Lyceum Movement. Several founders of the New Haven society worked in Brewster's shop, and the carriage maker eventually became a benefactor. In addition to helping its members out of minor legal trouble stemming from youthful pranks, Brewster payed for the group's rented room and supplied it with the nucleus of a scientific library. He also paid for lectures by Yale College professors, going so far as to provide the necessary laboratory apparatus. All the while he accrued trade distinctives, including building a carriage for newly elected President Jackson in 1829. Brewster likely sensed the irony, but he evidently considered a Democrat's money as good as any.[24]

Near the end of his life James Brewster informed an acquaintance that "copartnership" formation was one of his favored business strategies. Indeed, just two years after making him repository superintendent, Brewster elevated Lawrence to partner, changing the New York firm to Brewster, Lawrence & Company; in 1830 he invited his bookkeeper Collis to become his New Haven partner to form Brewster & Collis. Perhaps he conceived these to facilitate future expansion, for 1831 saw him planning an especially bold move. By now Brewster had become a seasoned contractor and developer, and he grafted his new expansion plans to a substantial real estate deal. In January 1832 Brewster purchased a large land tract called New Township, property that would eventually become part of New Haven itself. The site abutted the local steamboat landing, an important consideration in his plans, and on a nearby lot he erected a new shop on a new street named Wooster. The structure's size and style, with a fully equipped cost of around $20,000, made it more than a simple shop. By the time it was completed in May and occupied by about one hundred employees, it became apparent that James Brewster now owned a carriage factory.[25]

As with similar endeavors by James Goold and a few others later in the decade, however, Brewster's new enterprise was something of a transitional establishment. Retaining the old craft system's form and some of its substance, the facility embodied distinctly rationalized features. Neither quite the familiar craft shop nor yet a heavily mechanized factory, it embodied many of the ambiguities of a trade in transition. Given the scarcity of power and virtual nonexistence of production machinery, the core changes involved work organization. The changes aimed at increasing production to fill his repository, and a new facility allowed him to recast production methods along more effective lines. Brewster was doubtless aware of advances made by such American System manufacturers as his own state's clockmakers, and he adapted some of their techniques to his own enterprise. While his men were still woodworkers, blacksmiths, painters, and trimmers, their tasks were increasingly subdivided into simpler, rapidly executed segments. Even as he perpetuated the craft's time-honored forms, these same forms lost substance in their translation into a factory environment.

The Wooster Street carriage makers kept the familiar hours, including seasonal evening work, but those hours assumed an increasingly rationalized texture. More rigidly fixed starting and stopping times were posted throughout the factory and marked by bells, a significant departure from relaxed craft shop rhythms. If the men continued to use long-familiar hand tools, each employed them for more restricted tasks. Brewster not only retained his regular cash wage payments, he also began paying them to his apprentices as well as his jours. By the 1830s, then, Brewster engaged in the syncretic practices found in other trades experiencing similar structural shifts. Was Brewster now a master craftsman paying wages to his apprentices, or was he a factory owner whose wage earners bore antiquated titles? We have no record of how or even whether Brewster faced this pivotal question, and, in any case, the matter persisted long after his own retirement. He continued to take on apprentices and jours, even though he had less time to devote to them personally; hand work and craft methods remained in a more rationalized setting; and it would be twenty years before another New Haven firm truly mechanized a carriage factory and persuaded Brewster that steam-powered machinery represented the future. Thus it would be as much a mistake to portray Brewster a betrayer of his own craft heritage as it would be to assume his Wooster Street facility was a mechanized production plant. Both the man and his factory were very much works in progress, with all the contradictions the phrase suggests.[26]

Another symptomatic sign of change was Brewster's increasingly lengthy absences from the shop floor. Well before he opened the new factory he was already spending less time at the bench himself. The amount of financial wrangling, real estate speculation, construction, and renovation attached to his name required more attention than a fully employed tradesman could devote. Part of his rationale for making Lawrence and Collis partners was to free himself from some of his daily trade concerns. Though he still put in full days at what he called his "manufactory," he was as likely to be found at a desk in the office, for his were increasingly the cares of an influential developer and businessman. By the early 1830s Brewster had contributed mightily to New Haven's growth, and his prosperous carriage business and charitable work only enhanced his growing reputation. So it comes as no surprise that when local railroad interest reached fever pitch, the carriage maker found himself drawn in.[27]

Though Brewster later characterized his involvement with the Hartford & New Haven Railroad as his greatest mistake, at the time it seemed only a logical extension of his expanded business ventures. English experimentation with iron rails and self-propelled vehicles in the 1820s piqued American interest, and by the 1830s dozens of charters and a few operating railroads like the Mohawk & Hudson (1831) and the Boston & Worcester (1835) inspired others to follow suit. In 1833, James Brewster and several leading New Haven businessmen and investors formed an association to promote a New Haven to Hartford railroad and obtained a state charter. Though Brewster evidently saw himself as a promoter and financial backer, lack of initiative among his fellow speculators propelled him to the company's presidency, where he fielded correspondence, commissioned the route survey, and bankrolled more and more of the operation. Throughout the early phase of this project, Brewster managed to keep his hand in carriage making, though his profits increasingly went into the railroad. At the same time, his son Benjamin's sixteenth birthday compelled him to consider the boy's employment. In 1833 young Brewster commenced work in the paint department before eventually winding up in the office. No information survives about the degree to which Benjamin mastered the craft, but he certainly became familiar with the overall process. By this time the traditional craft apprenticeship was undergoing rapid change, and the mere existence of an office indicated how far removed Brewster's business had become from the little shop on the corner of Elm and High Streets.[28]

Circumstances intervened that materially affected not just the boy's plans but his father's as well. The year 1835 ended with a disastrous December fire in New York. Though not affecting Brewster directly, millions of dollars in destroyed and damaged property hurt many railroad stock subscribers and sent ominous shivers through the regional economy. A much more direct threat came during February with another fire, this one at Brewster's factory. An unusually cold winter had closed the harbor, and when the fire broke out the building contained some $60,000 in finished work awaiting shipment. Brewster carried insurance for only half that amount. While his workers and local fire fighters saved some of the vehicles, the loss still totaled around $17,000. Meanwhile the fragile structure of easy credit, paper money, and rampant speculation that had artificially stimulated the national economy began collapsing. James Brewster had the tremendous misfortune of facing a burned-out factory and a stalled railroad venture just as the nation plunged into the economic depression started by the Panic of 1837.[29]

Brewster pondered what to him seemed an unavoidable choice: factory or railroad? The factory had not been entirely consumed, and his insurance and financial reserves allowed immediately rebuilding, just as his New York repository remained a reliable sales outlet. For all its belabored progress, the railroad's success still seemed only a matter of time. The rapidly worsening panic, however, seemed to preclude moving ahead with the business and the railroad together. In the end, Brewster stepped back from both. In February 1837 he retired from the carriage trade by selling his interest to his partners, the New Haven factory becoming Collis & Lawrence and the New York repository Lawrence & Collis. The Hartford & New Haven Railroad had been plagued with political and financial problems from the very beginning, and by now persistent nonpayment of stock and perpetual fundraising efforts made the job of president an ulcerous one. When the board finally accepted his resignation, Brewster felt immensely relieved.[30]

The Family Trade

The period 1836–37 marked a distinct break in James Brewster's trade experience, as he coped with a disastrous fire and a nightmare speculative project. Meanwhile, Benjamin too left the trade for the dry goods business, but both soon returned to their accustomed work. In early 1838, James and Benjamin joined with New Yorker Jonathan W. Allen to form a pair of new carriage trade partnerships. The first was a New Haven shop opposite his former Wooster Street factory, an enterprise that went under the name James Brewster & Son; the second was a New York City repository called James Brewster & Company. This was also a new facility, further north on Canal, one of the many side streets off Broadway. This location offered cheaper rent but remained close to the better neighborhoods of his potential customers. James Brewster's role consisted primarily of lending his influential name and considerable finances, leaving Benjamin and Allen in charge of the business itself. An 1840 New Haven advertisement for Brewster & Allen suggests further reduction of James Brewster's involvement, but disaster intervened in 1842 with his son's bankruptcy. In later years, Benjamin lightly dismissed the episode as one of the unavoidable aftershocks of the 1837 panic; years later, an acquaintance claimed that Benjamin temporarily abandoned business and family, leaving his father and partner to face creditors. Brewster settled the outstanding debts, but doing so cost him money and no end of embarrassment. This would not be the last time Benjamin would experience severe financial setbacks, and it was only the beginning of James Brewster's frustrations with his sons.[31]

In the aftermath of this episode, Brewster sold his extensive eastern New Haven properties during 1842–44, including the East Street shop, his palatial home nearby, and several other pieces of real estate. While he and his wife accommodated their advancing age by moving to a smaller home, Brewster continued to earn an income through real estate speculation, and he still had before him the problem of his sons' employment. Henry appears not to have had even Benjamin's cursory trade apprenticeship; twenty years old in 1844, he had recently given up preparations for Yale for the sake of business. As son of a prominent carriage maker his choice seemed obvious, and perhaps—one imagines their father thinking—two brothers would succeed where one had not. So James Brewster entered the carriage trade once again, this time with both sons. Benjamin appears to have been nominally in charge, Henry serving as a clerk until attaining his majority. This time there would be no New Haven factory, or any factory at all. The new firm occupied the former Canal Street building in New York that had acted as a repository for carriages made by others but sold under the Brewster name. When Henry turned twenty-one he became the third partner in James Brewster & Sons, which in 1848 moved to a more advantageous Broadway location. It remained strictly a repository, buying "country made work" from firms like Bridgeport's J. Mott & Company (fig. 5.1).[32]

In 1849 the trio hired a young man named John W. Britton to act as bookkeeper, salesman, and clerk. Born the son of a Staten Island cobbler in 1823, Britton enjoyed a limited New York City public schooling before working in a Manhattan retail store. In 1847 he became a salesman in Isaac Mix's Broadway carriage repository and evidently entered the Brewsters' employ at Benjamin's suggestion. Despite limited formal education, Britton was an ambitious young man with an insatiable curiosity fed by omnivorous newspaper reading. Though sometimes impatient and exceptionally stubborn, his charismatic personality eventually carried a great deal of weight in the trade. James Brewster took an early liking to him, an affection that quickly proved mutual. While Britton evidently did not share the older man's religious fervor, they each cared deeply for the welfare of young working men. Just as he had done years ago with Solomon Collis and James Lawrence, the senior Brewster would eventually grant John Britton a substantial stake in the business. First, however, he sent him to New Orleans.[33]

In their ever-expanding search for additional markets, northern carriage makers continued to look southward, and by the 1850s many had found a bonanza in sales to the southern states. Some of the more ambitious firms established repositories there, and in 1851 the Brewsters elected to join them. New Orleans was a bustling, cosmopolitan city whose planters, merchants, and other upscale residents formed a tempting market for fine carriages. John Britton took charge of the New Orleans repository under the name Brewster & Company and immediately began sizing up the competition. Besides other out-of-town firms, their sharpest rival was local carriage maker John C. Denman. It was a risky endeavor given high transportation costs, communication delays, and even disease; yellow fever, dysentery, malaria, and other scourges killed many every year, and those from northern climates were especially vulnerable. While Britton remained healthy, his wife and two children repeatedly fell ill. Despite this, Britton radiated confidence as he exploited local demand for Brewster's heavy C-spring coaches, as well as lighter work made by others like the Cooks. Still, the Brewsters soon had second thoughts and in 1852 authorized Britton to sell the remaining inventory and return to New York. Any disappointment from this was quickly muted by more positive developments back home.[34]

Britton returned to a partnership in James Brewster & Sons through purchasing some of Benjamin's stock. Henry and his family now lived in Bridgeport, where he supervised the firm's newest venture. By the early 1850s the Brewsters sensed the disadvantages of lacking their own factory and remedied this by purchasing Bridgeport's struggling Union Carriage Company in 1852. Union Carriage had begun three years earlier as a joint stock company founded by striking carriage workers and was funded in part by showman and Bridgeport developer P. T. Barnum. The company floundered from its January 1852 commencement and, after a failed bid to lure Britton away from New Orleans, opted to sell the entire business to the Brewsters. Henry moved to Bridgeport in 1852 to oversee the operation, which henceforth supplied the New York repository with coaches and other heavy carriages. In order to meet growing demand for light work, the firm purchased a New York concern already experienced in this line. Over on Third Avenue, Lawrence & Townsend had been turning out light carriages and road wagons, and James Brewster & Sons bought them out, making Henry their agent and pairing him with James W. Lawrence to form Brewster & Lawrence. They retained the addition intact, and though Henry's name appeared on the paperwork, he remained in Bridgeport. Now James Brewster & Sons had both a heavy coach and a light carriage factory and seemed well positioned to meet all comers.[35]

They probably were, had not an even greater challenge come from inside their own organization. It began in the mid-1850s, when Benjamin suggested he and his brother divide ownership of the repository and factory. Though this would effectively split the family business, each firm would continue to receive James Brewster's advice and financial backing while Henry made carriages for Benjamin to sell. In July 1856, James Brewster & Sons and Brewster & Lawrence dissolved by mutual consent; Henry and Britton withdrew and with Lawrence formed Brewster & Company, recycling the name of the short-lived New Orleans branch. This new partnership purchased the stock and assets of both the Bridgeport factory and the Third Avenue facility and moved everything into newly leased quarters on Broome Street, another east-west street intersecting Broadway. Henry moved back to New York, and Brewster & Company opened for business in mid-July. Benjamin meanwhile combined James Brewster & Sons and Brewster & Lawrence into one firm, strictly a repository under the name James B. Brewster and operating out of the 396 Broadway property. Per the agreement, Henry would supply his vehicles as well as lower grade "country work" to Benjamin, who in turn would furnish a guaranteed sales outlet. Presumably, business would proceed as before.[36]

James Brewster had consented to all this against his better judgment. Dividing a relatively new enterprise between two inexperienced sons ran counter to everything he had learned about business. Still, Benjamin had been persuasive on the matter and had also showed signs of reassuring familiarity with the New York City business world. Brewster may have decided to test his oldest son's maturity, or he may have simply given him free rein for lack of a viable alternative. Henry objected to the confusingly similar firm names, but he also understood his role as family subaltern and went along with the scheme. Britton left no record of his opinion and, as both a minority partner and non–family member, perhaps remained silent. Benjamin's partners nearly all harbored nagging doubts about the venture. Benjamin justified it on the grounds that the reorganization would enable him to more easily disengage from the business; avowing his intent to remain only until Henry and Britton had become fully operational, Benjamin informed Henry that he planned on liquidating his assets and indulging in rural retirement.[37]

None of this sounds especially convincing; only thirty-eight in 1856, Benjamin had hardly accumulated a retirement-sized nest egg, and he still owed his father money from his bankruptcy. If Henry suspected a ruse, he may have been correct. Benjamin was animated not by visions of a bucolic future but by growing resentment toward his younger brother. Part of the animosity likely stemmed from the often blurred lines separating family and business. One of proprietary capitalism's underappreciated byproducts is its tendency to inflame family disputes. Benjamin relished his firstborn status and assumed this guaranteed his ascendancy to head of the business, yet his disastrous past complicated an otherwise logical conclusion and made him combative. Though his father sufficiently trusted him to manage the family business funds, Benjamin likely sensed a lingering undercurrent of mistrust.

For his part, Henry objected to having only indirect access to his father's money and, like the prodigal son's brother, resented his father's seeming willingness to overlook Benjamin's transgressions. Henry felt his contributions to the business had been muffled beneath his father's experience and his brother's belligerence. As he later recalled, Benjamin "always assumed the control of the business . . . and if I attempted to assume any of the rights supposed to belong to a partner in business, I was met with sneers and insults." An already fixed habit of inserting himself between his father and the other partners heightened Benjamin's abrasiveness and may itself account for the reorganization. If Benjamin could not master his younger brother, then perhaps he could at least segregate him. The extant correspondence suggests as much, but anything beyond this is speculative. Those outside the family remained close-mouthed, and sources speak mostly of subsequent developments. In time, the cumulative aggregation of slights and perceived slights, grudges and innuendo, charge and countercharge, transformed a private matter into a public display of embarrassingly bad taste. As the layers of personal, financial, and business disputes accumulated, so did its internal heat, until the rank midden finally exploded into open flame. Even while their father still lived, it became distressingly clear that the heirs to the Brewster carriage dynasty hated one another.[38]

Some of the bitterness came from business difficulties immediately following the 1856 split. Despite the enormous advantages of previously acquired assets and James Brewster's name, both firms faced daunting competition. Benjamin's was hardly the only city carriage repository, and more than one observer predicted the early demise of the impossibly large, six-story Broome Street factory. The factory itself faced the greatest challenge. While a repository required only minimal accommodations, a production facility necessitated much larger expenditures. James Brewster estimated that it took twice as much capital to operate a factory as it did a repository, an observation Henry soon validated. One indicator of the level of anxiety was the torrent of letters passing between New York and New Haven: Benjamin and Henry requested note endorsements while Brewster authorized loans and scribbled advice. Above all, he wanted his sons to realize their interests were "indissolubly connected" and must remain so to avert disaster. Yet Brewster became increasingly frustrated with both his sons, and by early 1857 he and Benjamin were exchanging sharply worded letters.[39]

Meanwhile Henry continued to spend money as if it grew on New Haven's elms. Life in a city famed for everything from elegant restaurants to pungent oyster stalls opened new gastronomic possibilities even as it expanded his waistband. "I feel exceedingly anxious to effect a change in Henry's habit of expenditure," his scandalized father wrote to Benjamin that summer, urging him and Britton to force Henry to curtail his spending. The specter of failure loomed over them all: the senior Brewster assumed greater liabilities than he felt prudent, and by the fall of 1857 he warned Henry that Brewster & Company's collapse would ruin him financially. "The fact is my age should excuse me from assuming large liabilities," he wrote, "for in spite of all my reasoning and efforts it depresses me exceedingly." He did take heart at funds from sale of the 396 Broadway property. It was another of his firstborn's ideas to which he had reluctantly agreed, but the profits helped them all immensely, while a cheaper repository site served just as well. Brewster happily credited Benjamin for his foresight, but no real estate transaction could heal the brotherly rift. Near the end of 1857 Brewster had to mediate a new business agreement between the firms, as the original pact broke down from continuous bickering. While they all managed to survive the year, the old man's bank account and his patience wore increasingly thin.[40]

The year 1858 brought more bad tidings in Brewster & Company's lack of profit and Henry's profligacy. By summer Brewster had all but given up on his youngest son. Not only had he refused parental advice, he did so with "great want of respect"; one of Henry's many letters had so upset the old man that he found himself unable to eat. Henry's appetite remained hearty as ever, even if to Brewster it seemed that "gormandizing had absorbed all his better feelings." Brewster sympathized with Britton and Lawrence, whose modest pay seemed inadequate recompense for having to put up with Henry and the danger of his dragging them to ruin. "Were I in their place," he wrote in July, "I should be discouraged!" Increased activity at the Brewster & Company factory during the fall lifted the gloom somewhat, and not just for its New Haven backer. "We have never had as many orders as this Fall," Henry wrote optimistically in September, "and I feel like confining myself closely to business, and try and make up the bad luck we have had of late years." He spoke of night work in the smith shop and rushing to meet orders. "We are working hard for a reputation, and our customers are beginning to appreciate it."[41]

While a variety of "start up" problems typically plague most new manufacturing concerns, Brewster & Company suffered from a more a fundamental malaise. In concentrating primarily on traditional heavy coach work, it fell victim to inflexibility in a changing market. Heavy carriages continued to sell, but light carriages formed the trade's real growth sector. James Brewster had seen the beginning of this trend back in the 1820s and 1830s, and subsequent development of buggies, rockaways, and other new styles only accelerated the changes. Light, handy, durable, and often requiring only a single horse, such vehicles made obvious sense in pavement-poor America. By midcentury a particular subtype arose from enthusiasm for fast driving on city streets. Carriage makers catered to demands for lightness and speed by building stripped-down, open buggies designed specifically for pleasure driving. A typical example consisted of a narrow, low-hung tray body equipped with a seat for a single occupant and mounted on a set of spidery patent wheels (fig. 5.2). Roughly analogous to the twentieth-century sports car, such "road wagons" or "speeding wagons" became increasingly popular in urban areas. They ended up saving Brewster & Company from ruin. After another gloomy annual inventory in early 1858, Henry, Britton, and Lawrence decided to shift their emphasis to light work, speeding wagons especially. Relying on Lawrence's considerable practical carriage-making skills, Brewster & Company made the necessary design and production changes to turn out road wagons. This accounted for the busy factory in the fall of 1858.[42]

The following year held promise as Brewster & Company's new product met ready sales even in the face of competitors touting the same goods. A spring 1859 advertisement announced the firm's new specialty, its "Light Weights" being especially intended for "Speeding," having "received the approbation of the best authorities on the road." While Benjamin ought to have been overjoyed at this eleventh-hour reprieve, it instead elicited more jealousy, something he even admitted to Henry in an unguarded moment. Benjamin continued to respond disingenuously; even as he avowed his hope of retirement, he lay the groundwork for his own factory. In the winter of 1859–60 Benjamin had carriage maker John C. Parker carefully measure a Brewster & Company speeding wagon with an eye toward duplicating it. Meanwhile, Benjamin opened negotiations with two potential business partners. On the very first day of 1860, he joined Theodore E. Baldwin in forming Brewster & Baldwin, carriage dealers, at 786 Broadway. This unannounced move startled Henry and his partners, none of whom cared for Baldwin or his reputation. What really provoked their ire, however, was Benjamin's second partnership, one specifically organized to manufacture speeding wagons (fig. 5.3).[43]

John C. Parker, younger brother of the volante-building Milne Parker, was an experienced carriage builder who had plied his trade both in New York City and nearby Yorkville; his Twenty-Fifth Street shop became Benjamin's new carriage factory when he became Benjamin's other partner. Parker had already established a reputation for his own light speeding wagons, thus Parker, Brewster & Baldwin featured the Parker Wagon as its signature product. This naturally angered the makers of the Brewster Wagon, who now faced unexpected competition. Whether or not Benjamin authorized any purloined measurements, he clearly intended to cash in on the trend. Once again the brother's business agreement collapsed. Not only had Benjamin acted surreptitiously, he had done so to usurp Brewster & Company as his primary supplier. Henry and Britton reminded Benjamin of his impending retirement, to which he retorted that "with regard to my business, both in view of retiring from or manufacturing in connection with it, I reserve to myself the right of determining, without any consultation on the subject." While Benjamin refused to state plainly his intentions, his actions suggested he would not be leaving the trade anytime soon. Several years later, Britton pinpointed the events of 1859–60 as precipitating open warfare between the Brewster brothers.[44]

James Brewster found solace investing his time and money elsewhere. In 1855 he realized a long-cherished goal in dedicating New Haven's new Orphan Asylum; two years later he hosted a reunion of some of his former apprentices. After a dinner at his home, they and the general public gathered to listen to a talk by the old man. Committed as ever to molding youth, Brewster addressed New Haven's young men by repeating the maxims and homilies so familiar to his former charges. This took place in Brewster's Hall, another and more spacious public auditorium Brewster built for the city in 1845. Not that the aging master craftsman had entirely abandoned his original occupation, for he remained an active investor in the local carriage trade. In 1858 he became a silent partner in New Haven's G. & D. Cook & Company. Established in the early 1850s, the firm produced large numbers of inexpensive carriages for the southern and western markets. The active partners were George and David Cook, the former one of John Cook's sons, as well as Hannibal I. Kimball of the well-known New England carriage-making family. Their factory boasted the latest in wood- and metalworking machinery driven by a powerful steam engine. Brewster openly admired George Cook for his business sense and his innovation. Brewster wrote to Benjamin that Cook's "*very ingenious* application of its power" had changed his mind, "as I was formerly averse to the use of steam." He thought it might be advantageous for Henry to follow suit.[45]

National events once more drew James Brewster into business complications. The outbreak of the Civil War threw many New England carriage manufacturers into chaos, and even substantial concerns suffered devastating blows by losing their southern markets. G. & D. Cook & Company failed in 1861 but with Brewster's assistance reorganized as a joint stock company—only to fail once more. Again Brewster faced the unpleasant choice of accepting his loss or making another risky attempt at recovery; once again he opted for the latter. With three other creditors he formed Hooker, Candee & Company to make inexpensive light work in the former Cook factory. During the middle of the war the firm became Henry Hooker & Company, with Brewster still a major stockholder. By this time he could no longer count on continued good health; 1864 found him suffering from heart trouble, though his spirits remained bright. Sustained by an abiding faith, he faced the future "as calm and peaceful in my mind as when I have laid me down to sleep till the morning." He remained so even in the grip of typhoid fever, which had him bedridden by the late fall of 1866. Slipping in and out of delirium, he softly repeated the lines from Alexander Pope's "Universal Prayer":

> Teach me to feel another's woe,
> To hide the fault I see:
> That mercy I to others show,
> That mercy show to me.

Not long after, the old carriage maker breathed his last.[46]

James Brewster died on November 22, 1866, a very wealthy man. His assets included more than $30,000 in Henry Hooker & Company stock, sizeable portions of which went to Benjamin and Henry, the former's less his remaining bankruptcy debt. Both sons inherited substantial fortunes, but they continued to lack the character their father so earnestly entreated them to acquire. Though the feud had already become public, James's death removed the last vestiges of restraint. Benjamin's business life became a flagrant attempt to match his brother's every move. He had been doing so at least since 1860, when he had issued a promotional lithograph mirroring one offered earlier by Brewster & Company. When Henry's company sold "Brewster Wagons," Benjamin used Parker's 1867 death to justify changing his "Parker Wagon" to a Brewster. His signboard disingenuously listed 1838 as his date of establishment, his vehicle labels sported "Old House of Brewster," and his *Times* and *Turf & Field* advertisements appeared next to Henry's. It was war, and war explains the failure of the one conciliatory overture stemming from James Brewster's death. In the spring of 1867 Benjamin approached Henry about Brewster & Company buying his business; he essentially offered to bow out of the trade in return for a handsome lump sum. Henry dismissed this as blackmail and turned him down flat.[47]

It was too late; there had been too many insulting remarks, too many scurrilous advertisements, and too many vitriolic letters penned in white-hot fury. The feud had taken on an inertia that only personal integrity and sacrificial character could overcome. Neither Benjamin nor Henry proved equal to the task. Benjamin's distaste for Brewster & Company grew to hideous proportions, as he derided it as a feckless aspirant to the Brewster mantle. Henry shared his brother's indiscretions by giving free rein to a partner who relished lashing back at what he regarded as a pathetic competitor. John Britton wrote most of Brewster & Company's advertising copy and voiced scorn for Benjamin all the while. "I am going to follow James B. Brewster to his grave," he allegedly declared in the summer of 1869, "and then I will evacuate upon it." Benjamin responded by accusing Britton of once saying Henry was not worth "shucks," being only valuable for his commercially useful last name. Benjamin may have fabricated this, but he may have told the truth. Britton naturally denied making either statement, yet they were entirely plausible for a man known for naked contempt of his enemies. Britton resented Benjamin's exclusive claim to the Brewster name and legacy as much as Benjamin resented Britton's warm relationship with his father. If Britton openly excoriated Benjamin, one wonders if he secretly loathed Henry as well.[48]

In what appears to have been a last-ditch effort to vindicate his side of the dispute, Benjamin brought suit against Henry, Britton, and Lawrence in New York's Superior Court on October 14, 1869. The lengthy complaint averred Benjamin's long standing in the trade, previous business with his father, and that Brewster & Company owed its very existence to Benjamin's "advice, counsel, and means." It accused the firm of "conspiring and confederating" to libel his reputation, having "wickedly threatened to vindictively pursue" him to destroy his business, and that its "unjust, wicked, and unlawful acts" had caused him "fifty thousand dollars damages." He cited two recent Brewster & Company advertisements as evidence and thoughtfully enclosed clippings for the judge's perusal. He wished to restrain Brewster & Company from publishing similar advertisements and to recover damages. Four days later, Henry, Britton, and Lawrence appeared with their counsel to fight back with characteristic vigor. The court adjourned after hearing arguments and, pending a decision, ordered Brewster & Company to refrain from further accusatory advertisements. Both sides produced dozens of affidavits and sworn statements, some touching on events as far back as the 1830s, while disputing everything from the degree to which James Brewster had once been Benjamin's partner to the nature and conduct of current businesses. Given the contested nature of nearly every significant event over a roughly thirty-year period, it comes as no surprise the judge required time to digest it all. In the final analysis, however, the suit boiled down to this: Benjamin claimed to be heir apparent to the Brewster name and business and as such could manufacture and sell the "Brewster Wagon" without interference.[49]

A dynastic battle over succession in a family business had become increasingly complicated by matters as disparate as factory production, ready-made parts, repository sales of outside work, and questions of who might commercially exploit a name denoting both product and family. Advertising copy highlighted many of these concerns; at the end of an otherwise conventional trade announcement in *The Spirit of the*

Times, Brewster & Company reminded readers it had "no connection whatever with the person of similar name, so long notorious as a dealer in Cheap, Country made Carriages, on Broadway, . . . impudently claiming the reputation which they alone have made for the 'Brewster Wagon.'" While technically accurate on the matter of Benjamin's product source, this claim also implied country work's inherent defectiveness even as it conveniently ignored James Brewster's past role as one of those same "out of town" manufacturers. The brand-name-cum-surname raised the stakes, as attested by a second Brewster & Company advertisement warning that "a person who has the good fortune to bear a name similar to our own" had been advertising "a cheap imitation" of their own work in revenge for their refusal to buy out his business. It went on to insist that Benjamin's claim to manufacture the "original Brewster Wagon" was as spurious as it was deliberately deceptive.[50]

In its desperate struggle for survival, Brewster & Company had stumbled on a new reality of the carriage trade: the ascendant value of brand name recognition. Increased commercialization since the 1840s generated sharper competition, and brand names and trademarks became useful means of navigating in this more complex and competitive environment. Brewster & Company's counsel cautioned that the applicable trademark legislation dated back only to the 1840s and failed to provide an exact precedent. He thought that Brewster & Company could make a legally defensible claim to the "Brewster Wagon" despite the appellation having evolved through informal public recognition rather than concerted advertising. Still, conflation of a de facto trademark and a shared last name clouded the matter of Benjamin's interference, even given his transparent attempts to confuse the public and benefit from Henry's new-found solvency. If Brewster & Company had the right to call its product what it wished, had not Benjamin the right to use his own last name?[51]

The dispute was further muddled by the simple fact that the vehicle type in question was in no way original to either party; it was merely another light buggy variant intended for recreational driving. Vehicles of this kind had recently rolled out of many different shops under a variety of names, few of them more imaginative or less proprietary than the Brewsters'. If anything, ease of manufacture through ready-made parts designed specifically for such vehicles prevented any one manufacturer from monopolizing the concept. By the 1860s, carriage makers chafing to join the road wagon craze needed to look no further than the catalogs of the various spring, shaft, gear, body, hardware, and wheel suppliers to get the necessary parts. If the business of carriage sales had the virtue of easy entry, so too did carriage manufacture inasmuch as the proprietors employed "store bought" parts. Although manufacturing still required more capital than sales did, heavy reliance on outsourced components dramatically decreased overhead in expensive items like woodworking machinery. Brewster & Company prided itself on in-house vehicle manufacture, but it hardly eschewed ready-made spokes, hubs, springs, and other goods. One of its 1859 road wagon ads proudly noted its use of English coach lace and trimming materials, as well as W. H. Saunders's axles, and these were hardly the only outsourced items in the firm's stock rooms. Both James Brewster and Benjamin had made even greater use of such goods prior to Brewster & Company's 1856 birth.[52]

What originated as a simple, if intense, family squabble came to embody many of the complexities of a craft becoming an industry. In their efforts to justify their respective positions on repositories versus factories, sales versus manufacturing, and the right to a contested trade name, the Brewster brothers demonstrated not only the perils of primogeniture in an industrial age but also the frustrations of proprietary capitalism in an environment where the rules constantly changed. The craft they inherited from James Brewster differed enormously from the one he learned in 1804, and the wracking strain they all felt in the 1850s and 1860s signified a greatly intensified commercial environment. In their efforts to defend their businesses first against outsiders and then from each other, the Brewsters appropriated many of the ambiguities of the industrialized carriage trade as weapons in a desperately bitter war. Patented suspension systems, ready-made parts, and steam powered woodworking machinery took their place alongside brand names, illustrated carriage catalogs, and creative advertising to form the arsenals of warring companies commanded by fratricidal leaders.

Brewster & Company's counsel wisely refused absolute prediction in surveying the proposed legal battle; he felt confident that Brewster & Company stood an excellent chance of refuting Benjamin while validating its own interpretation. Then he made the most useful suggestion of all. Since such cases "wherein brothers are parties, testing the rights of each to the use of the common surname as a trade mark, the plaintiff must make out a strong case for the Courts will otherwise lean on the side of the party attacked," he wrote to Henry. In other words, Brewster & Company retained benefit of the doubt unless Benjamin had the money, time, and legal skill to prove his point. Barring that, Brewster & Company had already won. No record survives of how Henry and his partners responded, but in any event they had little time to ponder; on

November 3, Benjamin walked into the court house and discontinued the suit. Perhaps he doubted his ability to sustain protracted litigation, or maybe he intended only to make a point without going beyond the opening salvo. Whatever the case, he dropped charges, settled costs, and opted to forgo legal satisfaction.[53]

Going Their Separate Ways

Continued business solvency may have influenced Benjamin's sudden reversal, for Parker's former Twenty-Fifth Street shop had since become a respectably sized carriage factory. Now a multistory building covering more than a dozen city lots, equipped with a large steam engine, elevator, and other up-to-date features, Benjamin's factory likely excited envy in at least some of his competitors. A heavy-handed promotional article in a Chicago newspaper from the previous year described a standard configuration: basement smithy, ground-floor drafting and woodworking departments, and upper-story painting and trimming shops. Every vehicle started out in the design room, where a draftsman employed floor-to-ceiling blackboards to render full-size elevations for each new design. In the basement the smiths worked the usual assortment of Scandinavian and domestic wrought iron as well as the Bessemer steel used for axles and tires. They had the advantage of centrally supplied air for the fires, though power made few other inroads there. One level above, shop hands operated the latest woodworking machinery, one estimate claiming mechanization enabled the firm's approximately one hundred employees to produce twice what they could by hand.[54]

Acknowledged as labor-saving, the machinery also permitted uniform parts fabrication with obvious assembly and repair benefits. The newspaper reporter felt that more exact shapes, tighter joints, and higher levels of finish all flowed from the repeatable accuracy of extensive mechanization. Each of the four main branches formed a distinct department, with its foreman familiar with all the work. Their combined supervisory and inspection roles reveal Benjamin's understanding of an important element of rationalized factory production. The writer noted rigid inspection of in-house and outsourced components as intrinsic to "Brewster quality" but might have specified its role as enforcing the uniformity vital to quantity production. The basement steam engine mostly powered woodworking machinery, forge blowers, and the freight elevator, though there may have been some other devices such as a belt-driven paint mill. Despite the newsman's obvious bias, he accurately surmised the benefits of the economy of scale when he wrote of Benjamin's ability to "sell a better carriage and sell at a smaller price, than any of his rivals who do not have equal facilities." An obvious dig at Henry, this nevertheless touched on the essential truth of machinery and rationalization.[55]

Trimming and painting entailed fewer novelties, but their upper-story location and custom finishing abilities indicated Benjamin followed common practice. As of summer 1869, the factory included a small repository, which, according to the article, sold only in-house work; Benjamin had already announced plans for a much larger repository adjacent to the factory. In another stab at his brother, Benjamin stated that he saved the cost of Broadway showroom rent in order to offer lower prices to his customers. His work included popular styles such as coaches, landaus, phaetons, coupes, light trotting wagons—anything likely to be used for city, country, or turf. Such luminaries as General Grant "of the late unpleasantness," Alexander T. Stewart of the New York dry goods house, and Collis P. Huntington of the Central Pacific Railroad owned J. B. Brewster carriages, and the firm shipped work as far away as Europe and Hong Kong. According to the piece, Brewster personally supervised the entire operation, Parker having died and Baldwin now back on his own. The factory belonged to Benjamin alone, and he clearly took pride in his achievement.[56]

Benjamin's repository plans took their cue not just from the need for additional sales space but also from his obsession with Brewster & Company. In late 1867 the Broome Street house opened a sumptuous Fifth Avenue repository in a renovated mansion in an upscale residential neighborhood. Delmonico's restaurant opposite, already a New York institution, was the only other business in the immediate vicinity. Ezra Stratton, charting its progress in *The New York Coach-Maker's Magazine*, described a Doric-columned edifice with a fashionable Mansard roof and an iron-fenced courtyard for outdoor displays. Inside, the ground floor had nineteen-foot ceilings paneled in expensive woods, and the staircase newels featured two four-foot coachmen carved from solid walnut. The steam-heated building provided ample space for up to 175 Brewster carriages and a hand-propelled freight elevator allowed easy stock movement. A clever set of reflectors provided nighttime illumination of the first floor display in an effect Stratton declared "charming and magical." When it opened on December 9, 1867, the $40,000 facility was easily the finest carriage repository in existence. Benjamin's, completed three years later, obviously strove for similar effect. His four-story Italianate building, featuring enameled brick facing, columns, and the now obligatory Mansard roof, shouted "class" in the accepted New York manner. The siblings clearly remained rivals.[57]

Benjamin likely timed this grand opening to coincide with yet another change. In January 1869 he had dissolved his partnership with Theodore Baldwin, leaving him the 786 Broadway repository, and changed his Twenty-Fifth Street firm's name to James B. Brewster, incorporating in 1870 as J. B. Brewster & Company. With this, each firm now assumed what would be its best-known form: Brewster & Company and J. B. Brewster & Company were the competing New York City houses that became so celebrated over the years. Often referred to by their locations long after subsequent moves, the Broome Street and Twenty-Fifth Street Brewsters remained fixed opponents whose feud became as well known as their quality carriages. Benjamin enhanced his name by developing a popular patent gear. To overcome the sidebar wagon's weak body-to-gear attachment, he added a second pair of half-springs hung from the bars to provide extra body support. It was a classic case of a simple solution to a widespread problem, one he patented in 1872. Known as the "Brewster Patent Combination Cross Spring," but popularly as the "Brewster Side Bar" or "Brewster Gear," it became one of the most popular buggy suspensions of all time. It was perhaps his single greatest triumph.[58]

Both firms underwent incremental expansions, which sometimes necessitated moving factories, repositories, or both. Episodic leapfrogging up Manhattan reflected the city's northern sprawl, as each firm sought real estate close to its prized upper-class markets. Brewster & Company made its most important change in 1874 when it abandoned Broome Street for the corner of Broadway and Forty-Seventh and a brand-new factory. The culmination of nearly twenty years' manufacturing experience and not a little dreaming, the new plant followed plans Britton had drawn up several years earlier. Like the Fifth Avenue repository, it was a model facility garnering its own media coverage. *The Hub*'s George Houghton had recently inaugurated a series on "Modern Carriage Factories," and he chose Brewster and Company's for the June 1874 issue. He cited three reasons for including this one: its great size, its efficient design, and the impressive amount of machinery and "modern conveniences." The building, which we would today classify as an industrial loft, was a five-story brick structure with two hundred feet of Broadway frontage. The floor plan revolved around a central courtyard surrounded on three sides by the roughly C-shaped building. As befitting its location on the city's premier shopping avenue, the exterior presented ranks of stately pillars, rows of windows, and a roof line crenellated with what seemed like dozens of small chimneys (fig. 5.4).[59]

Britton's design followed the general pattern of basement smithy and successive upper-floor departments but nearly all made innovative departures from standard practice. He had devoted special attention to the blacksmith shop, with the large number of chimneys (one for every two of the forty brick forges) providing better than average ventilation. He gave equal attention to lighting in the form of brightly whitewashed walls and cleverly situated street-level skylights. These last cast daylight on a row of filers' benches as well as on the main floor, where courtyard windows provided additional light. Excavated to the basement level, the courtyard served as a staging area for newly unloaded raw materials, where they remained until distribution via elevators to the various departments. The smithy's stock area was a model of neatness with 154 sizes and grades of iron and fifty-two sizes and grades of spring and tire steel in orderly, labeled racks and bins, giving the foreman exceptional inventory control and the workmen speedy access. A steam engine occupied a corner room; besides powering forge blowers, it also drove the adjacent heavy saws, planers, edgers, and other woodworking machinery. The convenience of a basement location for heavy equipment and its proximity to the power source evidently suggested this unorthodox arrangement.[60]

Though the firm retained its opulent Fifth Avenue repository, the factory had its own on the first floor, which also housed vehicles awaiting repair. This floor contained the office as well, which featured novelties such as speaking tubes leading to each department, a telegraph connection to the Fifth Avenue repository, and controls for the twin rooftop steam whistles regulating the workday. Besides a separate receiving and repair office, unloading ways leading to the courtyard, and a repair or "jobbing" department, the first floor also had entrances to the exterior elevators and three iron stairways, themselves designed for fire prevention and to preserve useable floor space. Trimming took place on the second story; this department featured a large stock room and cutting tables and shared the floor with a wheel-building department and silver plating room. The third floor housed the woodworking shop, which featured a steam-heated glue warmer, a steam box and wood bending jig, and a steam-heated lumber drying room. Nearby stood a bank of small power saws and "dressing up" machines (planers, joiners, and perhaps a shaper) powered from overhead lineshafting. Since Britton had elected to place the majority of the heavy woodworking equipment in the basement, these auxiliary machines spared woodworkers excessive trips downstairs. An innovative drafting department on the same floor rendered the latest European and New York designs on paper and full-size chalkboards.[61]

The final two stories were devoted to painting. The fourth-floor paint shop contained the space and equipment for sanding, priming, painting, striping, and varnishing carriage gears. Careful partitioning prevented finish coat contamination and the separate varnish room boasted tightly fitted doors for additional protection. The fifth floor contained the body painting department, along with a rubbing deck for intermediate wet sanding and a pair of body varnish rooms. It featured a body storage area for work in the white; having nearly completed bodies on hand enabled the firm to swiftly fill orders for the popular styles while retaining the ability to provide customized paint and trim. Like all the other floors, the uppermost had its own water closet, elevator, and stairway access, as well as buckets, hoses, water tanks, and Babcock extinguishers. These last, plus the exterior elevator and stair towers and interior iron doors, aided in preventing and containing that bane of so many nineteenth-century manufacturing establishments.[62]

Houghton came away deeply impressed. What impacted him most was Britton's overarching commitment to rationalized factory production. The floor plan promoted regular work movement without undue "backtracking," and raw materials flowed smoothly from delivery wagon to elevator to storeroom to application. While hardly a production line's linear work flow, the arrangement nevertheless marked a significant attempt to treat the entire facility as an efficient machine. Houghton felt the factory contained "one of the finest assortments of machinery that has ever been brought together for the purpose of coach-building." Most of it came from the city's own Wood & Light Machine Company, though other firms supplied the wheel-making machines. The machinery alone cost $25,000, and the cost of the fully equipped factory totaled more than $500,000. As of May 1874, the firm employed just over 400 men, including 101 blacksmiths and helpers and 84 painters, with the remainder spread among the other departments. Since its occupation in January of that year the company had been turning out a vast quantity of work, 180 carriages (both new and refurbished) in April alone. Brewster & Company's new factory stood as veritable showcase of nineteenth-century industrial production.[63]

Brewster & Company and J. B. Brewster & Company were at their respective pinnacles of productive popularity during the 1870s and 1880s. In the early 1870s Henry and Benjamin popularized "spring openings," open house sales events modeled after those in the dry goods, jewelry, and clothing trades. In 1878 Brewster & Company received international acclaim for its Paris Exposition display; the firm carried off a fistful of medals while Henry's lapel glinted with the French government's Decoration of the Legion of Honor. Additional recognition came in early 1879 in the form of a Tiffany gold commemorative tablet, as well as a scroll and autograph book containing the signatures of hundreds of subscribers, many direct rivals, grateful for the advancement of their collective international reputation. The carriage trade remained competitive, but it also retained a large measure of gentlemanly decorum.

Brewster & Company celebrated its twenty-fifth anniversary in July 1881 with a dinner for 539 employees and more than 70 guests, the kind of lavish celebrations the company increasingly hosted during these years. Benjamin made a strategic move in 1882 when he relocated his repository up Fifth Avenue to Forty-Second Street. In 1885, Henry's son William, who had formally joined the firm a few years earlier, sailed to France to attend the Parisian Dupont school. His nine months of study added to the company's ability to exploit the enormous American interest in European-inspired carriage design. Brewster & Company had long since become carriage makers to the elite, the firm from which the Astors, Vanderbilts, Morgans, Rockefellers, and the rest of the fabled "Four Hundred" bought their carriages. Cultivation of an exclusive clientele actually allowed the firm to raise its prices despite the dominant industry trend in the opposite direction. Henry and his partners recognized the glint of coin and made the most of it while they were able.[64]

Two of them were not able much longer. John Britton began experiencing puzzling weakness during 1884 and after a frustrating round of ineffective treatments, his doctors could only point to his driving personality and work habits. In 1886 Britton reluctantly sold his New York residence, rented his summer home, and prepared for what he hoped would be a healing foreign tour. As a final token of respect for the firm's 450 workmen, he surprised them with a predeparture announcement of reduced daily hours, making the new standard nine hours daily, Monday through Saturday. This proved his final such talk, for sixty-two-year-old John W. Britton died from a massive stroke in Austria a few months later. This was a particularly sore blow to the man who had depended on him for so long. Henry, allegedly known as "the best-dressed fat man in New York City," found his own health declining in the face of his lifestyle and eating habits. Still, it surprised nearly everyone when sixty-four-year-old Henry Brewster died in September 1887. Benjamin swallowed his pride long enough to attend the funeral, though he likely fumed over Brewster & Company's continued good heath. The firm remained in the competent hands of a second generation made possible by Lawrence's retirement and the recent deaths. The new partners were William Brewster, Channing Britton, and Charles J. Richter, each having considerable company experience.[65]

Brewster & Company had long since become the dominant of the rival firms. Besides its smaller size and output, J. B. Brewster & Company also failed to attain similar acclaim or even financial stability. Benjamin's reactionary obsessions and impulsive managerial style certainly hurt, as did the near-constant moves. While he achieved notable real estate successes in and around the city, Benjamin evidently lacked that elemental streak of business genius so conspicuous in his father. Not that Henry had exactly overflowed with it either, but he at least had partners who filled the gap. Britton had been especially effective; Benjamin had even once attempted to persuade him to become his own partner. Henry deferred to Britton in most business matters, and Brewster & Company's prosperity followed suit. It continued long after both men died. The 1890s brought continued profit and acclaim, including medals at the Chicago World's Fair in 1893, further accentuating its reputation as the best-known American luxury carriage builder.[66]

Time was less kind to Benjamin, who in 1895 suffered bankruptcy once again. If Henry had long enjoyed the benefit of wise partnership, Benjamin's twilight years found him increasingly surrounded by charlatans. Though he managed to reorganize as James B. Brewster & Company, he could not regain the old momentum and in 1897 transferred ownership to a trio of employees. The Brewster name still contained too much magic to disappear, and Cairn Cross Downey, Henry M. Duncan, and George M. White named the joint stock concern J. B. Brewster & Company. The owners had ambition, if not tact, which explains their moving the repository to Broadway and Forty-Ninth, provocatively near Brewster & Company's factory. Bringing their own plant to East Forty-Fourth Street, the triumvirate aggressively promoted their Brewster product while Benjamin retreated further into the background. He survived to see the new century, and died in March 1902 at the age of eighty-five. He at least enjoyed the satisfaction of outliving his old combatants and knowing that a firm bearing his name would in turn outlast him.[67]

Brewster & Company president William Brewster remained the sole family member to carry the third generation of carriage makers into the twentieth century. It had all started with his grandfather, whose 1809 stage coach stopover turned into a lifelong New Haven residency. Trained as a traditional carriage maker, James Brewster worked long enough to experience the first tentative signs of the revolution that would so transform his trade. Meanwhile, his own driving ambition, entrepreneurial streak, and charitable concerns led him far afield even as he continued to ply his traditional craft tools. As he built and remodeled, bought and sold land, and helped found an orphanage and a railroad, James Brewster's reputation rose higher than that of many craftsmen. Perhaps the central irony of his life was that his reputation owed more to his nontrade achievements and personal integrity than to any carriages he ever built. To his frustration and despair, however, Brewster could not pass on to his sons the character he most desired them to have. He once wrote that "knowledge and moral worth will always obtain a deference and respect which wealth cannot give"; his final years saw this validated in the negative by two of his own children. Indeed, it is hard to avoid concluding that having inherited a package of valuable trade skills, a fine family business, and a respected local name, Benjamin and Henry Brewster squandered much of their legacy by repeatedly making asses of themselves. Not that they lost money doing so; by the 1860s each had achieved a remarkable degree of prosperity. Still, the paradox of their own lives was that their fame derived in large part from their great wealth, not from the outstanding character their own pervasive deficiencies denied them.[68]

While their fundamental distemper overshadowed Benjamin and Henry Brewster's lives, their encounter with a trade in transformation proves more enlightening than yellowing letters written in the heat of some long-forgotten passion. The Brewsters acted out their vendetta during a time when the carriage trade shuddered from industrialization's relentless onslaught, and a degree of their angst stemmed from being proprietary capitalists during a period of enormous structural upheaval. Even as each established rival factories and repositories, the trade had already started to resemble something far different from the one their father once knew. In 1856 an aging James Brewster advised Benjamin of his unwillingness to once again superintend a carriage factory employing a hundred men. What would he have thought had he foreseen his sons employing many times that number? The first and second generations of carriage-making Brewsters were separated by a chasm called the Industrial Revolution, but the second and third generations experienced a different discontinuity. Little more than ten years after joining the firm, William Brewster saw an omen of the future in a tiny display of six motor cars at the Chicago World's Fair. In a few short years, his entire industry would view the automobile with growing alarm as a threat far surpassing any that had come before. The American carriage trade had survived the passenger train, the streetcar, and the bicycle, but how would it fare against this gasoline-propelled herald of the twentieth century? As head of an especially venerated carriage company, William Brewster would be among those soon to find out. So, too, would the leaders of a very different but equally venerable firm. The Studebaker Brothers Manufacturing Company had enjoyed commercial success arguably greater than the Brewsters', and its own long history reveals much about a trade shared by both.[69]

[6]

A Wagon Every Six Minutes

The Studebaker Brothers Manufacturing Company

The truth of the saying that there is no royal road to learning holds equally good with respect to the trip from the anvil to the office. The way is toilsome to any commanding position, but one who does not shrink from dust and heat, and is willing to plod patiently will in due time receive his reward.

Clement Studebaker, ca. 1886

The wagon maker cannot make a $35 vehicle look like a $75 one in his paint shop. He has to stand on genuine merits . . . for the merits of his work are more readily discernible, and defects likewise.

The Carriage Monthly, April 1904

In extolling the many virtues of the Hoosier state and its enterprising citizens, DeWitt Goodrich's *Illustrated History of the State of Indiana* lingered over a rapidly expanding city near the Indiana-Michigan line. Just a few decades previous to the book's 1875 publication, South Bend had been little more than an embryonic settlement on the St. Joseph River. If the city owed its name to the river's great southern bow cutting through the rich black soil, its reputation stemmed from its most famous residents and their industrious activities. As 1870s America already knew, South Bend was home to the Studebaker brothers, and the Studebaker Brothers Manufacturing Company was the largest wagon manufacturer in the country. In keeping with Gilded Age passion for industrial statistics, Goodrich proudly ticked off the firm's many superlatives: the two factories consisted of two-and-a-half acres of buildings, eight acres of floor space, and seven acres of timber sheds, not to mention two large steam engines driving over a hundred machines. Nearly six hundred employees drew Studebaker paychecks, and the firm indirectly contributed to the support of some two thousand people, more than 10 percent of South Bend's population. Perhaps even more astonishing was that one of these facilities had only just been built as a replacement for a factory gutted by an 1872 fire. Rapid recovery from being burned out twice in three years only added to the firm's already impressive reputation. The Studebakers' family business was a classic nineteenth-century industrial success story, and the Studebaker brothers' transfiguration from sons of Vulcan to captains of industry remains a signal chapter in the history of American wagon and carriage manufacturing.[1]

The Studebakers originated in western Germany's Rhine valley, where generations of them had practiced subsistence farming, wheelwrighting, and blacksmithing in and around the cities of Baden, Solingen, and Cologne. Bitter religious and political warfare had always generated internal migration, but by the eighteenth century many disenchanted Europeans looked further afield for relief. Whether desiring more land, greater religious freedom, or escape from compulsory military service, many began moving to the American colonies. Among them were thirty-eight-year-old Peter Studebaker, his younger brother Clement, their spouses, and another relative named Heinrich, all of whom arrived in Philadelphia in September 1736. Despite Pennsylvania's reputation as a haven for German-speaking people, local officials still fumbled with the language and left "Staudenbeckers," "Studenbeckers," and "Studibakers" scattered over the record books. In time "Studebaker" became the accepted American version. Peter Studebaker was a wagonmaker, and his son Peter Junior followed him, both plying their trade in southeastern Pennsylvania, where Clement and his wife also lived. Clement and Anna Catherina Studebaker prospered in York County, and Clement's death saw the family dividing a considerable amount of land and other property. Only four years old when his father died, Clement Junior eventually inherited his portion, married, and raised a family of his own. Widowed late in the century, Clement remarried to a woman who bore him a son named John Clement. We know little of his youth save that he became a blacksmith and married Rebecca Mohler in 1820.[2]

Like many of their north German countrymen, the Studebakers were Protestants, but of a decidedly independent stripe. John and Rebecca were members of the German Baptist Brethren, popularly called "Dunkers" or "Dunkards" (from the German *tunken*, "to dip") for their commitment to adult baptism by immersion. Part of the larger, loosely constituted Anabaptist movement stretching back to the Reformation, the German Baptists favored lay interpretation of the Scriptures, adult or "believer's" baptism (hence the pejorative Anabaptist, or "rebaptizers," conferred by their enemies), pacifism, and separation from state-supported churches and politics. The movement spawned a host of European denominations including the Mennonites, Hutterites, and the Amish; an overarching emphasis on experiential faith and conduct, an impulse known as pietism, marked these sects that settled in the middle colonies. Rebecca's family members were stalwarts of an extreme branch in Ephrata, Pennsylvania. Founded by Johann Conrad Beissel, a mystic who led a splinter group out of the larger Dunkard denomination in the late 1720s, Ephrata espoused a form of monastic Christianity. The group also included "outdoor" members, nearby families like the Mohlers who eschewed celibacy and communalism while embracing the faith's larger tenets. They evidently welcomed the attention paid their daughter by the young man from a closely related group. The Studebakers had been part of a general German Baptist migration to Pennsylvania that began in 1719, and that colony gradually took on a distinctive pietistic flavor.[3]

The young couple made their home back in Pinetown, near Gettysburg. John had recently completed his blacksmithing apprenticeship and hoped to start his own business. He purchased a hundred-acre farm abutting Conewago Creek, where he spent most of his time in the log smithy shoeing horses, repairing tools, and attending to the myriad tasks common to a village blacksmith. Besides resetting tires and repairing carts and wagons, John also engaged in wagonbuilding with a fellow craftsman named George Gardner, who performed the woodwork. Meanwhile, Rebecca cared for what would become a very large family. Of the thirteen children she bore John, ten survived infancy: Sally, Nancy, Henry, and Elizabeth were all born in the 1820s, Clement, John (later called J. M.), Peter, and Rebecca in the 1830s, and Maria and Jacob in the 1840s. John soon found himself head of a growing family, which he struggled to support. Despite the farm, shop, and comfortable home, John Studebaker continued to suffer financial stress that only grew in intensity.[4]

For all his virtues as husband, father, and churchman, John Studebaker made a poor businessman. The demands of independent craftsmanship frequently bested those with greater gifts, but his problem appears to have been primarily temperamental. A phlegmatic and trusting personality, perhaps buttressed by Dunkard convictions of human perfectibility, seemed to guarantee perpetual financial hardship. Not only did John find it impossible to compel payment from well-meaning but delinquent clients, such customers seemed attracted to his shop like bees to a flower. Some managed to get him to act as surety for much larger debts. Family tradition relays "Owe No Man Anything but to Love One

Another" as a favored John Studebaker motto, but his cosignatory role on defaulted loans forfeited his property, which is how the Studebakers came to Ohio. They left in 1836, the same year Rebecca gave birth to Peter. John had just sold the farm and built what young Clement remembered as a "huge Pennsylvania wagon," a heavy, four-horse Conestoga transporting the family's remaining possessions. They had already chosen north-central Ohio's Ashland County, which had become something of a pietist stronghold by the 1830s. There John took possession of a large farm on credit, but after just two years he had to sell out to settle his remaining Pennsylvania debt. A subsequent attempt to renovate an ancient mill proved equally disastrous. The Panic of 1837 only made the situation worse, and the family finally settled for renting a small plot just east of Ashland, where John would once more work as a blacksmith.[5]

"Pleasant Ridge" proved a considerable step down from Pennsylvania. A painting the brothers commissioned years later shows a ramshackle log cabin and shop whose thatched roofs suggest endemic poverty as much as hard work. There the family fell into a pattern of subsistence farming and blacksmithing. The older children had long since taken up tasks in kitchen, field, and shop. John taught his sons ironwork, some starting as early as age eight, and J. M. learned woodworking from another tradesman. Henry had been nearly ten, Clement five, and J. M. just a toddler when the family moved, but by the early 1840s they were all vital to the family economy. These three, eventually joined by Peter, brought in money as paid hands on outlying farms, the extra cash helping enormously.[6]

Plenty of work existed right on the home place. The Studebakers had located at a busy crossroads, and local farmers and westbound migrants alike brought business to the shop, where John shoed horses, repaired travel-damaged carts, and forged farm implements. As he had back in Pinetown, he cooperated with a local woodworker to make wagons. George Myers was a farmer who worked part time as a carpenter and general woodworker, and the two craftsmen routinely collaborated on wagon repair and construction. Still, this accounted for only a portion of the work passing across the Studebaker anvil; one contemporary of those years estimated the number of complete wagon ironing jobs at only five or six annually. Customers kept John busy with dulled plowshares, broken tools, and draft horse shoeing, and the old headaches soon followed. Hardscrabble farmers were unable to pay their bills, which forced John to purchase supplies on credit, with disastrous results. As one resident later recalled, "it was understood in the neighborhood that the old gentleman Studebaker owed nearly everyone in that part of the country, and every merchant in the county seat whom he could induce to trust him." While some of these businesses perhaps shared John's forgiving nature, others resorted to the law, and the hapless smith found himself sued for debt annually. Little but the scenery had changed for the Studebaker family.[7]

Despite the austerity, some of the children acquired rudimentary education. Clement in particular possessed a keen mind and a taste for books and learning, and he frequently assisted the harried teacher at the nearby district school. Many years later J. M. recalled of his older brother that "Father never had as much trouble with him as he did with the rest of us boys." Clement's abiding desire for schooling benefited from his brothers' equally ardent preference for work, inasmuch as Henry and J. M. frequently took on more labor to avoid the schoolhouse. Not that their lives precluded all levity; besides those chores they found pleasurable, the family took part in the activities of the local Dunkard community. Itinerant preachers provided another diversion (particularly for Elizabeth, who married one in 1849) and the family frequently boarded travelers. The oldest girl, Sally, also called Sarah, preceded Elizabeth in matrimony when she married a cooper and moved to Indiana.[8]

Even with additional child labor, the Studebakers continued to struggle to scratch out an existence from Ohio's soil and John from the pinched confines of his customers' wallets. The late 1840s found John once more considering migration, with Chicago particularly capturing his attention. In 1849 he made a solitary horseback reconnaissance trip through northern Indiana, stopping in South Bend to visit relatives and fellow church members who had been urging him to move. Their enthusiasm likely countered his oldest son's negative opinion: about two years earlier Henry had moved to Goshen, Indiana, where he sold his horse and secured a job with a local blacksmith. Yet several months of hard work netted him little profit; neither this nor his three-hundred-mile walk home favorably disposed him to northern Indiana. John thought the risks worthwhile, however, and decided to move west once more.[9]

Before he turned plan into reality, more of his children preceded him. Sally and her new husband moved in 1849, followed by Henry and Clement in 1850. In October of that year the brothers journeyed northwest to Sandusky on Lake Erie, where they took the Detroit boat, picked up the westbound train, and bounced into South Bend on a stage three days after leaving Ashland. Staying with Sally and her husband, now toll gate keepers near South Bend, Henry took a job in a local blacksmith shop while Clement began teaching. Sally had arranged for Clement to teach the district school, and his success in doing so in the fall of 1850 prompted his staying through the following spring. Fresh from the farm himself, corduroy-suited Clement looked like a local farmer slicked up for church, and he got along with the students famously. Despite this, however, he remained wedded to practicing the trade he had followed since age sixteen, so the spring of 1851 found him blacksmithing for the Eliakim Briggs Threshing Machine Works. His pay, less board, amounted to fifty cents per day, or about the same he received as a teacher. Meanwhile, John, Rebecca, and the remaining children had finally arrived by wagon that fall, and Clement helped his father clear a small homestead about three miles south of South Bend.[10]

His father's lack of success may have given Clement pause about starting his own blacksmith shop, but then his habitual caution usually impelled him to carefully consider all aspects of a situation. He likely had greater confidence in his long-practiced ironworking abilities than in his pedagogy, and the prospects of independent tradesmanship still held great appeal. South Bend was a growing community well able to support another blacksmith, and by the beginning of 1852 Clement was ready. So was Henry, whose ambitions ran along similar lines. When not working for another local blacksmith, he spent his time courting a young Dunkard woman from nearby Goshen. Henry's 1852 marriage to Sally Studybaker had particular bearing on his livelihood; during the winter of 1851–52, Henry and Clement were actively planning their own shop but lacked capital. Fortunately Henry's fiancée possessed above-average business sense and a modest nest egg she willingly loaned her suitor. Accordingly, Clement quit his job at the "thrasher" works and joined Henry who purchased Henry Warden's blacksmith shop. The Michigan Street building housed their entire assets: two forges, two sets of tools, and $68—forty of them Sally's. H. & C. Studebaker opened for business in February 1852. While later accounts often indicated this was a wagon shop, the brothers were general-purpose blacksmiths. Their first job consisted of shoeing a horse for twenty-five cents, and while they also engaged in wagonbuilding, their total output consisted of just two by year's end.[11]

While H. & C. Studebaker was a modest operation with no power or machinery, Henry and Clement had plenty of experience with hand tools and wagon fabrication. They built their first wagon in seven days—testament to rapid workmanship and probable use of at least some ready-made parts. As the brothers would soon discover, however, finding and retaining customers was as important as building vehicles. They sold the first wagon, a typical farm model with plain oak box, for $175.00 to a Mr. George Earl. The bright green-and-red paint scheme followed the norm in farm wagon aesthetics, as did the big yellow letters "S-T-U-D-E-B-A-K-E-R" down each side. Fewer details survive about the second wagon, but it likely followed the form and finish of its predecessor. The year's work almost certainly included tire resetting, gear and body repair, perhaps some repainting—the usual vehicle repair jobs. While the labor force consisted mostly of Henry and Clement, they evidently hired two or three others not long after commencing. They may have been blacksmith's helpers, an additional smith, or perhaps wagon or carriage makers able to contribute woodworking, painting, or trimming skills. Soon nineteen-year-old J. M. joined them to help with the wagons, as he previously had worked for a wagonmaker after his 1851 arrival in South Bend.[12]

J. M.'s restless nature drew him away almost from the start. Everyone knew of the 1848 northern California gold strike, and passing wagon trains were top-heavy with prospectors hoping to strike it rich. Captivated by the idea of travel and adventure, J. M. persuaded his brothers to trade a new wagon for his passage with a California-bound wagon train in the spring of 1853. Five months later, J. M. arrived in a Sierra Nevada mining town, a ragged slash of log cabins and unpainted frame buildings called Hangtown. Originally known as Old Dry Diggings, the town's new name reflected its criminal squalor. The whirlwind of drinking, gambling, prostitution, theft, and murder—the last usually followed by a wave of recriminatory lynchings—made Hangtown an especially rough place. J. M. ended up accepting a job offer from the town's wagonmaker instead of panning for gold. Hugh L. Hinds was a canny, pipe-smoking tradesman who wanted to exploit the strong local market for wheelbarrows, and he had been searching for a subcontractor to build them while he continued to repair wagons and work iron. Hinds wanted twenty-five to start with, at ten dollars each, and J. M. accepted on the spot. Wheelbarrow construction presented no problem for an experienced woodworker, it paid well, and if the rough cabin with its coffee sack bunks and cheap sheet-iron stove appeared primitive, it may not have seemed much different from Ashland.[13]

As he recalled the event many years later, J. M. came in for a shock on his first day of work. "The tools," he said, "were the worst you ever saw, and the only material was pitch-pine lumber." Two days' assiduous labor saw the first one finished; skeptically examining this initial effort, Hinds asked what he called it. "I call it a wheelbarrow," J. M. retorted. "Hell of a wheelbarrow," Hinds replied. True, it had a slightly crooked wheel and probably oversize dimensions to compensate for the soft wood, but it sold regardless. Soon J. M. settled down to a pace of roughly one wheelbarrow per day, acquiring the sobriquet "Wheelbarrow Studebaker" or "Wheelbarrow Johnny" in the process. Hinds's little shop was a beehive of activity, both men frequently working all night to make miners' picks and repair stage coaches in time for their early morning departures. Through all of this, the young man discovered the secret of the California gold fields: "I soon found that hundreds and thousands of the pioneers who tried the mines never made a cent, but those who stuck to steady jobs at good wages and saved their money were doing well." This included Hinds and Studebaker, as well as the local butcher, a New Yorker named Philip D. Armour. Armour's success bankrolled his later entry in the midwestern meat industry, where he founded one of the nation's leading packing houses. Over the course of his long lifetime, J. M. Studebaker would meet several men who traced their wealth back to entrepreneurial ventures on the Pacific slope.[14]

Back in South Bend, his older brothers worked to firmly establish their own enterprise. They pulled in a steady amount of general blacksmithing, wagon repairs, and wagon building, and within the first year or two purchased a neighbor's smokehouse for additional shop space. In 1854 they produced five wagons, and they also attempted to acquire nonfamily business partners, though neither Henry Smyser nor Israel Hoge elected to remain for long. The firm returned to the familiar H. & C. Studebaker in 1855, and the brothers erected a more substantial brick shop during the period 1856–58. Despite these positive developments, however, they lacked capital, making it difficult to expand seriously. Much of the pay came in livestock, crops, and notes, which kept the enterprise cash-poor. The year 1857 brought temporary relief in a subcontracting job from a neighboring wagonmaker named George Milburn. Owner of a wagon works in nearby Mishawaka, Milburn had secured a government contract for several hundred U. S. Army supply wagons. Like so many successful bidders, however, he found himself unable to meet the deadline, and chose to sublet a portion of the work. Henry and Clement jumped at the chance, but they too found the job stretched them to the breaking point. Their portion totaled one hundred wagons, to be completed in six months' time, which necessitated some drastic changes. They erected a hastily contrived lumber kiln to obtain sufficient seasoned wood, outsourced some parts, and hired extra hands to whom they gave ruthlessly abbreviated trade instruction during the punishingly long hours.[15]

H. & C. Studebaker made the deadline with time to spare. The crash program of hiring, training, and expansion paid off as they fulfilled their subcontract in just ninety days. The influx of federal money proved an eye-opening experience, even as it likely saved their business from failure during the Panic of 1857. Still, a single contract could not provide the capital required for significant growth; Henry and Clement were rediscovering the validity of the old axiom about money being required to make money. By this time, however, J. M. had accumulated between $8,000 and $10,000, and when he returned for a visit in 1858, found himself equipped to render meaningful assistance. J. M. evidently intended to return to California and its seemingly limitless possibilities, but the timing of his visit dictated differently. Not only did H. & C. Studebaker require a capital influx, but brother Henry wanted out of the partnership.[16]

Pacifism, War, and Profit

Most accounts of those years merely state Henry's disillusion with business and desire to return to farming, but his prime motivation lay in religion. As the oldest brother and the one with the closest Dunkard affiliation, Henry questioned the propriety of supplying wagons to the U.S. military. The rumbles of sectional strife could be heard across the country, and while it would be too much to suggest that he prophesied general warfare, the relation of the firm's fortune to the spoils of war, even if only sporadic skirmishes with western Indians, nagged at his pietistic conscience. Part of the reason for the recent wagon contract lay in heightened tensions between the federal government and Brigham Young's followers. The Mormon's brutal 1857 slaughter of 140 unarmed migrants in a remote Utah valley seemed to promise worse to come, never mind the larger sectional tensions threatening to tear the nation asunder. Whatever the depth of his own convictions, Henry also had to contend with the opinion of others he could not safely ignore: the local Dunkard fellowship was exerting the social pressure that forms the pietistic sects' chief disciplinary tool. These same factors may also have accounted for John's conspicuous distance from the business. Henry responded by withdrawing from the partnership in 1858.[17]

While Clement and J. M. had at least nominal ties to the same spiritual heritage, their own interpretation of pacifism, historically open to somewhat varied application even within Anabaptist circles, evidently allowed for government contracts. Whether the result of reasoned deliberation or hasty rationalization, neither Clement nor J. M. felt he was violating his conscience. Henry parted on good terms and continued to loan his brothers money and advice, and J. M.'s arrival suggested an obvious replacement. Clement convinced his younger brother to remain, and J. M. purchased Henry's share in the fall of 1858. If C. & J. M. Studebaker seemed similar to H. & C. Studebaker, the similarity ended there, for the new partnership boasted assets of approximately $10,000, much of it in sorely needed hard cash. Moreover, J. M. brought with him more than just gold and woodworking expertise; he also came equipped with hard-won entrepreneurial experience. J. M.'s willingness to take risks counterweighted Clement's habitual conservatism, and he injected new ideas, as well as new money, into the family business.[18]

Not that Henry and Clement had lacked insight and initiative; in 1857 they had expanded into passenger transportation by manufacturing their first carriage. A phaeton produced in time for display at the St. Joseph County Fair, the vehicle marked the birth of a company sideline to their primary role as wagonmakers. J. M.'s West Coast experience had shown him how lucrative the wagon trade could be. Westbound migrants frequently stopped for repairs in South Bend, and in some cases they purchased new and more appropriate wagons for the harsher terrain ahead. Local groups staging for the trek sometimes bought several wagons simultaneously, and the Studebakers were well aware of this market. They also realized that the migrants' arrival in California might also trigger additional sales, as the travelers exchanged heavy covered freighters for farm wagons. Western wagonmakers able to ship their goods to the point of demand stood to reap handsome profits. J. M. had seen the West and saw in it a chance to link his own fortune to one of nineteenth-century America's largest demographic shifts.[19]

The new Michigan Street shop turned out both wagons and carriages by the end of the 1850s. A period newspaper advertisement, written by J. M. and allegedly the firm's first, exhorted loyal Hoosiers to "ENCOURAGE HOME INDUSTRY" by patronizing the "NORTHERN INDIANA CARRIAGE AND WAGON FACTORY" of "H. & C. STUDEBAKER." Written after the new building went up but before Henry's departure, the advertisement underscored the firm's product variety while claiming that it was the only area vehicle manufacturer to guarantee its products. The assurance that "blacksmithing, painting, trimming, custom work, and repair" could be done on short notice attested to the continued need to supplement wagon manufacture; two years later, the firm's annual vehicle production stood at only twenty. By now the brothers employed just over a dozen hands in a small brick factory divided into the familiar blacksmithing, woodworking, painting, and trimming departments. During the early years they sold some of their output by the then-common practice of hitching several wagons in a row and hawking them about the countryside. Recognizing the need to explore new markets, J. M. turned to his younger brother Peter, now a successful store owner in Goshen, twenty-five miles to the southeast. In the spring of 1860, J. M. persuaded Peter to build a small addition for a Studebaker repository.[20]

Though only in his early twenties, Peter already possessed respectable retail experience. An outgoing personality combined with abiding business interest had early pushed him toward commerce; he had minimal craft training and started his working life as a seventeen-year-old store clerk in 1853. According to family lore, Peter shared J. M.'s adventurous tendencies and would have followed him to California had he not been steadily employed. Slow accumulation of modest wages eventually permitted him to become an itinerant peddler. Raised in the same hardscrabble environment as his siblings, Peter knew hardship well: "I know what it is to live week in and week out on mush and milk," he would say many years later, and "it makes my blood boil to hear any man say one word that will tend to encourage a man in waste of time and idleness." He spent the next few years selling needles and thread, pots and pans, and similar items as he traveled about the region. Marriage ended his itinerancy, and around 1857, in partnership with his sister Sally's husband, he established his own dry goods store in Goshen. His arrangement with J. M. marked the first regular sale of Studebaker vehicles outside South Bend.[21]

Peter's vehicle dealership coincided with the outbreak of war and its disastrous impact on New England carriage makers. The trade generally suffered from diminished demand for what remained for many a luxury item. Yet war quickened some forms of commerce and increased the demand for vehicles capable of conveying goods. Many wagonmakers found themselves busier than ever, especially those fortunate enough to capitalize on a house specialty during a time when the federal government needed all kinds of wagons. Besides the rugged supply wagons already in western use, the Civil War's shockingly high casualty rate made ambulances a high-demand item, and the conflict's mobility required additional wheeled conveyances ranging from portable kitchen wagons and forge carts to gun carriages and baggage wagons. Naturally the federal government appealed to American wagon and carriage makers, and the Studebakers found themselves ideally suited to the task. The

Figure 1.1. Perhaps the ultimate early American expression of conspicuous consumption, heavy coaches—like this example from the 1770s—epitomized the coach maker's art. (George W. W. Houghton, *Coaches of Colonial New-York*, 1890)

FIGURE 1.2. The ubiquitous farm wagon changed relatively little over time. This single-horse model dates from the 1880s. (*The Hub*, August 1881, courtesy Dover Publications, Inc.)

FIGURE 1.3. Prepared for an 1877 issue of *Harper's Weekly*, this engraving features some of the many varieties of mid-nineteenth-century American wagon and carriage types. (Collection of the Long Island Museum of American Art, History & Carriages)

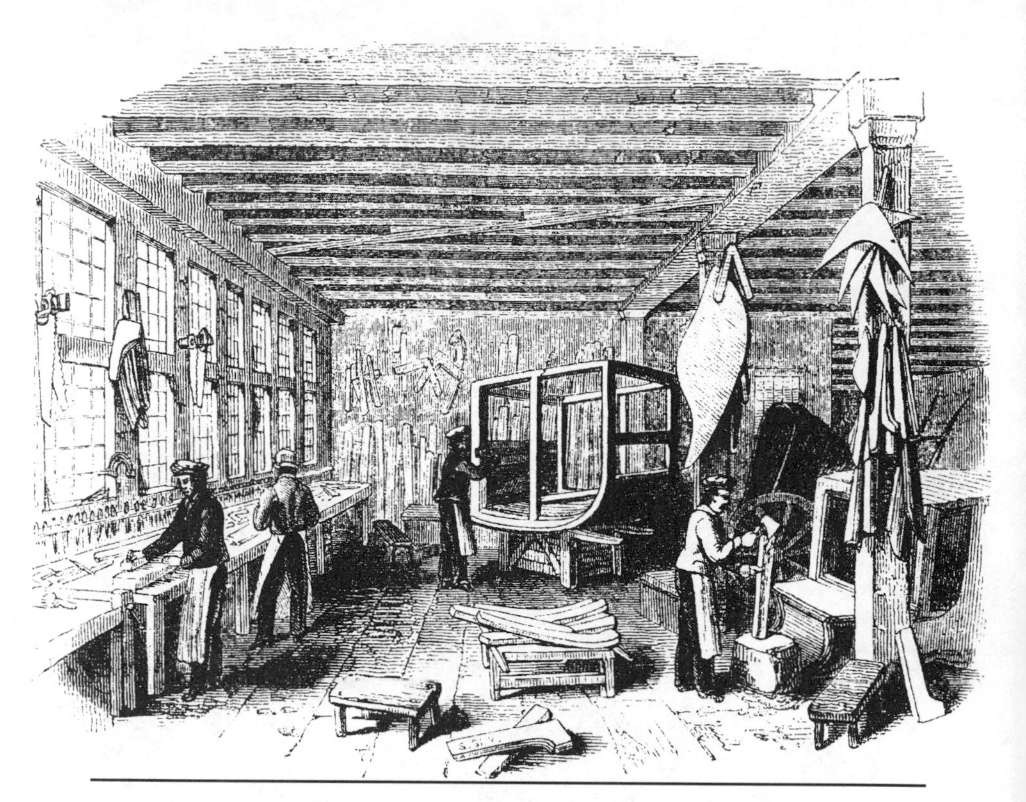

FIGURE 2.1. This view of an 1840s English coach shop reveals a scene familiar to carriage makers on both sides of the Atlantic. The woodworker on the right hews a curved piece to shape, while his shopmates at the benches "shove the jack plane," and a fourth assembles a coach body resting on a trestle. Note the handsaws and chisels above the benches, as well as the large number of patterns adorning the posts and far wall. ("A Day at a Coach-Factory," *The Penny Magazine*, 1841)

FIGURE 2.2. This woodcut of an 1840s wheelwright shop shows three steps in wheel making: the man seated at the great wheel lathe (propelled by a sturdy apprentice) turns a hub while another uses his drawknife to shape a felloe. The mallet-wielding tradesman at center drives spokes into a hub securely mounted in a wheel pit. (Edward Hazen, *Popular Technology*, 1846)

FIGURE 2.3. Heat-shrinking an iron tire onto a wooden wheel required group effort; here two apprentices pour water over the hot iron tire while the blacksmith and his helper keep it centered with hammer blows. This wheel platform is equipped with a retaining screw, but many shops used plain stone or brick platforms and relied on the wheel's own weight to hold it in place. ("A Day at a Coach-Factory," *The Penny Magazine*, 1841)

FIGURE 2.4. Simple newspaper and city directory advertisements like this one for Cleveland's Jacob Lowman brought wagon and carriage makers' wares to the attention of the buying public. The copy not reproduced noted that Lowman had wagons on hand but would also make them to order, as did nearly any craft shop. (Cleveland City Directory, 1845–46)

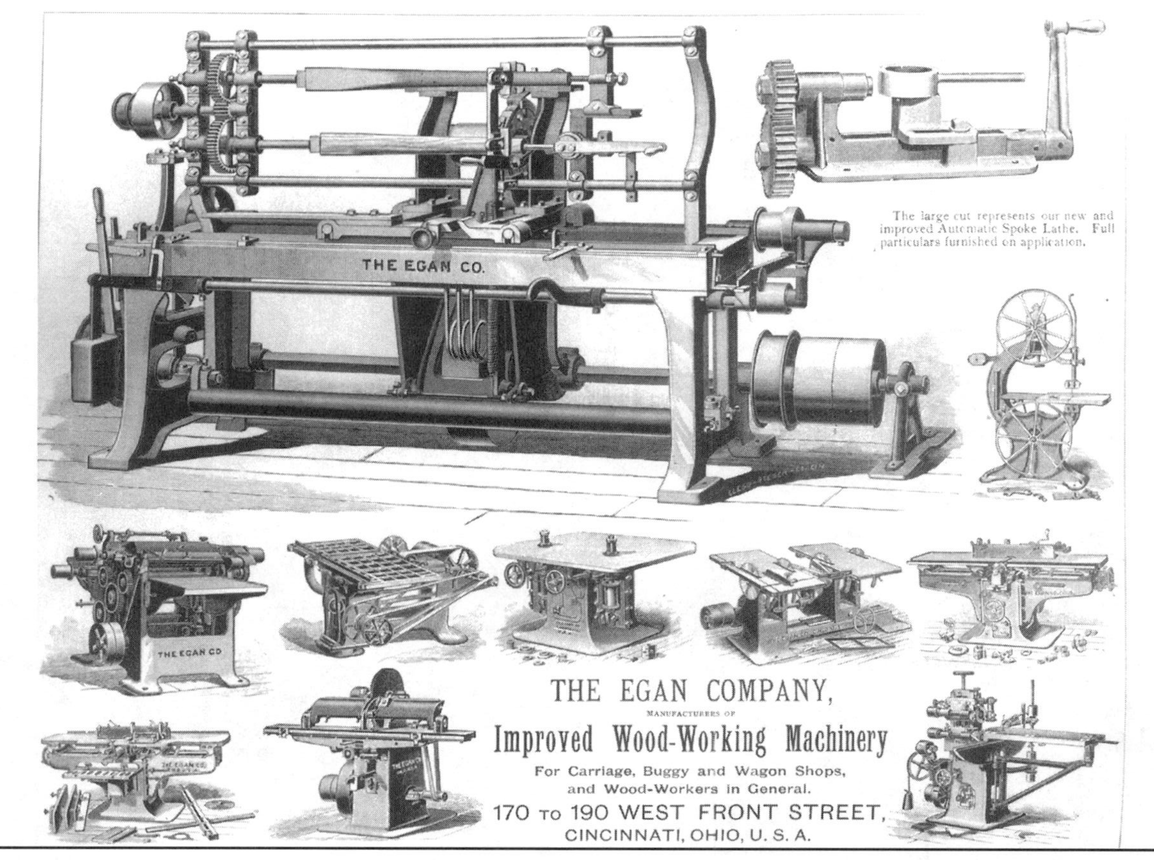

FIGURE 3.1. By the 1870s machine builders such as Cincinnati's Egan Company were producing an increasing variety of general- and special-purpose machinery for woodworking industries such as railroad and street railway car building, barrel making, and the wagon and carriage industry. The "Automatic Spoke Lathe" pictured here operated on the same principle as the Blanchard copying lathe. (*The Carriage Monthly*, November 1886, collection of the Long Island Museum of American Art, History & Carriages)

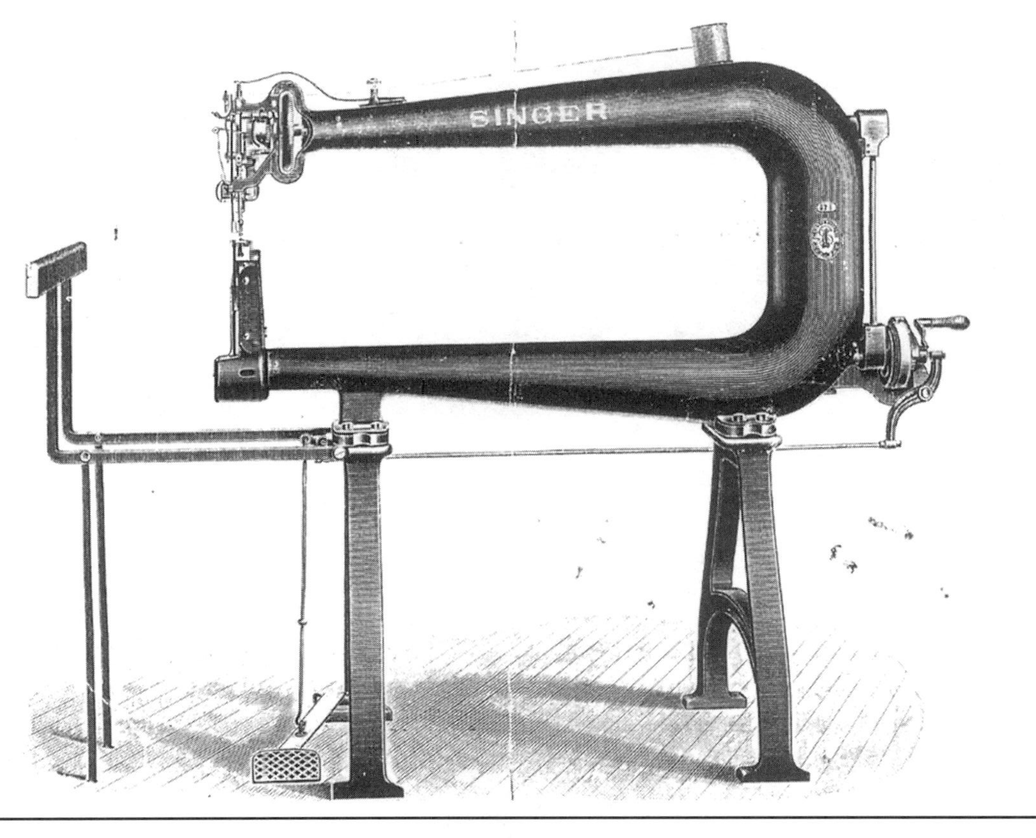

FIGURE 3.2. The sewing machine was easily the single most important machine in the carriage trimming department. The earliest models, for home and commercial use, were suited to only light fabrics. Soon, however, leading manufacturers like Singer produced heavy-duty, crane-necked models specifically for the heavy leather and sturdy fabric used in carriage dash and fender work. (*Class 67 For Stitching Carriage Dashes and Fenders*, 1904, courtesy of Singer Sewing Company)

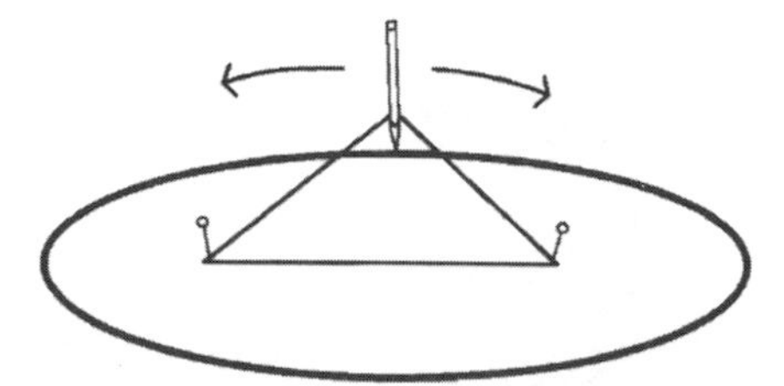

One of the oldest methods of scribing an oval consisted of marking the length and width, placing a pair of pins equidistant from the center, and using a loop of string and a pencil to draw the outline. Provided the tradesman held the pencil consistently and the string did not stretch, he could obtain satisfactory results, though it might require more than one attempt.

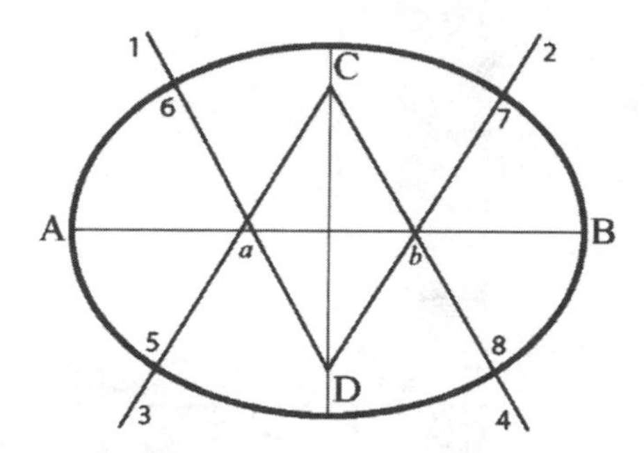

With the French Rule the tradesman used pencil, straight edge, compass, and elementary geometry to obtain an exact form on the first try. For an oval with a width half its length (other proportions could be employed) he would draw line AB bisected by CD and divide AB into three equal parts (Aa, ab, and bB). He then would mark on CD from aC or aD the same distance as from Aa and draw the lines DI, D2, C3, and C4. Placing the compass point at a, he next scribed arc 5–6, and with the point at b, arc 7–8. Then setting the point at the intersection of AB and CD he scribed arcs 6–7 and 5–8 to complete a perfect oval.

FIGURE 3.3. This simple comparison of two oval scribing methods contrasts the old "rule of thumb" with the new geometrical drafting methods the American carriage trade began adopting in the 1860s. (Adapted from M. T. Richardson, *Practical Carriage Building* [New York, 1892], 2:56–59)

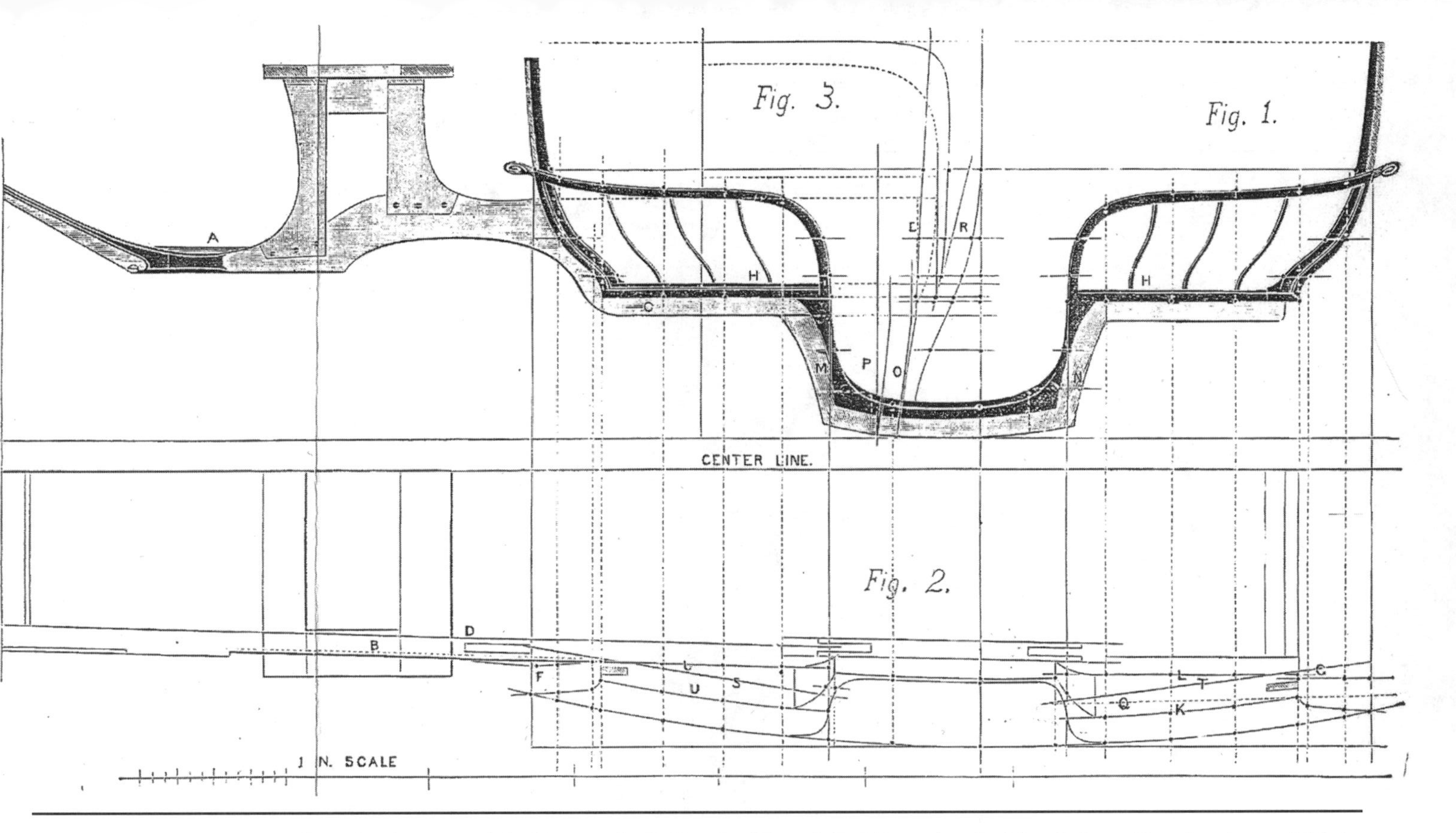

FIGURE 3.4. One important benefit of improved drafting methods like the "French Rule" consisted of detailed working drawings, which by the 1870s were disseminated to the trade via the trade press. Twenty years later, simplified working drafts dramatically improved American carriage bodies. They also permitted American carriage makers to outproduce the Europeans who had taught them the system in the first place. ("Working Draft of Light Vis-a-Vis, without Doors," *The Carriage Monthly*, March 1891).

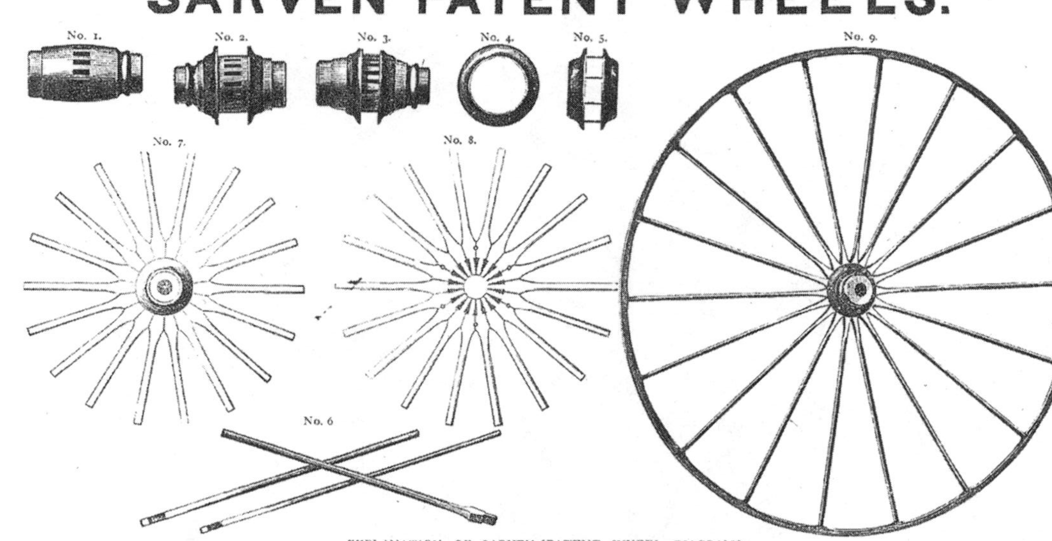

FIGURE 4.1. Wheels were one of the earliest horse-drawn vehicle components available outside the wagon and carriage shop. The New Haven Wheel Company was an early proponent of patent wheels, and versions such as the Sarven proved exceedingly popular with carriage builders. (*The Hub*, October 1872)

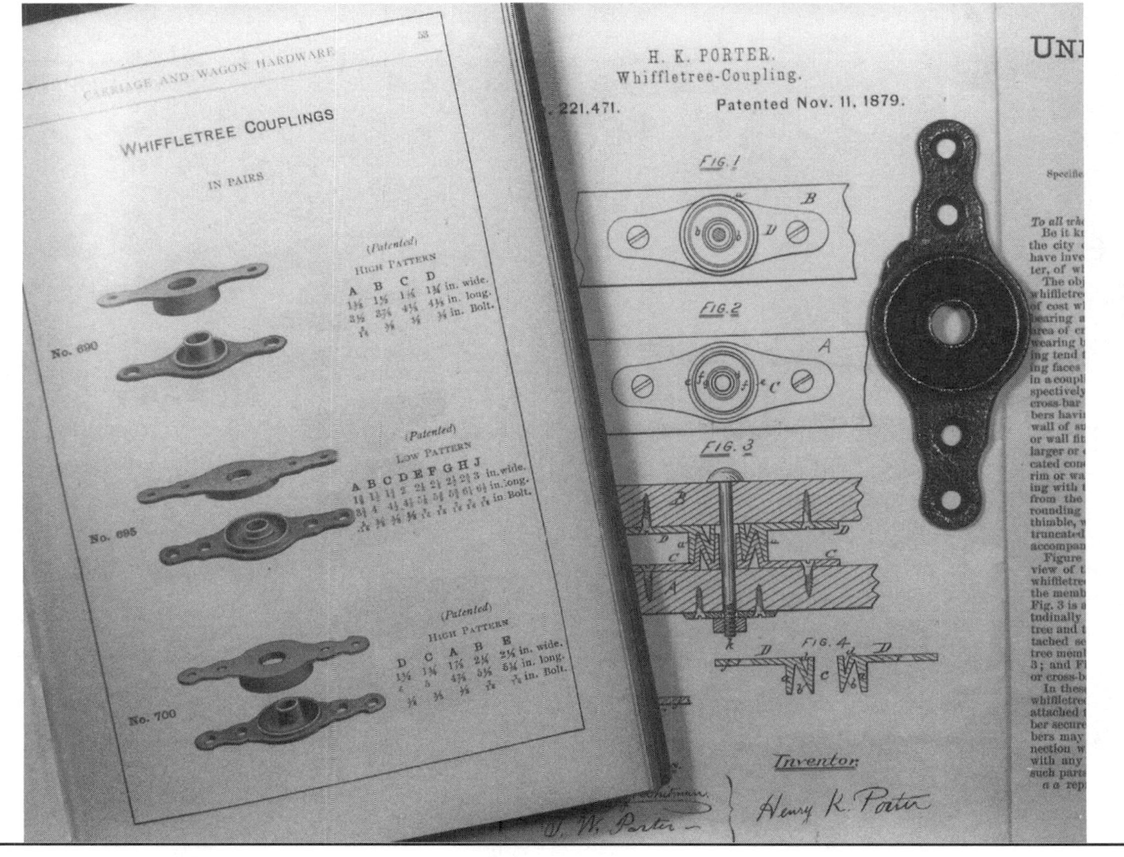

FIGURE 4.2. One of the thousands of varieties of malleable iron castings, its U.S. patent, and the catalog that sold it to the trade. This example of Eberhard's no. 695, Patent Low Pattern Whiffletree Coupling (bottom half) came from the factory of a former wagon gear builder in Akron, Ohio, while the patent and catalog reside in the archives of the Eberhard Manufacturing Company in Strongsville, Ohio. (Author's collection; patent and catalog courtesy Eberhard Manufacturing Company, division of the Eastern Company)

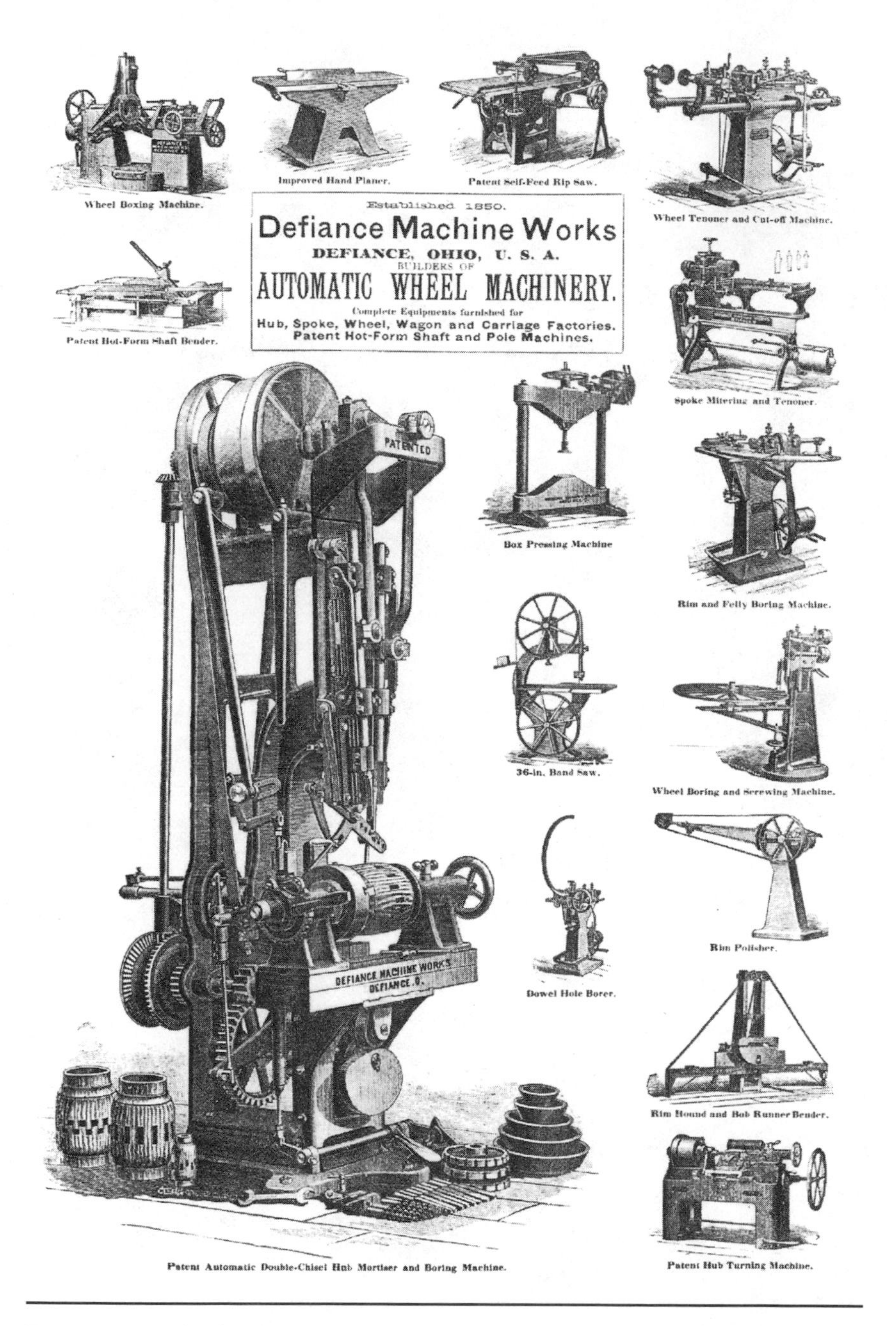

FIGURE 4.3. Wheel making had been heavily mechanized for over forty years when this advertisement appeared in the 1880s trade press. Firms such as Ohio's Defiance Machine Works supplied the majority of the machinery in use by this time. (*The Hub*, March 1888)

FIGURE 5.1. A coach epitomizing the kind of heavy carriage work typical of James Brewster's shop. This example from the late 1840s or 1850s left Brewster's repository at 396 Broadway wearing the family name, though it had almost certainly been built elsewhere. At this time neither James nor his sons were manufacturing carriages, instead stocking their New York repository with vehicles made in New York and Bridgeport—a perfectly legal and common practice in the nineteenth-century carriage trade. (Collection of the Long Island Museum of American Art, History & Carriages, gift of the Brooklyn Museum, 1961)

FIGURE 5.2. Though built around 1880 by another firm in nearby Brooklyn, this road wagon displays the same characteristics of the famed Brewster versions of the 1860s. Benjamin Brewster had patented this form of side-bar suspension in 1872. (Collection of the Long Island Museum of American Art, History & Carriages, museum purchase, 1953)

FIGURE 5.3. This detail of the wheel from the road wagon in fig. 5.2 shows the extremes to which some makers went in paring weight to an absolute minimum. Such spidery wheels had to be of the finest hardwood, usually hickory, to withstand the strain. To a swell racing down Fifth Avenue or in Central Park, however, the risks were well worth the thrill of speed. (Photograph by the author)

FIGURE 5.4. When in 1879 *Scientific American* ran a feature article on Brewster & Company's new factory, it took pains to illustrate its interior wonders, including the heavy bending press, drafting room with chalkboards, the body and varnish shops, and the unusually well-lit blacksmithing department. The men at upper right perform an informal but effective test on a freshly hung carriage body before sending it upstairs to be painted and trimmed. (Hamilton S. Wicks, "The Manufacture of Pleasure Carriages," *Scientific American*, February 1879)

FIGURE 6.1. The "celebrated Studebaker wagon" was in reality little different from most others. The simple, robust design ideally suited rigorous agricultural use, and the demand for such an essential device seemed limitless. (Albert Russel Erskine, *History of the Studebaker Corporation*, 1918)

Figure 6.2. No other wagon maker hawked its wares like Studebaker. Innovative advertising had been a company hallmark from the start, and by the 1880s the firm included music as part of its advertising budget. This jingle came out in 1892–93. (Courtesy of Frank Frost, Hastings, Minnesota)

Figure 6.3. In the wagon division's cavernous blacksmith shop Studebaker smiths performed tasks intimately familiar to the firm's founders—but on a scale far greater than anything they had known in years past. (*Studebaker Souvenir,* 1893, collection of the Long Island Museum of American Art, History & Carriages)

Figure 7.1. Craftsmen, assemblers, or machine tenders? Woodworkers in these illustrations operate some of the machinery that revolutionized wheel making in the late 1840s and 1850s. Though primitive by later standards, such machinery reveals that portions of the trade had already become heavily mechanized well before the Civil War. (*G. & D. Cook & Co.'s Illustrated Catalogue of Carriages and Special Business Advertiser* [New Haven, 1860], courtesy Dover Publications, Inc.)

FIGURE 7.2. By the end of the nineteenth century sophisticated special-purpose equipment like this automatic double-chisel hub mortiser combined speed and precision to an astonishing degree. Ohio's Defiance Machine Works announced that "the jigging, spacing, feeding, etc. are entirely automatic" and that a single operator running a pair of them could turn out four hundred hubs in a ten-hour shift. (*The Hub*, September 1908)

FIGURE 7.3. The nineteenth-century equivalent of the twentieth-century automotive repair garage, small shops like this combination blacksmithing and carriage-painting business employed skilled craftsmen to repair and refurbish horse-drawn vehicles of all kinds. (Mark L. Gardner Collection, used by permission)

Figure 8.1. This advertisement proudly noted the ongoing partnership between the equally venerable firms Valentine & Company and Brewster & Company, as the latter launched "The Brewster" automobile. (*The Vehicle Monthly*, March 1917)

Figure 8.2. By the summer of 1927, John Crowley had been working at Cleveland Hardware for fifty-seven years—since 1879. He had already become an expert at forging automobile parts. (Courtesy Cleveland Hardware and Forging Company)

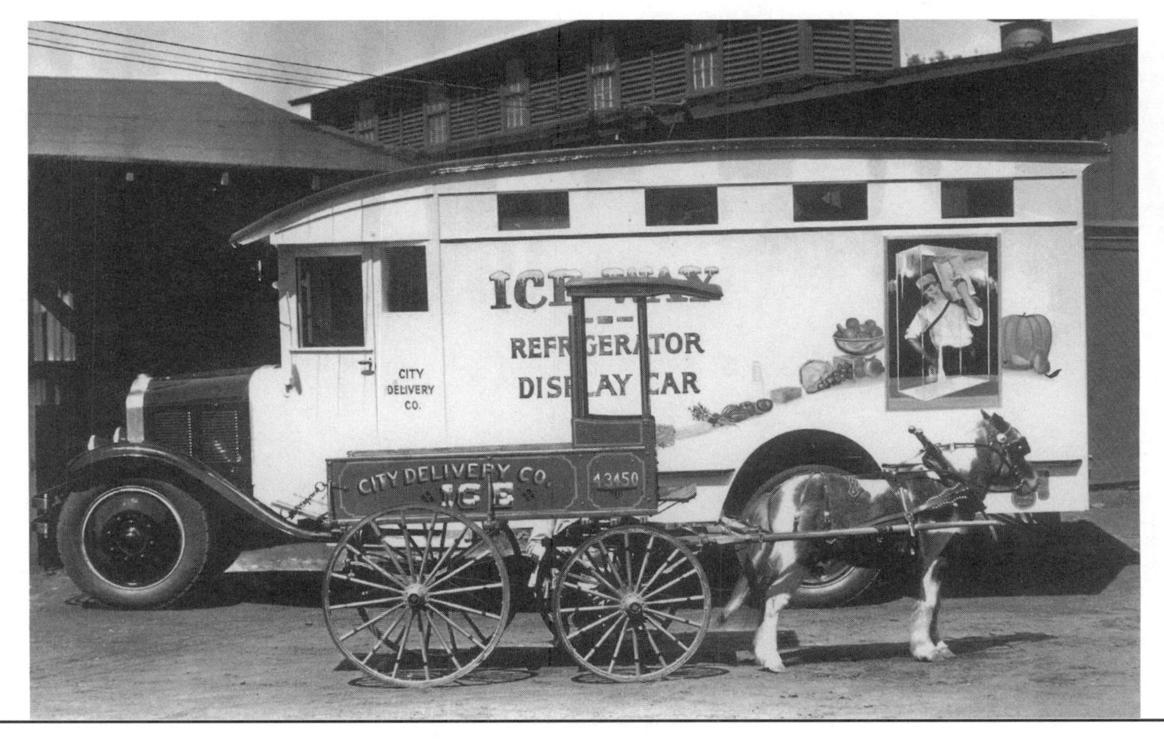

FIGURE 8.3. As late as the 1920s, horse-drawn wagons continued to deliver commodities like ice, even as motor trucks became increasingly common. (Courtesy National Museum of American History, Smithsonian Institution)

government's rigid product specifications were already familiar to the brothers, and they proceeded with confidence. As early as 1861 they turned out an average of eight wagons a week, or just over four hundred annually.[22]

Federal demand for wheeled vehicles only increased during the conflict. Despite innovative reliance on railroads, the Army of the Potomac still required vast numbers of horse-drawn vehicles to wage war effectively. Moving from its initial advocacy of ambulance carts—dubbed "avalanches" by wounded men who cursed their agonizingly jarring motion—the government turned to four-wheel ambulances of various designs and produced by numerous contractors. Supply wagons formed the real transportation backbone of both Union and Confederate forces. In praising the quartermaster corps' organized conduct during 1864, Lieutenant General Ulysses S. Grant characterized its wagon train as stretching "from the Rapidan to Richmond," each bearing ingenious markings denoting, corps, division, brigade, and contents. Hundreds of them crowded the roads before, during, and after every campaign, conveying everything from hardtack to ammunition. The Confederacy also employed them whenever possible; the wagon train carrying Confederate wounded from Gettysburg extended for seventeen miles. Some were doubtless captured Union wagons, for the South lacked the North's abundant supplies, and firms like Studebaker sold their goods only to the federal army. Ambulances and supply wagons switched sides as Union and Confederate forces raided one another's supply trains; one such action after the battle of Spotsylvania netted the Confederates twenty-five or thirty Federal supply wagons, which Union forces soon recaptured. Likewise Major General Philip Sheridan's burning of 180 Confederate supply wagons during the Appomattox campaign created more hardship for the already beleaguered Army of Northern Virginia. Besides capture and destruction, loss by hard use ensured continual demand for more.[23]

Government contract money permitted the Studebakers to undertake an ambitious expansion. In response to pressing production demands, Peter and his family moved to South Bend in 1863 to concentrate on sales. Although government payments proved an invaluable benison, the experience it bequeathed had even greater value: the lessons the brothers learned about quantity production opened entirely new vistas. Timely infusion of federal funds permitted increases in the number of employees, the size of the factory, and the pace of production. The pressure of fulfilling successive orders for impatient government agents forced the brothers to pay continual attention to speed and efficiency while maintaining high quality standards. The mandate to dramatically increase production and the capital to make it reality flowed directly from Union wagon contracts, and the results extended far beyond 1865. The Civil War left the Studebaker family business in far better shape than it had been five years earlier, and the challenge for the brothers after Appomattox lay in successful transition back to civilian production. Fortunately their products lent themselves well to many nonmilitary applications. Henry and Clement founded the enterprise on the premise that farmers, merchants, and many other people required wagons, and this remained as true in 1865 as it had been in 1852.

New market challenges prompted Peter's move to St. Joseph, Missouri, in 1865. Though the overall number of westbound migrants had declined from a peak of some 55,000 in 1850, traffic remained steady on the Oregon, Santa Fe, Mormon, and other trails. Many wagon trains staged along the Missouri River; Independence had been favored since the late 1820s, but upriver competitors like Fort Leavenworth in Kansas and St. Joseph in Missouri had since become serious rivals. Wagonmakers in Independence and nearby Westport had long provisioned emigrants, but St. Louis became the premier Missouri wagonmaking city. St. Louis builders like John Cook and Louis Espenschied had vast trade experience and had also filled Union army contracts. Still, an aggressive "out of towner" might horn in on the market, and Peter Studebaker became the company's western spokesman, remaining in St. Joseph until 1871. By then the transcontinental railroad had severely diminished the number of caravans, and Peter's talents could be better used elsewhere.[24]

In 1868 the brothers decided their operation warranted something more than a simple partnership, so on March 26 they incorporated under Indiana law as the Studebaker Brothers Manufacturing Company. Clement became president, J. M. treasurer, and Peter secretary, their titles reflecting new administrative realities. Even before incorporation, the brothers found themselves confronting that paradox so familiar to the successful craftsman. In the early days Clement had rounded out a full day at the anvil with an evening's bookkeeping, but business growth had long since demanded his full-time office presence. J. M. made a similar transition when he left the workbench to become general manager. He wisely preferred close contact with the factory floor and continued to improve wagon design and fabrication. Peter continued as advertising specialist and spent part of his time visiting established repositories and agencies, as well as creating new ones. He effectively promoted Studebaker wagons through an aggressive and innovative publicity campaign. The same held true for twenty-four-year-old Jacob, who in 1868 replaced Peter as secretary (the latter becoming treasurer in place of J. M., who became vice president) and also ran the carriage division. Perhaps taking advantage of an old Ashland acquaintance, John had arranged for Jacob's return to Ohio to apprentice under Peter Van Ness, a Tiffin carriage maker. Here he learned the carriage trimming skills he put to good use in South Bend. Jacob completed what Clement would refer to as the "chain of five brothers." Later company literature bristled with a virtual tribe of bearded Teutonic wagonmakers and announced that theirs was a proprietary endeavor indeed.[25]

Family businesses bring to mind family feuds, and the wagon and carriage industry had its share. Whether the Topliff patent battle or the Brewster's marathon rivalry, family ties and capitalistic enterprise did not always mix well. Yet for every example of spectacular disharmony, there were many others that by their quiet efficiency passed through the era with hardly an outward ripple of discord. Studebaker appears to have been one of them. While business strains and family matters generated episodic disagreements, little spread beyond the office doors. "Our success in the past, as I look upon it," Clement wrote Peter in 1890, "has been largely due to the fact that as brothers we have had a brotherly feeling for each other." Despite the occasional "sharp word," this bond always prevailed in the end. Perhaps some of this came from their austere childhood, where each quickly became vital to the domestic economy. Members of large families often shine at collective endeavor, if only through years of practice. It certainly owed something to their uniform upbringing and their similar world views, just as their collective drive likely stemmed partially from witnessing their father's repeated failures. Certainly the business called upon their individual talents; one striking aspect of Studebaker managerial style lay in an efficient division of labor. This sometimes manifested itself in surprising ways, as with delegating religious charitable requests according to each brother's denominational affiliation. Clement handled the Methodists, J. M. the Presbyterians, Peter the Episcopalians, and Jacob the Baptists, with requests by others receiving a collective hearing. It was a characteristically efficient solution to one of many business challenges.[26]

All this developed gradually; in 1868 the Studebakers had still to refine their organization at all levels. Their $75,000 in capital and nearly $250,000 in assets reveal their practice of plowing much of their profit back into the business. A period advertising cut reveals just how large this had become: the Michigan Street operation still retained the one-story office and two-story repository, but they were now dwarfed by commodious new additions erected during 1866–68. One had a modest smokestack, which bore witness to the firm's increased mechanization, and a similar structure housing the blacksmithing operations occupied the other side of the lot. The courtyard between acted as an erection department where workmen assembled wheels, gears, and boxes into complete wagons, and a three-story building at the rear contained the carriage works. By this time Studebaker employed between 150 to 190 skilled and unskilled hands, the number depending upon the season, and 1868's production totaled nearly four thousand vehicles. The breakdown is instructive: 150 sleighs, 205 bob-sleds, and 300 carriages and buggies testified to continuing variety, but 3,300 wagons announced the house specialty.[27]

By the late 1860s, the firm looked to South Bend's southern outskirts as a place for possible expansion. In 1867 the brothers purchased a large frame building about a half-mile south, at Lafayette and South Streets, and moved their woodworking machinery there. This department now had much larger quarters, and its location on the Lake Shore & Michigan Southern Railroad enhanced shipping efficiency, an important consideration for an enterprise that had long since ceased to be local. Their purchase of more property south of the line in 1871 allowed the brothers to build what would become their showcase wagon factory, while carriage production remained at the still-expanding complex in central South Bend. Work on the new facility temporarily halted in June 1872 when a fire in the woodworking shop consumed the entire building and its contents. In Colorado at the time, J. M. received word a few days later and wired his brothers with typical Studebaker flair: "I start home tonight. Kill the cow that kicked the lamp over." The loss totaled between $70,000 and $80,000, a serious blow especially so soon after incorporation and major expansion, and the firm's insurance covered less than half. Fortunately, adequate financial reserves permitted immediate reconstruction, and J. M. appointed his old California boss Hugh Hinds as construction superintendent.[28]

The new works consisted of a three-story I-pattern main building, with a forty-forge blacksmithing department projecting rearward. A centrally located powerhouse and chimney contained the steam engine that supplied air to the forges and drove the lineshafting propelling the woodworking machinery. Open air lumber piles separated these two structures from the woodworking department at the rear, and the ten-acre lot was enclosed on two sides by a serpentine timber shed hundreds of feet long. Three hundred employees worked here, two hundred more in the carriage works, and together they had an estimated output of an unprecedented twelve thousand vehicles. This of course did not take into account the inevitable market fluctuations, equipment breakdown, and other contingencies, but by the end of 1873 the firm had turned out just over ten thousand vehicles. These were impressive figures, but this was a state-of-the-art vehicle factory, one of the most substantial in the entire American trade, and it positioned Studebaker to make bold strides in the national wagon market. To the shock and horror of all, fire again broke out, and the August 24, 1874, conflagration came much closer to being a total loss. The new building was a smoking ruin and several other structures suffered serious damage. Vast quantities of seasoned timber fed the blaze and added to a loss totaling more than the cost of the entire 1872 renovations, only a third of it insured. Predictions ran rampant that this surely spelled the end, but Studebaker bounced back once more. This time the firm elected to take advantage of recent improvements in factory architecture and rebuilt with brick instead of timber.[29]

When completed in 1875, the new main building ran for five hundred feet along Lafayette, with a wing seventy-five feet longer fronting the rail line. Most of the other buildings were now also brick, with the primary exception of the timber sheds. These snaked down the property for over a thousand yards, comprising about seven acres of floor space. Such huge stocks of wood reflected not only Studebaker's commitment to air-cured lumber but also the tremendous quantities necessary to supply the enlarged facility. The firm's two-hundred-horsepower C. H. Brown steam engine powered forge blowers and wood- and metal-working machinery, as well as the factory's numerous elevators. The carriage works back at Michigan and Jefferson now contained a multistory repository (erected in 1874) and factory. Another steam engine supplied air to thirty-five forges and the wood shop's saws, planers, and molding machines, and the complex included a high-ceiling repository and an office where Jacob Studebaker presided over it all.[30]

A Studebaker wagon originated in the company's timber sheds where workmen selected the lumber to be sent into the woodworking department. The wagon began with the wheels, made in the firm's own wheel department. In this heavily mechanized area, workers employed power lathes to turn and bore hubs before others chucked them up in the mortising machine in preparation for spokes. These required their own series of shaping operations, with machines jigged to turn out uniform spokes with precisely sized tenons. A power spoke driver (a trip hammer version seating each with a few sharp blows) installed them before workmen attached the felloes, also made in house. An 1875 description notes that a wheel, complete save for the later-applied tire, passed through fourteen machines and the hands of twenty-three men in its construction. Ready-made hub boxes and precut strapping simplified the blacksmiths' boxing, banding, and tiring, and the smiths heat-shrank tires via a collapsible platform that dropped the wheels into a water tank.[31]

Wagon gear and box proceeded similarly, each beginning with seasoned lumber sawed and planed to specific dimensions. This stock preparation required the usual general-purpose planers, rip and cut off (swing) saws, molding machines, mortising machines, and tenoning machines, besides special-purpose equipment like the lathes used to turn axle ends. The reduced amount of forging in the blacksmithing department suggests the use of ready-made metal hardware, but painting remained a time-consuming process. The firm managed some streamlining by substituting dipping for some brush work. After each wheel received a preparatory coat of boiling linseed oil, shop hands ran it through a paint-filled trough before leveling each of three such coats with a brush. Though pinstriping still required striping pen and good coordination, and drying still occupied a great deal of time, these innovations reduced the man-hours required to finish each wheel. Farm wagons required almost nothing in the way of trimming, but the carriage-making division employed sewing machines to speed the work along. Little of this resembled vehicle making as the Studebakers had learned it, and machinery had a great deal to do with the difference. One 1870s observer remarked that at Studebaker "manual labor is in every possible way economized by the steam engine and machinery," and he counted no less than 106 machines at the wagon works and 24 in the carriage division.[32]

Mechanization formed only one of the operation's three organizational distinctives. Rationalization and interchangeability were equally intrinsic to the Studebaker production ethos. Mechanization lent impetus to the latter even as it accelerated production, as noted by a description from the 1870s: "the axles and all other appliances are turned, shaped and fitted by machinery, one part being the exact counterpart of thousands of others." Jigs stopped cutters at specified points to ensure that tenons fit mortises and boxes fit hubs, just two examples of efficient assembly. The raw material and its application allowed for greater tolerances than a piece of intricate, all-metal machinery, and hand-fitting still remained. A much-republished late photograph of the wheel department shows a young man using chisel and mallet to touch up machine-mortised hubs, and hand work continued elsewhere as well. Fine-tuning machine processes to render predictable results aided uniformity and interchangeability, as did design simplification and task specialization. Mechanization worked best with rationalization, and the two abetted interchangeability; together, all three drove sustained quantity production. Very early in its existence the Studebaker Brothers Manufacturing Company mass-produced horse-drawn vehicles.[33]

This classification suggests features conspicuously absent from actual practice. The biggest error would be to assume, as dozens of writers have, that there was a Studebaker wagon assembly line. None existed, for assembly took place throughout the plant. In the case of subcomponents such as wheels, gears, and boxes, workers assembled them in the departments that fabricated their constituent parts. Combining these discrete units into complete vehicles also took place in different parts of the plant; blacksmiths often hung bodies in or near the smithy, tops went on in the trim shop, and the shipping department disassembled vehicles to fit them into packing crates. Misinterpreted late-nineteenth-century factory photographs continue to engender confusion. An example from around 1900 frequently identified as a production line shows three workmen examining the first in a short row of finished carriages. The logic of such interpretation rests on the understandable tendency to project later developments onto earlier situations; the photograph actually depicts what appears to be the spring vehicle trim shop, and in any case it hardly resembles a production line. Studebaker's practice of translating production figures into a specific number of vehicles per day, hour, or minute only reinforces the impression that finished vehicles rolled off an assembly line. Such advertising ploys served to underscore production figures, not provide literal descriptions of manufacturing practice. Finished vehicles exited the plant in quantity but hardly in the manner of products coming off assembly lines. While the tremendous output made possible by Studebaker's mechanization, rationalization, and interchangeability qualify it as mass production, historians should refrain from assigning it nonexistent attributes. Linear assembly and related elements of Fordist mass production were no more part of Studebaker mass production than was the principle of the Model T body drop to wagon box installation.[34]

Ready-made parts contributed to production efficiency in nearly every wagon and carriage factory, but Studebaker also made many of its own vehicle components in-house. Outsourcing remained an important strategy, but as the firm sought to better regulate product specifications, it found that in-house parts fabrication made great sense. Studebaker-made wheels were one of the earliest such examples, but during the

1880s and 1890s the company advanced into manufacturing technologies beyond the capabilities of most horse-drawn vehicle makers. By the early 1890s, Studebaker boasted its own metalworking and machine shops crowded with sophisticated production equipment. Banks of punches, shears, and drop forges turned out everything from reinforcing rods to steel leaf springs, all items coming to most other firms ready made. This surge of integration gave Studebaker a highly efficient internal supply structure, while outside sales of surplus parts recouped a portion of the investment in additional machinery, labor, and raw materials.[35]

Studebaker was by no means the first to apply these principles to horse-drawn vehicle manufacture, but it was perhaps the first to implement them on so great a scale, to such efficient effect, and for goods with seemingly limitless consumer appeal. Despite the great many variations produced over the years, Studebaker's signature product and perennial best-seller remained the humble farm wagon. Though incremental design and construction improvements altered the product over time, the essence of the Studebaker farm wagon remained the same for decades. As company advertising correctly noted, wheels formed a wagon's literal foundation. Studebaker wagon wheels were heavy-duty versions with iron-hooped hubs and hardwood spokes, the latter featuring the firm's "sloped shoulder" tenons, which "fit into the hubs as though originally grown to their place." Each wheel had its rugged tire, the larger versions featuring those made from three-quarter-inch thick iron. Though the firm eventually offered a small assortment of patent wheels, its farm wagons retained the simple but durable plain-hub variety. Studebaker placed great faith in boiled linseed oil, thoroughly immersing each wheel in an oil bath to preserve and prepare the surfaces for paint and varnish. Front and rear wheels differed only slightly in size: the necessity of a low-mounted box with maximum interior space dictated the absence of a front wheel cut-under, though slightly smaller front wheels reduced the turning radius (fig. 6.1).[36]

The farm wagon's expected use dictated a dead-axle gear suited to heavy loads and uneven terrain. While lack of suspension made for a bone-jarring ride, it also enabled the user to load the wagon to, or even beyond, its rated capacity. In other words, farm wagon design eschewed comfort for strength, an acceptable trade-off for most purchasers. The basic wagon gear consisted of a pair of massive, iron-banded hardwood axles, each terminating in tapered axle arms. Studebaker capped these with thimble skeins, sheet-metal sockets threaded to accept the wheel-retaining axle nut. Each axle supported a bolster, a parallel timber sporting a pair of upright stakes to contain the wagon box. A coupling pole, the farm wagon equivalent of the coach's perch, supplemented on both ends by splayed wooden braces called hounds, connected these axle/bolster units. The front axle pivoted on a center-mounted, vertical iron king pin and was stabilized in its traverse by the iron fifth wheel above. Axles, bolsters, coupling pole, and hounds comprised the major wooden components, each supplemented by appropriate metal hardware. Some of the heavier models featured iron rod-braced gears, others had axles reinforced with strap iron. Like the wheels, the gear featured a boiled linseed base coat below successive layers of paint, pinstriping, and varnish.[37]

The standard wagon body consisted of a long rectangular box resting on bolsters and retained by bolster stakes. Hardwood framed and sheathed, it featured front- and rear-hinged endgates and a slanted front footboard. Most included capacity-increasing detachable sideboards held in place by sheet-iron sockets. The basic wagon came without a seat, but one mounted on half- or full-elliptic springs, a rare concession to operator comfort, formed a popular and inexpensive option. Like the gear, the box featured a variety of iron bracing to maintain structural integrity under heavy loads. A common paint scheme consisted of red gear and green box pinstriped and emblazoned with "The Studebaker" down the sides. The basic wagon's purchase price included the necessary accessories for a team. Besides the tongue (drop or stiff), they included a heavy double tree (evener) bolted and chained to the front of the gear, and two single trees attached to this, one for each horse. Neck yoke or tongue chains and an axle nut wrench completed the equipage. The tongue type was purely a matter of personal preference; the stiff version remained above and parallel to the ground, while the neck yoke or tongue chains provided two different means of connecting the pole tip to each horse's harness. The purchaser required only team and harness to have a ready-to-work outfit.[38]

There was also the matter of options and accessories, however. In the early years, buyers determined what they wanted through informal discussion with one of the proprietors, but Studebaker's explosive growth and increasingly uniform product specifications meant that customer-derived options came from a prescribed list. Still, the choices were quite extensive and allowed for a remarkable degree of customization. For example, in 1883 a Studebaker farm wagon purchaser could, for small additional fees, add any of the following to his new wagon: spring seats, brake ("with Patent Ratchet"), feed boxes, lock chains (to prevent the wagon from rolling when not in use), bows for a cloth top, and replacement parts, which a prudent farmer might want to have in case of breakage.

Perhaps the most significant option related to capacity; by the early 1880s Studebaker's basic farm wagon came in six different sizes, each suited to heavier loads. Besides incremental increases in the dimensions of axles, wheels, tires, and boxes, the fundamental differences lay in strength of construction and vehicle weight. The smallest standard farm wagon's box was six inches shorter than the largest, but the vehicles weighed 485 pounds and 1,500 pounds, respectively. Likewise, the smallest had a 1,200-pound capacity while the largest could accommodate 5,000 pounds. Surprisingly, the prices varied little, with the small wagon selling for $80 (in 1883) and the largest for $110. Studebaker handled the thorny matter of track (which it measured from tire centers, on the ground) by offering a wide version at five feet and another six inches narrower. These presumably accommodated most wagon ruts nationwide. In time, customers desiring allegedly more durable iron-axle versions could choose from five additional wagons. Studebaker clearly catered to customer choice.[39]

In the final decades of the nineteenth century, huge numbers of customers chose Studebaker over other wagon brands. Not that the brothers let them come unbidden; one of the most significant distinctives about the South Bend firm lay in its extensive promotion and distribution schemes. Though not originating the repository or agency concepts, Studebaker developed both to a greater degree than any other wagon or carriage manufacturer. Repositories, or the more dignified-sounding "branch houses," consisted of buildings Studebaker erected especially for the purpose in strategically located cities. The earliest, and the majority, were in the Midwest and far West, with later ones in New York City and a few other eastern cities, but all served the same purpose. Beyond providing a sales outlet for the Studebaker wagons and carriages kept constantly on hand, however, some light manufacturing sometimes took place there as well. A late-1860s company advertisement referred to the St. Joseph branch as a repository and a "branch factory." Production outside South Bend consisted primarily of acquisition or fabrication and installation of accessories such as seats, as well as trimming and painting—in other words, processes permitting customization of vehicles for the consumer. Repositories also repaired vehicles, ordered replacement parts and accessories, and provided greater visibility for the company name. Studebaker's first substantial repository in St. Joseph was eventually followed by others in Kansas City, Chicago, Salt Lake City, San Francisco, and New York, among others.[40]

Studebaker supplemented its repositories with a much larger number of agencies, established businesses selling Studebaker products in return for a percentage of the profits. Some agencies also supplied components such as seats and brakes, while others accepted repair jobs—it all depended on the agent's primary business and facilities. Many Studebaker agencies simply displayed a selection of the firm's fully finished products, and most could order any vehicle from the factory. From the company's perspective, agencies required a far smaller investment in return for sales in areas often beyond the reach of the nearest branch house. From the agent's perspective, the addition of a brand-name item with proven customer appeal promised greater profit. The arrangement benefited both and grew accordingly: as early as 1868 Studebaker already had at least sixteen agencies spread across Iowa, Texas, Tennessee, Kansas, Missouri, and Minnesota, with more being added all the time. Eventually the firm had a few European agencies, though the majority (as many as 1,500 by 1889) remained in the United States. Branch houses and agencies formed intrinsic parts of the strategy propelling Studebaker to national leadership: rationalized and expanding plants produced the vehicles, an advertising campaign kept the name before the public, and an unparalleled sales network brought the goods to consumers nationwide.[41]

The Power of Promotion

Astonishingly enough, Studebaker's production had increased despite the two devastating fires; while 1872 merely matched the previous year's production of just under seven thousand wagons and carriages, 1874 totaled an incredible 11,050 horse-drawn vehicles (Table 6.1). Two separate factories partially accounted for this, as did the abundant capital reserves permitting rapid rebuilding. Demand continued to escalate all the while, stimulated by the firm's aggressive and far-reaching marketing. Peter continued to direct most of it personally, making an 1872 promotional trip to Texas, creating a company advertising sheet ("The Wagon") in 1874, and continuing to print catalogs and place advertisements. In later years, some company literature suggested the firm's tremendous postwar popularity owed to publicity from its Civil War supply wagons and ambulances. Yet Studebaker was only one of several federal wagon suppliers, makers' names often did not appear prominently on vehicles in military livery, and veteran-farmers had a wide range of choices when choosing a wagon for private use. The war's greatest benefit to the company consisted of capital influx; sales received far greater stimulus from another source. What distinguished Stude-

TABLE 6.1. Studebaker Wagon and Carriage Production, 1852–1876

	Number of Vehicles	Number of Employees	Amount of Sales ($)
1852	2–3	2–3	—
1853	—	—	—
1854	5	—	—
1855	—	—	—
1856	—	—	—
1857	—	—	—
1858	—	—	—
1859	—	—	—
1860	20	—	—
1861	417[a]	—	—
1862	417[a]	—	—
1863	417[a]	—	—
1864	—	—	—
1865	—	—	—
1866	—	—	—
1867	—	140	—
1868	3,955	190	380,000
1869	5,115	220	412,000
1870	6,505	260	573,000
1871	6,835	285	625,000
1872	6,950	325	691,000
1873	10,280	455	896,000
1874	11,050	550	1,000,000
1875	15,000	600	1,250,000
1876	15,500	650	1,400,000

Source: Adapted from M. L. Milligan, "Outline History of Studebaker, 1851–1921," 1–4 and 8 (Studebaker scrapbook #1:43–44, 141; #2:15; #15:49; #38:38, 282, 297).

Notes: Fires, sporadic recordkeeping, and constant expansion make Studebaker production figures difficult to tabulate. These are the best available approximates from company records. The number of employees generally excludes office and sales staff, while production figures are in some cases projections or estimates rather than actual counts.

[a]Estimated.

baker from equally reliable but less well-remembered firms like Milburn, Moline, Espenschied, and others was persistent and innovative publicity. Peter Studebaker's wagon business debut coincided with the trade's awakening to the possibilities of advertising.[42]

The Studebakers had already learned the intricacies of quantity production, but with that came the mandate to increase sales accordingly. Capitalism required markets, and good salesmanship found and exploited them to the fullest. A May 1867 newspaper advertisement announced "THE GREAT FAILURE AT SOUTH BEND, INDIANA OF STUDEBAKER BROS." Interested readers discovered that the firm had "failed . . . heretofore to supply the people of Northern Indiana and Southern Michigan with their superior Carriages, Buggies, and Farm Wagons" but that its newly expanded facilities would prevent a recurrence of this regrettable tragedy. Clement evidently penned this one, but Peter proved the true media master. He employed the age's positive progressivism with mottos like "Never to late to learn, and learning, improve," while also mining his family's faith and the Bible. Thus another advertisement asked customers, in a phrase lifted from the Book of Proverbs, to "let our works praise us; we ask no other endorsement." The brothers loved slogans and squeezed maximum mileage from them; an 1868 advertisement included no less than three: the biblical-sounding "by their works ye shall know them," the scriptural paraphrase "let our works praise us," and a promise to vindicate their motto (cribbed from Virgil): "*Labor Omnia Vincit*"—roughly, "labor conquers all things." Perhaps Clement delved into his schoolmasterly Latin for this last.[43]

Studebaker did not feel restricted to newspaper announcements and used a variety of media to advertise its wares. By the 1880s the wagon and carriage industry turned out richly illustrated price lists and catalogs, and Studebaker naturally followed suit. Its 1883 "Catalogue and Price List" typified the genre: a handy, pocket-sized booklet of eighty pages illustrating Studebaker wagons and carriages. Besides the obligatory aerial factory view, lithograph of the Ashland homestead, purple prose ("the wheels of the Studebaker Wagon are perfection itself"), and improbable claims ("Five hundred thousand people in the United States are using the Studebaker Wagon, and all are happy"), the booklet included detailed, illustrated descriptions of farm, freight, and special-purpose wagons, as well as passenger carriages.

While ever-larger and more lavish catalogs followed, the company continued to branch out into formats like music. Early use of another slogan, "Wait for the Wagon," which more than one wagon manufacturer evidently used, became something of a hit song during the 1880s.

Purchasing the copyright to music previously used to promote Austin, Tomlinson & Webster's "Jackson" wagons, Studebaker distributed thousands of copies of sheet music intended for singing, piano, and parlor organ beginning in 1883. The musical endorsements continued with "Richard and his Sweetheart Nell" (1893), the "Studebaker Grand March" (1899; eventually available on piano rolls), then a later revival of "Wait for the Wagon." Its musical promotions continued into the 1920s. There were also decorative stamps for sealing envelopes, chromolithographs, souvenir booklets, and more—much of it surviving today in museums, antique shops, and among collectors who avidly search out Studebaker advertising ephemera (fig. 6.2).[44]

This stands in stark contrast to eastern carriage makers like the Brewsters. Henry and Benjamin Brewster employed advertising for more than just internecine sniping, but Henry at least found that family reputation alone carried sufficient weight to form a de facto advertising campaign. Studebaker's reputation owed something to this same quality, but much more to an extensive, innovative, and relentless publicity drive. Much of it bore the stamp of horsy nineteenth-century humor, like the "locomotive/farm wagon collision" poster. Here a stunned engineer in a wrecked 4-4-0 gazes in astonishment at an unscathed Studebaker wagon driven by a leering farm boy. "The Secret of Strength of the Studebaker Wagon," the copy says, "Lies in the Scientific Curing of the Timber Employed in its Construction." Of course no one took the illustration literally, but it did suggest that sound timber had its virtues. Such down-home whimsy charmed the public for its very lack of sophistication, much like a kindly old man whose quaintness wins out over his recycled stories. Whether or not 1903 *Carriage Monthly* readers really believed there was a "Sultan of Sulu" offering two wives for two Studebaker carts, such "news items" certainly kept the company name at the fore.[45]

Tall-tale advertising aside, just how did Studebaker wagons compare with others? An 1880 government contract dispute sheds light on this very issue: in March of that year the Department of the Interior's Office of Indian Affairs solicited bids for 331 farm wagons and received many replies. Some, like the Austin, Tomlinson & Webster Manufacturing Company (Jackson, Michigan), the Moline Wagon Company (Moline, Illinois), the Studebaker Brothers Manufacturing Company, and the Kansas Manufacturing Company (Fort Leavenworth, Kansas), qualified as large manufacturers, but the bidders included many smaller partnerships and individual shops around the country. The department's choice of Austin, Tomlinson & Webster and Moline raised the ire of

Alexander Caldwell, president of the Kansas Manufacturing Company. Caldwell wrote a stinging letter to commissioner of Indian affairs, R. E. Trowbridge, demanding an explanation for the rejection of his lower bid. Defending the decision on the basis of best quality for the price, Trowbridge cited an earlier purchase of Kansas wagons and their disappointingly brief service life. He enclosed the opinion of E. L. Cooper, the department's wagon inspector, who found Kansas wagons inferior in several respects. Trowbridge requested additional information, and Cooper provided a detailed analysis of more than two dozen wagon brands.[46]

Cooper had impressive qualifications, having spent much of a forty-year hardware career selling iron wagon hardware to manufacturers, and he also had experience purchasing farm wagons, carts, and trucks in various quantities. Prior to his Office of Indian Affairs appointment, he had supervised large shipments of wagons from various manufacturers to the government. As an inspector he had become acquainted with an even larger variety of wagons, making detailed observations of the Studebaker, Milburn, Jackson, and Moline factories as well as a more cursory examination of those of Espenschied, Lukin, Gestring & Becker, Weber & Damme, and Luedinghaus. On the basis of this experience, Cooper provided illuminating commentary on the leading manufacturers' relative merits. Dismissing out of hand advertising's value as a means of judging quality ("each of the makers claim some advantage over the others"), Cooper instead relied on his own observations of product and manufacturing process. Citing the lack of interchangeable parts, poor quality timber, and indifferent workmanship, Cooper recommended the government avoid St. Louis wagons. He considered Kansas wagons of only fair quality, and in his opinion the firm's convict labor (federal prisoners from the Fort Leavenworth penitentiary) made the workmanship suspect. This last informed his dim view of Austin, Tomlinson & Webster ("Jackson wagons"), which also used convict labor, though he admitted the workmanship appeared superior to the Kansas and St. Louis products. He had a much better opinion of Moline and Milburn, both firms producing a strong and serviceable wagon, though the Milburn featured better iron hardware.[47]

Cooper reported "Studebaker's best wagon as good as any," having previously visited the factory on another assignment. While in South Bend he toured the works to ascertain the quality of raw materials, observe the workmanship, and randomly select wagons for inspection. He liked what he saw. Based on his observation of the extensive covered lumber sheds, general manufacturing process, and the material, ironing, and workmanship of the individual vehicles, Cooper surmised that Studebaker produced stout, serviceable wagons. Assured that the contract wagons comprised the company's finest product, he approved them for shipment; later, however, he reported that "when put to use some of them are said to have failed." He offered no further details, but his comment about Studebaker's two product grades suggests a possible explanation. "I knew at that time from them that they made two grades of wagons," he recalled, "one for the West and one for the South, the Carolinas, Georgia, Alabama, & c., the southern wagons being lighter ironed." Though Cooper never says explicitly, this implies Studebaker sold the lighter-duty version to the government. The inspector suggested that future orders specify the western model. So long as they did so, he felt Studebaker wagons ranked with the best.[48]

Cooper recognized the inevitable subjectivity of public opinion, noting that customers sometimes preferred local brands over superior offerings by the larger firms. Despite relatively minor differences in manufacturing and a few ancillary matters ("About the only fault that I have ever found with Moline is that they are dilatory in getting wagons ready for shipment"), Cooper felt that a cluster of wagon brands compared favorably, a cohort that included Studebaker. Whatever the cause of its wagons' western shortcomings, Studebaker continued to land government contracts. Like many of the other firms the inspector discussed, Studebaker grasped the importance of a workable balance between quantity and quality. Despite the compelling logic of the economy of scale, a manufacturer courted disaster when carrying it too far. Caldwell's hot scorn derived in part from his shock at being outbid by makers of more expensive goods. His use of prison labor (in 1874, 200 of his 233 employees were minimally paid convicts) permitted him to routinely underbid competitors while maintaining impressive production numbers. This in turn generated his baffled rage; he angrily informed Trowbridge that he sold more wagons for western use than Moline and Austin combined. "This is not an unusual claim to be made by a manufacturer," Trowbridge wrote to his superior, "the experience of this office is that the same assertion has been made by other leading manufacturers of the same article." When defending his products on the basis of sales figures, the hapless wagon magnate found himself dismissed for hubris. This, combined with Cooper's dispassionate assessment, proved that contracts ceased when quantity outstripped quality.[49]

Studebaker Brothers Manufacturing Company did not sacrifice quality for quantity, at least not often. Federal contract work placed manufacturers under enormous pressure, often exacerbated by strict deadlines enforced by hovering government officials. One justification for federal inspectors lay in their motivational quality; Cooper frequently traveled to factories solely to ensure suppliers met deadlines. Harried manufacturers faced terrible dilemmas, and some cut corners to make their cut-off date. For wagon manufacturers, such expediency might include building with improperly seasoned wood. Indeed, the government's rejection of Caldwell's wagons stemmed partially from brittle timber, a classic symptom of hurried seasoning. Most wagon and carriage makers agreed that slow air-curing made the soundest timber, yet many resorted to kiln drying to overcome this potential supply bottleneck. Henry and Clement completed their first government contract ahead of schedule partially by using a kiln to "push" green timber into a useable state. While such ploys might defeat a time constraint, overreliance could also mean a disastrous loss of reputation, declining sales—or worse. Studebaker's extensive lumber yards permitted the firm to avoid this disaster, and its spiraling production during 1865–1900 came at no cost in product quality. However much Studebaker owed its enviably high profile to creative advertising, no clever newspaper ads, parade floats, or promotional trinkets could compensate for inferior quality. Lacking carriages' fashion component and the potentially camouflaging coach lace and cushions, wagons and their makers rose or fell on serviceability, not appearance. If anything can be said with certainty these many years later, Studebaker wagons remained eminently serviceable.

By the 1870s Studebaker enjoyed an almost logarithmic increase in general recognition. When the nation's manufacturers displayed their goods at Philadelphia's 1876 Centennial Exposition, the Studebaker Brothers Manufacturing Company was present, and not just in the farm wagon building. Over in the Main Annex, a Studebaker top buggy stood in a long row of carriages by such luminaries as Rufus Stivers, the Brewsters, and William D. Rogers, and the South Bend firm came away with awards. Commemorating twenty-five years in business the following year presented the firm with another publicity-generating opportunity, closely followed by the Paris Exposition of 1878. Since the firm had more domestic business than it could handle, Clement had only moderate enthusiasm for foreign exposure, but he did end up displaying a Studebaker military wagon. The firm presented this to France's minister of war at the end of the event, handily saving the cost of return freight. Well aware of the value of endorsement from on high, Studebaker put its vehicles at the disposal of several American presidents. When Ulysses S. Grant returned from his 1880 post-presidential world tour, he rolled through Chicago in a Studebaker landau. Studebaker prepared a similar carriage for President James Garfield before his assassination ended the project. When it furnished carriages for President Benjamin Harrison in 1889, the firm gleefully noted this as the first instance of a western firm supplying vehicles for official White House use. This surely irritated eastern carriage manufacturers, which was precisely the intent.[50]

The 1880s found the Studebaker Brothers Manufacturing Company thriving beyond its founders' wildest dreams. Company records show that as of January 1883 it employed 1,200 hands (900 of them in the wagon plant), collectively paying out over $40,000 in monthly wages. These employees produced more than thirty thousand wagons the previous year, with carriage production taking the total even higher. Company promotional material touted the statistics of Studebaker's success: total plant acreage (eighty), floor acreage (twenty), number of electric lamps (fifty-eight), average letters received daily (five hundred), annual stationery expenditure ($2,500), and the secret behind its one-day payroll processing ("ten swift accountants"). Its product line fell into several distinct categories, each available in a range of sizes and variations. Besides the several wooden- and iron-axle farm wagons, the catalog listed low-wheel versions "particularly suited to Macadamized roads," California and Oregon wagons (larger capacity, heavier construction, brake-equipped) for large western farms, and one-horse wagons for the small farmer. The freight wagon line consisted of a civilian model as well as the Four Mule U. S. Government Wagon and Six Mule U. S. Army and Transportation Wagons, all with the familiar canvas-covered tops.[51]

By now Studebaker also had a line of light, spring-equipped delivery wagons and offered its Three-Spring Express and Business Wagons ("for the farmer, planter and businessman"), the Scroll Spring "Diamond" Wagon, and models suitable for grocers and baggage-handling. The carriage line tended toward versatile family models, such as platform (spring) wagons, light vehicles whose twin removable seats accommodated several passengers. These came in versions seating four, six, and more, as well as heavy-duty mountain wagons with brakes, others with tops, plus at least one example of the lowly but enduring buckboard. Special-purpose vehicles included log wagons, lumber trucks, and hand carts. Purchasers furnishing their own boxes could choose from the Nevada Wagon Gear ("adapted for heavy freighting in all mining countries") or the California Header Wagon Gears ("especially adapted to the Great Wheat Fields of California"). Besides accessories such as seats, tool boxes, and top bows, one could also order replacement parts as complete as entire wagon boxes or as simple as a singletree ("finished and painted but not ironed"). This catalog billed Studebaker as manufacturers of "Farm, Freight, Plantation, Platform and Spring Wagons," as well as "Fine Carriages and Light Buggies." The impression, an entirely intentional one, was that the company specialized in one-stop shopping regardless of location or need.[52]

As the hard work of the 1850s and 1860s gave way to the successes of the 1870s and 1880s, the Studebaker brothers found themselves becoming wealthy and influential citizens. They remained intensely loyal to their parents; late-nineteenth-century photographs show old John and Rebecca enjoying their golden years in unaccustomed luxury. Before John died in 1877 and Rebecca a decade later, they had witnessed their sons' business triumphs. Each had married and established his own household, Clement and J. M. in adjacent South Bend estates "Tippecanoe Place" and "Sunnyside." Clement enjoyed tradewide recognition through his 1886 presidency of the Carriage Builders National Association. He had been one of several espousing its formation, and Brewster & Company's John Britton nominated him as temporary chairman of the 1872 inaugural meeting. The Studebaker Brothers Manufacturing Company held charter membership, and Clement remained active as a vice president until his 1886 election to president. Presiding over the 1887 meeting in Washington, he may have reflected that one reason for this honor lay in his alleviation of east-west regional tension within the organization. Indeed, there were likely those objecting to a Studebaker at the podium. Was this man not a *wagonmaker*, the kind that daubed his name down the sides of his own wagon boxes? Bold hucksterism allegedly grated on eastern carriage makers' sensibilities, even though they doubtless envied Studebaker's annual sales figures. If the Brewsters found money in taste, the Studebakers found even more of it in mass production. Size counted in the CBNA, which routinely rewarded large factory proprietors with its presidency. Despite its title, the association theoretically served all horse-drawn vehicle manufacturers, and most members objected less to wagonmakers than they did to hardware magnates who built no vehicles at all.[53]

The CBNA may also have factored in Clement's prominence in its favorite political party. By the end of the 1870s he had become increasingly important to the Republican Party in and outside Indiana. Some of this influence derived from a widening circle of friends, including the oleaginous South Bend publisher and industrialist Schuyler "Smiler" Colfax, promoted from Congress to serve as vice president under Ulysses S. Grant. Clement had been a delegate to the 1880 and 1888

Republican National Conventions, and entertained Presidents Grant, Harrison, and McKinley at his South Bend home. While he shied away from political office himself, he continued to lend his influence to movements he regarded as worthwhile. This accounted for his role in forming the National Wagon Manufacturers' Association. Established in Chicago in 1879 by several leading wagonmakers, the group emulated the CBNA by promoting the trade, curtailing price wars, and providing opportunities for socializing. Clement became its first and long-standing president, despite his disenchantment with the perpetual internal squabbling that prevented it from achieving notable influence.[54]

However much the Studebaker brothers prospered, they clearly shared their wealth with their many employees; even a cursory glance at wages credits them as generous employers. The very earliest Studebaker employees often had to accept scrip instead of cash; Henry and Clement themselves took livestock and produce in payment for work. This practice pleased no one; in the early years, Clement's credit wore thin at the grocer's, but he managed to obtain sufficient cash to keep the merchant happy and his men in food. With more abundant hard money, general reduction in barter, and creation of their own cash reserves, the brothers began paying cash wages, usually twice per month. One of their first employees received $12 monthly, but this amounted to less than Clement's wages at the thresher works. Higher-skilled workmen commanded better wages. In 1880 the average pay for a Studebaker factory worker was $1.50 per day, up to $1.80 by the early 1890s. This figure was actually above average for the industry, at least according to the U.S. Census Bureau, which calculated that the average unskilled factory hand made about sixty cents less per day. Intensive mechanization meant that many of the jobs consisted of repetitive machine operation, and Studebaker employed many young men with little or no trade experience. Still, the factory required skilled workmen at various posts, and they received better pay. In 1895 Studebaker's skilled woodworkers typically earned anywhere from just under $2.00 to as much as $3.50 for a ten-hour day, significantly higher than the $1.80 trade average. Foremen and supervisors received more, sometimes considerably more, for their years of practical experience. Overall it appears that the Studebaker Brothers Manufacturing Company paid its workmen well.[55]

Good pay continued to mandate long hours: over the course of the firm's post-1860s history, the ten-hour day and sixty-hour week became the norm, this mirroring common industrial practice. Many city carriage shops had reduced their hours to this level as early as the late 1830s and early 1840s, though there were plenty of firms wedded to different standards. Studebaker resisted reducing its hours despite sporadic agitation in and outside the trade. In 1893 the Chicago branch suffered a brief strike over hours, among other things, but the company retained the ten-hour day still in use by many of its competitors. Brewster & Company held to ten hours until 1886, and plenty of its own contemporaries genuinely felt reduction to nine or eight a disastrous innovation. Indeed, the ten-hour day remained common in the trade into the early 1890s, and as late as 1910 the majority of wagon- and carriage-factory hands put in fifty-four- to sixty-hour weeks. Even this could be exceeded in times of great demand, as when Studebaker temporarily mandated twelve-and-a-half-hour days in 1897. The seasonal demand cycle and occasional government contract ensured uneven production schedules, and increased hours remained a favored accommodation. The brothers' own experience with twelve-, fourteen-, and sixteen-hour days doubtless convinced them that longer hours were sometimes necessary: for years an office sign reminded passersby that "Success Means Sacrifice." For nonsalaried employees, it at least meant larger paychecks.[56]

The prosperous 1880s witnessed at least one disaster for the company when Jacob died in December 1887. Loss of their forty-three-year-old brother proved a sore blow to the others. This "break in the chain" saddened them all, and not just as a familial loss. As Clement remarked, "we naturally looked to have him left for us to lean upon," and his death meant that continued family control would require someone from outside the founder's generation. Jacob Studebaker had at least lived to see the carriage division mature splendidly. Though never so large as the wagon department, the Studebaker carriage works turned out an impressive variety. Though it emphasized practicality, the firm's carriage line by no means restricted itself to buggies and surreys. The very first Studebaker carriage had been a fashionable phaeton, and long after Jacob's death, Studebaker continued to advertise its broughams, coaches, and opera buses. By the 1890s the company boasted its "Izzer" buggy line—the name differentiated them from the competitors' "wuzzers," another of Peter's corny cognomens—as well as another group of carriages. Studebaker's 1898 purchase of the optimistically named World Buggy Company gave it a lower-priced market presence.[57]

The 1890s commenced with other important developments, beginning with a new factory on South Bend's Tutt and Lafayette Streets. Carriage production continued at Michigan and Jefferson, but change came here with the incandescent lighting that brightened dark winter mornings for that factory's 250 workers. Ever-watchful for promotion opportunities, Studebaker participated in Chicago's 1893 Columbian Exposition with a spectacular show vehicle. Though following ordinary farm wagon design, the custom display model featured a holly-inlaid rosewood body embossed with over thirty medals the company had won over the years. The most novel touch consisted of highly polished hardware made from an exotic new alloy called "aluminum." If this celebrated the past, so too did the brothers' 1897 return to Ashland for a reunion. There Clement, J. M., and Peter revisited the scene of their childhoods, though they particularly missed Henry, who had died two years earlier. Nearly seventy years old and a prosperous northern Indiana farmer, Henry had long since broken with the Dunkards, but in fathering ten children perpetuated at least one other family tradition. The Ashland reunion proved especially meaningful for Clement, though J. M. and Peter also carried away memories they cherished for the remainder of their lives—which in Peter's case consisted of only a few more days. Sixty-one-year-old Peter Studebaker died shortly after his return to South Bend, and Clement once more spoke of a fractured fraternal bond. Perhaps work provided a measure of solace; Clement and J. M. remained very much full-time firm members, and, to use another metaphor favored by both, had expressed their wish to die in harness (fig. 6.3).[58]

Work certainly continued as much as ever. The year 1899 dawned with the opportunity to build yet another presidential carriage, this time a park phaeton for William McKinley's personal use. Clement greatly esteemed the man and welcomed this opportunity even as his firm rushed to fulfill another government contract, this time for a foreign entity. War had broken out between the British and Boers in South Africa, and the British Army found that irregular warfare required large numbers of durable supply wagons. Unable to obtain sufficient numbers from its own manufacturers, the British turned to Studebaker. Three hundred Studebaker wagons left the factory in December 1899, with more to follow. Commenting on his choice of an American product in his later report to Parliament, British commander Frederick S. Roberts defended Studebaker wagons as superior to any obtainable in England or Cape Town. One suspects Studebaker's proven military contract record also informed Lord Roberts's choice. The South Bend company had certainly created something of a niche for itself in the specialized world of military transport. A 1901 brochure "In Six Wars" listed the firm's supply of wagons for the Indian Wars, the American Civil War, the Spanish-American War, the Boxer Rebellion, and the Boer War. Pa-

TABLE 6.2. Studebaker Wagon and Carriage Production, 1877–1900

	Number of Vehicles	Number of Employees	Amount of Sales ($)
1877	18,200	700	1,525,000
1878	19,000	700	—
1879	20,000	800	2,000,000
1880	22,000	700	—
1881	25–28,000	1,100	2,000,000
1882	30,000[a]	1,200	—
1883	30,000[a]	1,200	—
1884	—	1,300	—
1885	40,000	—	—
1886	—	1,300	—
1887	—	—	—
1888	—	—	—
1889	40,000	—	—
1890	40–50,000[a]	1,500	—
1891	60,000	1,500+	2,500,000
1892	65,000	—	—
1893	50,000	1,860	—
1894	—	—	—
1895	—	—	—
1896	—	—	—
1897	—	1,600	—
1898	75,000	1,800	—
1899	75,000	2,000	—
1900	100,000	2,500	3,000,000[a]

Source: Adapted from M. L. Milligan, "Outline History of Studebaker, 1851–1921," 8–19 (Studebaker scrapbook #38:145, 282, 297; #4:17, 48, 75; #2:53; #6:13, 39, 51, 67; #7:8; #8:49, 56, 94, 128, 216).
[a]Estimated.

triotism and profits comfortably coexisted at the Studebaker Brothers Manufacturing Company (Table 6.2).[59]

The twentieth century found Clement and J. M. commanding a firm with a grand history, one neatly symbolized by an advertising poster from around this period. Illustrations of twenty-seven vehicles roll across the poster board, each towed by tractable horses or mules with characteristic Studebaker efficiency. Curtain- and glass-sided school buses transporting school children flank a park wagon carrying several suspiciously familiar bearded men, while an empty lumber truck returns for another load near two heavily laden log wagons. Delivery wagons bearing dry goods, furniture, milk, ice, and coal to elm-shaded homes rumble down cobbled lanes scrubbed clean by a Studebaker street sweeper, while a road sprinkler and oiler care for the dirt lanes nearby. A sewer flusher and garbage wagon assist the underappreciated souls consigned to their use, and a courting couple passes in their Izzer buggy, preceded by a family out for a drive in a surrey with the obligatory fringed top. Yet work remains the dominant theme, as the three-seated platform and business wagons appear to hurry. The former waits for an auto-seat buggy to crest a hill, while the latter stalls behind a grocery wagon pulled by a blinkered truck horse. Its pace owes less to the heavy load than to the slow but steady progress of the farm wagon preceding it, and which in a larger sense, preceded them all.[60]

Clement ended up preceding J. M. in death. Business had often pressed heavily on him over the years, particularly during the depressions of the late nineteenth century. While he successfully conned the company through them all, each took an increasingly physical toll. "I would much rather undertake the work of building up a large business," he confessed, "than to take care of it and keep it intact after it has become established." His health had become much harder to maintain. In early 1901 a lingering cold wore away at him until he and his wife fled to Europe for relief. While the mild climate helped, a wrenching fall on the New York docks on his return seemed to cancel much of the gain. Even when confined to his bed, Clement's thoughts revealed his ongoing loyalties: "This is an ideal room to be sick in," he told his wife. "I can see the boys go to the office, and I can see the children and hear their voices." He could also see the end coming and so put his affairs in order. Clement Studebaker died on November 27, 1901, the fourth of the five brothers to pass away.[61]

J. M. remained the sole link in the fraternal chain, and he too felt the passage of time. A second generation that included Clement's sons Clement Jr. and George, and his own J. M. Jr. continued the family presence in the business, but talented outsiders like son-in-law Frederick S. Fish seemed poised to lead the firm in new directions. As early as 1895, Fish had pressed for Studebaker to consider automobile manufacture. Clement regarded autos as a fad to be endured rather than a goal to be pursued. He did sanction experiments that included an 1897 prototype electric runabout, and the company built wooden bodies for various automobile manufacturers. Still, Clement knew wagons and the wood from which they came; the automobile seemed to require a greater leap of faith than he was able to make. In that sense his death removed an obstacle, and J. M. appeared more willing to consider the matter. Studebaker would indeed make the plunge into automobile manufacture, and it distinguished itself above all other wagon and carriage makers by this successful transition. Before exploring that triumph, however, we need first to consider those whose hands guided the machinery at Studebaker and throughout the American wagon and carriage industry. For as much as the carriage to automobile shift instructs those willing to examine it, wagon and carriage workers had already been subject to a transition of another sort entirely.[62]

[7]

From Craftsman to Assembler?

Industrialization and the Wagon and Carriage Worker

An educated man finds out the value of machinery and desires to use and improve it. Instead of fearing its rivalry he welcomes it; he remembers that all tools, even the saw and the hammer, are machines, and that the hand, the human hand that guides those tools is but a perfect machine obeying the guidance of the brain more quickly and in a more varied manner than any man-made machine. The American workman therefore uses machines more and more.

George A. Thrupp, 1877

The late Phineas Jones, in the course of a speech before a body of carriage makers, perpetrated a witticism that now savors more of fact than fancy. He asked himself the question, "What does the carriage builder do?" and after a short pause said, "Why, he paints the carriage."

The Hub, February 1902

In February 1895, *The Hub* carried a provocative discussion of the relative skills of past and present carriage woodworkers. "Reflections by an Old-Time Wood Worker" expounded on the talents of the traditional carriage maker before machinery, working drafts, and factory organization so thoroughly altered his routine. The anonymous author admitted such innovations simplified the work and made it less tiresome, yet he lamented that contemporary workmen lacked the skill of their predecessors. He recalled the pre–Civil War years, when body makers were guided by little more than rough sketches and their own wits, making patterns as the work progressed and using their eyes to lay out curves and sweeps. In his view, "cut and try" produced superior workmen and excellent work: "There are a few today who outrank in skill and technical knowledge those of fifty years ago," he admitted, "but they who do the bulk of the work are not skilled as were the former workmen." Whatever readers thought of his contention then, his observation cuts to the heart of a debate about industrialization that continues to animate historians more than a century later. The central issue is whether or to what degree industrialization turned skilled craftsmen into unskilled machine tenders. As the "Old-Time Wood Worker" pointed out, skill lay at the heart of the workman's world, so it seems particularly appropriate to assess industrialization's impact in this light. Additionally, by examining mechanization, rationalization, ready-made parts, and workers' reactions in letters, protests, and unionization, we also limn the essential contours of a process that had repercussions far beyond the confines of the carriage trade.[1]

As this piece appeared in *The Hub*, a far more systematic examination was taking shape at the hands of a federal employee with an abiding interest in American manufacturing. Union army veteran, attorney, and statistician, Carroll D. Wright had worked for the Massachusetts bureau of labor statistics in the early 1870s and subsequently became an authority on American industry. An 1882 book on the factory system led to his appointment as the first commissioner of labor in the U.S. Department of the Interior two years later, a post he would occupy for twenty years. Here he studied various facets of American industrialization, publishing reports on convict labor (1887) and economics and crime (1893), before turning to the relationship between machinery and skill. His 1898 annual report, "Hand and Machine Labor," compared hand and mechanized work in eighty-four different industries. In the introduction, Wright noted that he had prepared the report at a time when intimate knowledge of traditional hand methods remained in individual and collective memories, even as they rapidly faded from use. Thankful he had been able to gather such knowledge before it disappeared completely, he confidently predicted his report would stand as the "largest possible contribution of facts for the consideration and discussion of the great machinery question." The question was machinery's impact on the American worker, and Wright hoped to provide hard data for this hotly contested debate.[2]

Like craftsmen in dozens of other trades, the horse-drawn vehicle builder's skill with hand tools formed the basis of his reputation and livelihood. As much as Ezra Stratton hated his jack plane, his ability with that tool and dozens of others enabled him to transform rough planks into finely wrought carriages. Each of the trade's four main branches had its own armamentarium: the carriage woodworker used hand saws, chisels, common and specialized planes, and the drawknife, as well as hammer and nails, screwdriver and screws, and animal hide glue. The blacksmith worked with his forge and anvil, hammers, punches, fullers, hot and cold chisels, files, and perhaps a grindstone or emery wheel.

Trimmers wielded awl, needles, thread, pins and tacks, while painters employed body, gear, and striping brushes, and pigment grinding equipment. Normally the tradesman's personal property, most of these tools were not unique to wagon and carriage making; the genius of the skilled carriage maker lay in his ability to turn general-purpose implements to highly specific tasks. The joiner and carriage maker owned similar hand planes and manipulated them in similar fashion, yet one fashioned cupboards, the other a hearse; therein lay the essence of craft skill. Traditional craftsmanship stressed the hand-eye coordination necessary to follow a minimally detailed plan, as well as the ability to deviate from it when required. The craftsman's hard-won skills were his most valuable possession.[3]

Machinery changed all this. Not instantly, of course, or comprehensively, inasmuch as machinery failed to penetrate every corner of the trade, yet mechanization had profound consequences nevertheless. Wright's report provides an unparalleled contrast between old and new; in most cases the hand work he describes was common during the period 1840–65, while the machine methods were employed in the mid-1890s, at the time of the survey. Compiled from observation and workers' memories at several firms, the document provides a unique assessment of the major differences between hand and machine work in vehicle and parts manufacture. In the first category, Wright surveyed construction of a buggy, three sleighs (a set of team bobs and two Portland cutters), and two wagons (farm and road). The spring-equipped, folding-top piano-box buggy was built in 1865, a typical product of a typical small shop. The woodworker made the gear, body, and seat from planks cut and planed to shape. Wheels appear to have been purchased partly finished, though Wright's table does not explicitly state this. Likewise, the blacksmith produced the hardware (springs excepted) from bar iron. Again, no mention is made of whether the springs were made in house, but it is virtually certain that they were purchased ready made. Painters used their brushes and pots to apply color and varnish coats, while the trimmer employed patterns, knives, and shears to cut material to size, and awl, needle, and thread to trim the carriage.[4]

Thirty years later the manufacturing process for the same vehicle had changed considerably. The factory took its power from a water wheel or turbine, though steam had replaced water in most places long before 1895. Power had become necessary to drive the equipment used for each process, beginning with basic stock preparation. For sawing out the wood for body, seat, and gear, the woodworkers employed powered rip (table) and crosscut (swing) saws; for squaring and smoothing they turned to an assortment of other machines: the buzz planer (jointer) smoothed and trued edges, while the sticker (outside molding machine) cut moldings. These machines and operations created the various parts, while tenoning machines and gainers (mortisers) created the joints required for assembly. Other equipment complemented the already considerable list: band saw, scroll saw, and shaper created curved surfaces, while a boring machine (drill press) drilled screw holes, and drum or belt sanders smoothed most of the woodwork. While the scroll saw's mechanical limitations necessitated hand finishing head blocks and spring bars, this was an exception to the rule. Factory woodworkers used general-purpose woodworking machinery for the overwhelming majority of their tasks.[5]

Once the blacksmiths had welded, fitted, and set the axles, the woodworkers assembled the gear, body, and seat using glue pot and brush, a flexible shaft drill, screwdrivers and wrenches. Assembly required no machinery other than the power drill and a few common hand tools. This particular factory purchased partially completed wheels that required only boxing and tiring. Those steps necessitated a tire bender (water-powered), drill, a hand-powered tire setter, a "flexible shaft nut runner" (power ratchet), wheel borer, and wheel boxer. Ironing the gear and body still required the usual blacksmithing hand tools and techniques, but the smith also used a trip hammer, drill press, and mechanical punch. It seems likely that paint and some of the hardware came from an outside supplier. The steps listed for axle fabrication indicate use of ready-made axle ends joined to properly sized bar iron, but by contrast this company produced its own dashboards even though ready-made models had been available for decades. Painters employed no machinery, and trimmers required only sewing machines and mechanical eyelet presses, all water powered.[6]

Wagon and carriage hardware traditionally came from the blacksmith, and the earliest specialty manufacturers relied on his proven tools and techniques. Such was the case with leather-covered bow socket fabrication in 1865. Though Wright listed only "blacksmith's tools" on the chart, the individual steps reveal specific implements: hot and cold chisels to cut slat irons, a breast or post drill to bore rivet holes, and a hammer and cold chisel (or shears) to cut the sheet iron. Carriage trimmers used patterns and sharp knives to cut out the leather covering, stitching it in place before handing the subassembly back to the blacksmith for forge welding and riveting. Thirty years later the process had become extensively mechanized: workmen used drop hammer, trimming press, and tumbler to forge and finish the pivot at the unit's base. The sockets began as sheet steel shapes cut out by steam shears and tube cutter, then passing through three machines that formed the lock (the joint holding the tube together), bent them into tubes, and pressed tight a permanent joint. Pivot and tube connection required the blacksmith's forge-welding skills, while finishing consisted of a mix of hand and machine work: plain bow sockets received a hand-brushed japanned finish before oven curing, while leather-covered versions required a steam-powered sewing machine and seam trimmer. Leather cleaning and socket blacking were still done by hand.[7]

Wooden parts fabrication involved similar contrasts. Though the hand and machine methods of hub fabrication required a heavy-duty lathe, the other operations bore little resemblance to one another. Cutting the billet to length in 1879 required a handsaw, but the 1895 wheel factory used a belt-driven crosscut saw. Rather than hand mortise the hub to accept the spokes, the wheel factory used a mortising machine and a hub-boring machine replaced the auger for boring out the center. It was the same with spoke manufacture: in 1848 the craftsman used axe, saw, drawshave, and spoke shave, whereas in 1895 the job required a spoke-turning machine, power saws, throating and tenoning machines, and a belt sander. Wheel making was vastly transformed: where once the lathe was the only machine required—and a human-powered one at that—wheel makers came to rely on some of the most sophisticated special-purpose machinery in the industry (fig. 7.1). Even so, hand work remained. In the 1895 example, two nonmechanized processes were spoke driving (performed with a maul) and hand fitting each spoke into the rim. Though mechanical spoke drivers had been available for some time, not all firms used them, and the latter operation largely defied mechanization. The report did not include painting, but wheels were often wholesaled in the white. Whoever painted them, however, usually did so by hand, even if with ready-mixed paint.[8]

What does all this tell us about machinery's effect on worker skill? First, it is important to note that even relatively late factory production still required many hand tools and processes. A number of them, such as rubbing down leather bow covers, applying japanned finish, and some assembly procedures, required little skill. Others were moderately difficult, including hand finishing head blocks and spring bars, spoke driving, and gear assembly; still others required the craftsman's practiced eye and hand. No unskilled factory worker could weld pivots to sockets or hand stripe wheels and gears. Mechanization had nevertheless permeated most production sequences; how did this affect worker skill? Conventional historical wisdom has long held that mechanization transformed skilled craftsmen into semi- or unskilled machine tenders. While certainly the case in some trades, this generalization often obscures a more complex reality. As a field, the history of technology has moved beyond many formerly widespread truisms about mechanization: one perceptive look at a nineteenth-century federal armory demonstrated the surprising persistence of skilled workmen. Mechanized firearms production did not entirely displace craftsmen partially because many machines were incapable of reproducing fine hand work—a pattern repeated in many other industries.[9]

Close examination of the equipment itself yields sometimes surprising information, for machinery was by no means all of a kind, nor were its effects on the laborer exactly the same. Broadly speaking, most machine processes fell into one of four distinct categories. The first consisted of mechanizing tasks impossible to perform by hand, as with cylinder turning. This required a skilled craftsman but also a lathe, as no hand could perfectly duplicate that device's circular motion. This explains the great wheel lathe's centrality to even primitive wheelwright and wagon shops. A second machine application involved reducing the effort of a physically taxing or repetitive task, but in a skill-dependent manner. The blacksmith's power hammer saved hours of hand hammering, but the smith still manipulated the hot metal between blows and had to understand the correct procedure to forge a particular piece of hardware. Open-face dies considerably lowered the skill quotient, but blacksmiths did not invariably employ them. Planers became enormously popular wood shop fixtures for their elimination of the joint-wracking hand planing formerly required for stock preparation. The planer required less talent than a jack plane, but the tremendous savings in time and effort made this a loss many craftsmen willingly sustained. It also freed woodworkers to spend more time fabricating and assembling. In general, these two machine categories complemented rather than supplanted craft skill.[10]

The third machine application consisted of "self-acting" equipment that performed a single task without constant operator attention, essentially an elementary form of automation. A semiskilled workman operating a self-acting lathe could chuck up the workpiece, activate the mechanism, and allow the machine to take over from there. This proved a more significant departure by reducing the workman's role to transferring raw material and finished product in and out of the machine, even though the items often required subsequent finish work. Textile factories boasted some of the earliest American self-acting machinery, but devices employing similar principles began turning up elsewhere, as with the Blanchard lathe, used to turn gunstocks, axe handles, spokes, and other irregular wooden objects. The fourth and final category consisted of multiple-task machines capable of shuttling the workpiece through a series of discrete operations. Whether by physically moving the work or by applying multiple tools, this extension of the self-acting principle came much closer to true automation. Here the workman had only to feed in raw materials and remove a product that required little or no finishing. A good nineteenth-century example would be the Howe pin-making machine, a highly complex mechanism that took steel blanks through a series of steps to create common sewing pins. These last two machine categories required far less operator skill, and this, combined with higher volume output, lent them particularly well to factory use.[11]

Mechanization permeated vehicle and accessory manufacture alike, but each branch of the industry relied on distinct varieties. Machinery common to wagon and carriage manufacture generally fell into the first three classifications, particularly the first two, as vehicle production relied most heavily on the general-purpose wood- and metalworking machinery rather than self-contained, multiple-task equipment. Whether table saw, planer, or power hammer, the efficient flexibility of general-purpose machine ideally suited it for the widely varied demands of horse-drawn vehicle manufacture. By contrast, the accessory industry made maximum use of special-purpose machinery, equipment capable of repeating a sequence of closely related steps with little or no operator guidance. Special-purpose machinery allowed for quantity production of uniform and interchangeable parts, which became the specialty trade's hallmark. Stock preparation and other preliminaries required some general-purpose machinery, such as the drop hammer and steam shears in the bow socket factory in Wright's report. Special-purpose machinery turned raw material into finished product, and in cases such as wheel manufacture, constituted the overwhelming majority of a factory's equipment. Technically, much of this fell into the "self-acting" category, though in most cases the machine only performed a single or pair of closely related tasks. Few special-purpose production machines were as fully automatic as a Howe pin machine or a Fourdrinier paper maker.

General-purpose machinery required the most skill. Just as chisel, handsaw, or fuller could be turned to different tasks, so too could table saw, planer, or power hammer. The selectivity necessary to make hand tools produce a wagon box instead of a desk remained the same in turning general-purpose machines to vehicle building rather than furniture production. Mechanized buggy manufacture in the report reveals the primacy of these general-purpose machines: crosscut and rip saws, planers, mortisers and tenoners, belt sanders, post drill, trip hammer, and sewing machines were all part of the process. Their use predicated an intimate understanding of each raw material, its strengths, weaknesses, reaction to being worked, and specific application. While a woodworker might no longer spend hours pushing his jack plane, planing remained fundamental. Although a blacksmith used his sledge hammer less and his drop hammer more, hammering still shaped hot iron. In most instances the new corpus of general-purpose machinery acted as an adjunct to traditional craft skills. So long as the workman could relate to it as he had to his hand tools, machinery alone could not consign him to the ranks of unskilled labor.

The accessory industry's embrace of special-purpose machinery had more insidious effects. Because of its ability to perform a single task largely unassisted, such equipment restricted its operators to far more attenuated duties and skills. Take the triad of machines that bent and crimped sheet steel into bow sockets: little more was required of the operator than to feed precut pieces into the first machine, transfer them between each successive machine, and remove the completed sockets from the last one. This routine would have been wearily familiar to the spoke factory worker spending his time chucking fresh blanks in the spoke lathe and tossing finished ones into a bin. Wheel factory workers followed similar rituals as they attended some of the most highly developed special-purpose machinery in the industry. Though most of these machines possessed features ensuring precise and uniform work, they still demanded certain operator qualities, among them alertness. Dull minds around high-speed, unguarded industrial equipment often resulted in missing fingers, or worse. Equally important, and more a matter of real skill, proper machine set up, adjustment, and operation were all necessary for successful use. One participant in an 1890s wheelmaking debate put it this way: "I am frank to say that no man ever lived who could mortice the fourteen holes in a hub with a bit, chisel, and mallet as true as a well constructed mortising machine can do, if a skillful man conducts its operations." Still, the majority of special-purpose machinery required only moderate skill in use, and this necessarily affected the nature of the specialty factory labor force. Reduced skill requirements and constant economizing combined to mandate hiring cheaper, less skilled industrial workers (fig. 7.2).[12]

The relationship between mechanization and skill naturally raises the question of the basic motivation behind mechanization. A favored economic history truism long held that American labor scarcity prompted mechanization, in contrast to England, where abundant and inexpensive labor dampened mechanical innovation. Despite its seeming plausibility, however, this theory has repeatedly broken down in the face of detailed case studies, including a brief but enlightening look at the Cincinnati carriage trade. In the United States the desire to maximize production, rather than a perceived need to minimize use of skilled tradesmen, prompted factory owners to pursue mechanization aggressively, a conclusion Wright amply supported. The desire for greater productivity motivated proprietors to invest in mechanization, though owner and worker alike hotly debated its impact. Machinery's impact on skill was determined to a great extent by the basic nature of the devices themselves. The general-purpose wood- and metalworking machinery in vehicle manufacture demanded from its operatives skill and judgment very much in keeping with traditional craft skill, where the accessory industry's special-purpose machinery relegated its workers to a more passive role. Wagon and carriage factory workmen could still legitimately claim membership in one of the trade's four traditional branches, but those in the accessory industry found themselves increasingly identified as dedicated machine operators. Brewster & Company's carriage mechanics still conceived of themselves as gear makers, body men, and blacksmiths, while the Eberhard Manufacturing Company's machine tenders had more in common with general industrial labor. The machinery had a great deal to do with that difference.[13]

Rationalization, Ready-Made Parts, and the Small Shop

Even so powerful an agent as mechanization acted in conjunction with others that had an equally significant impact. The reason modest amounts of machinery could seriously alter established work patterns stemmed partly from the nature of the equipment, which suggested new and more productive methods. In other words, aspects of the factory system tended to crop up wherever machinery made its appearance. Whether referred to as specialization, rationalization, or something else entirely, it had everything to do with a fundamental reordering of the division of labor. The trade had long embraced the basic specialties of woodworking, blacksmithing, painting, and trimming; traditionally the craftsman in each could perform all the work necessary to move a job through the department. General-purpose machinery disturbed this order the least, so long as the men followed the work through the various processes as they had once done with hand tools. In such a setting, workers related to the machines as they once had to the hand tools, which still continued to supplement machinery.

Nevertheless, the craftsman-machine relationship changed as more machines crowded the shop floor. The factory system's productive triumph owed much to subjecting production to rigid systematization, a process economist Max Weber called rationalization. This consisted of reducing a complex process into a series of discrete and simplified steps, a move that greatly elaborated the division of labor. Rationalization simplified tasks, sped their rate of execution, and materially reduced the requisite skills, all to achieve often dramatic productive gains. It shifted the division of labor principle from macro to micro, forcing rationalization into the very structure of work in each of the four traditional departments. The very earliest significant experiments in marrying rationalization to carriage production, like those tentative experiments by Brewster and Goold in the 1830s and 1840s, owed much to this scheme. None of these large shops or primitive factories had much in the way of machinery; little of it was even available, and Brewster remained skeptical of its applicability in the first place. Their productive innovations were solely organizational and correspondingly modest. When the Cooks, the Studebakers, and others began linking the concept to intensive mechanization in the 1850s and 1860s, however, the advances became truly dramatic. This is when the process began to impact worker skill markedly.[14]

Aggressive exploitation of the rationalization-mechanization connection by 1850s New Haven carriage makers changed James Brewster's opinion of machinery. The Cooks and Hannibal Kimball divided the production process by machine type, assigning an operator to each. If the results lacked handcrafted durability and finish, they nevertheless suited the southern market, which seemed an ideal outlet for the variable quality products of these early carriage factories. This machine-operator pairing featured most prominently in the accessory industry. Wright's investigation of bow socket manufacture provides a case in point: where once a single blacksmith produced one set at a time, a series of dedicated machine operators now turned them out in vast quantities. No longer could the bow socket factory worker be considered a blacksmith; as Wright's tables confirm, he had instead become a punch press operator, a shearman, or a seam roller. Likewise, wheelwrights had become lathe operators, wheel-box press men, or spoke polishers, while hardware blacksmiths gave way to molders, furnace tenders, or annealers.[15]

Though general-purpose machinery lent itself less readily to such practices, rationalization still suggested dedicated operation for maximum efficiency. Restriction to a single machine, even one requiring some talent in use, consigned the workman to the role of machine tender rather than skilled craftsman. Again, this comes across clearly in the 1898 report. While workers' titles in the 1865 carriage shop reflected their craftsman status (blacksmith, woodworker, painter, trimmer), those in the 1895 buggy factory bore names eerily similar to the machines they manned. Here men were stickers, shapers, scrollers, welders, or any of numerous similar titles. In this environment the foremen remained the retainers of the old time craft training, a quality leading one trade journalist to refer to wholesale factories as "those formidable and relentless enemies of skilled labor." This development contrasted not only with previous American practice but also with the trade as carried on in Europe. There vehicle manufacturers tended to retain older work practices and eschewed ready-made parts, though these found increasing use during the late nineteenth century. Yet the contrast remained, prompting one visiting foreign coach maker to declare that "an American coach manufactory should properly be called a mill!"[16]

Many of the workmen in such establishments could lay claim to only a fraction of their predecessors' abilities. A woodworker spending most of his days operating a table saw could not acquire or maintain a comprehensive set of woodworking skills. An 1895 correspondent to *The Hub* lamented that in the case of wheel factory workers "each one can soon be proficient in his particular branch, and yet never become a wheelwright." He broadened his scope to include quality when he challenged anyone using the new methods to best him in making sturdy wheels. The editor took up the gauntlet in a respectful but telling manner: "It is possible for a specially skillful workman to make a very superior wheel, 'by hand,' but who does? No man is foolish enough in these days to hew out and round up his spokes, cut his tenons, bore and mortise his hubs, or bore his felloes, nor can he do these things better than they can be done by machinery." The editor took the position that the man, not the tool or machine, determined product quality: a skilled hand tool workman could surely produce a better wheel than a careless machine tender. Any truth to machine-made wheels' alleged inferiority owed less to machinery than to its adjustment, operation, or the kind of raw material being used. These fell under the heading "skill," not "machinery." The journal editor felt the matter came down to this: though it was still possible to make good wheels by hand, doing so was an act of economic suicide.[17]

In a sense both tradesman and editor were correct. The wheel maker stated a fact when he insisted that rationalization failed to produce competent wheelwrights. Yet the goal of a detailed division of labor and special-purpose machinery was to produce wheels, not wheelwrights. The editor rightly identified indifferent workmanship rather than the degree of mechanization as chief product quality determinant. Yet he too missed the point. Machinery, though not intrinsically harmful to the finished product, was capable of great harm to the worker. The wheel maker and the editor shot past one another; the real issue was not machinery's efficiency but what machinery and rationalization together did to those subject to them in the workplace. This was a thornier question, though few shied from attempting an answer. "What skill or knowledge does it require," one contributor asked, "to feed trees into one end of a machine, and watch alleged wagons come out the other end?" Others were more blunt: one scornfully noted that a recent factory-trained applicant for work at his firm "knew no more about the art of carriage-painting than a Fiji Islander." While the culprit may have varied, the crime, as it were, remained the same: the deskilling of the craftsman.[18]

Not only did the skilled craftsman have to face atrophy of his hard-won skills, but he also had to contend with those who had never won them in the first place. As the twin imperatives of mechanization and rationalization permeated the wagon and carriage trade, a variety of consequences both intentional and otherwise manifested themselves. One took the form of the unskilled worker. As routine production and sophisticated machinery eroded the skill quotient necessary for adequate performance, those with no trade pedigree at all began replacing traditional craftsmen on the shop floor. For the employer, this trend flowed from compelling economic logic. For example, a skilled blacksmith could "cut and nut" thirty bolts in ten hours by hand, but a bolt-making machine tended by a boy could complete twice that number in a single hour—and the boy might tend three machines simultaneously. If the equipment featured sufficient mechanical refinements, the same boy might tend up to six of them, doing so for the same low wage. Even J.L.H. Mosier, himself a highly talented blacksmith when he penned this comparison, conceded that raw economics favored the boy. Despite his own long service for Brewster & Company, a firm that definitely valued its many skilled employees, Mosier was a perceptive observer of the trade at large. Rationalization combined with machinery vastly increased output even as it suggested tempting economies through employing unskilled labor.[19]

This had been a bone of contention even before widespread mechanization. In 1868 an incensed New York carriage trimmer published the exposé "Castle Garden in the Trimming Shop," lambasting certain New York carriage firms for "farming out" their trim work to contractors employing unskilled foreigners. "B. Ilinover" decried it as a "disgraceful system of semi-slavery" and a "sweat system," but the practice was formally known as "inside contracting." This happened when a manufacturer placed a discrete portion of his production process in the hands of an individual who used the factory's space, power, and equipment but hired his own workmen and managed them as he saw fit. Originating in antebellum New England factories and a distinguishing feature of many American System manufacturers, inside contracting spread throughout many of the metal-fabricating trades in subsequent decades. Its popularity derived from its mutual benefits: the employer freed himself from a portion of the cares of production, while the contractor stood to make a great deal of money. Indeed, the practice in arms manufacture and other metal trades gave rise to numerous technological innovations as ambitious inside contractors sought to expedite their work. Yet this same motive also lent itself to abuse when contractors opted to save money by employing substandard help. According to the author, these operators applied to the Emigrant Aid Society (Castle Garden served as the immigration entry point in the years before Ellis Island) for harness makers, mattress makers, and cobblers (even a "superannuated upholsterer") to trim carriages for piddling wages. Not only did this result in substandard work, he charged, but it would greatly damage the trade. Should the practice expand, he predicted, skilled New York carriage trimmers would become as rare "as a no-wife-adult mormon is in Salt Lake City."[20]

B. Ilinover perhaps overstated his case for dramatic effect; an upholsterer, however superannuated, did not qualify as unskilled, and even mediocre carriage trimming required at least moderate ability with needle and thread. His complaint showed signs of a turf war, a boundary dispute over who should be permitted to work in American carriage shops. Despite its appearance here and in other places at various times, inside contracting never became the fixture in wagon and carriage manufacture that it did in the making of firearms, sewing machines, and other precision products. Still, the trimmer's complaint had some validity: in a system where profit overrode all other concerns, the temptation to substitute unskilled laborers for skilled craftsmen became too much for some employers to resist. The timing of his complaint suggests that this substitution might take place even in the absence of machinery; the trimming departments of 1860s urban carriage shops were hardly beehives of mechanization. If employers had already started to realize the possibilities of unskilled labor at that time, how much more apparent had it become when machinery formed an intrinsic aspect of production? The answer, of course, was that it became very apparent indeed. The combination of mechanization and rationalization presented employers with the opportunity to realize substantial wage savings even as they harnessed new machinery and organizing principles to push through productive barriers. Increased productivity, however, owed something to yet another factor, one that impacted even the small shop tradesman: the advent of ready-made parts.

In the course of interpreting his data, Wright made special note of the importance of "parts of the article [bought] ready-made." The organization of his survey reflected the fundamental division between the makers of horse-drawn vehicles and the makers of parts and accessories. By the time the report thudded down on library shelves and government office desks, ready-made parts had been a feature of the trade for more than fifty years. Purchase rather than fabrication of a complex component eliminated potential production bottlenecks and pushed throughput to new levels. Ready-made parts eliminated the need for costly special-purpose machinery, and their often impressive range of types and sizes allowed easy response to market trends and customer desires. Such advantages ensured their use in vehicle firms everywhere. Ohio's Walborn & Riker epitomized the practice in its enthusiastic use of Sheldon axles, Timken springs, Bookwalter wheels, Sherwin-Williams paint, Valentine's varnish, Goodyear rubber tires, and McKinnon dashes. Many smaller firms adopted ready-made parts for the same reasons. The most unexpected effect of "store bought" parts, however, was their contribution to the survival of the small shop.[21]

As industrialization reached full stride, it appeared that the small wagon and carriage shop would be forced out of business. During the 1870s and 1880s the wholesale vehicle factories of Amesbury, Columbus, and Cincinnati seemed poised to sweep away all challengers; most injured by this development were the small builders who attempted to match the big factories' low prices. The wholesaler's price advantage stemmed from the economies of scale: prodigious and efficient output permitted a reduced profit per vehicle. While a small builder might require an average gain of \$40–50 per vehicle in order to prosper, a large factory with a daily output exceeding the shop's annual production often required as little as \$5 per unit to turn a handsome profit.

This seemingly impregnable advantage gave rise to widespread predictions of the small shop's demise, a prophecy seemingly validated by industrialization's impact on many other trades. Not everyone agreed, however. While *The Hub's* editor George Houghton perceived that "the two classes conflict seriously," he saw a number of possibilities for small builders. Besides counseling them to avoid price wars and to uphold product quality, he advocated concentrating on specialized markets and repairing. Grant H. Burrows, president of Cincinnati's Standard Wagon Company, charitably suggested another strategy: small shops should be willing to sell the work of the wholesale factories alongside their own. Citing the experience of a small Ohio carriage builder whose business eroded through fruitless price competition, Burrows urged small firms not to dismiss wholesale as a temporary phenomenon. Doing so courted disaster; if the small shop refused to sell these goods, the itinerant trader, the general storekeeper, and the agricultural implement dealer would fill the gap, as many had already done. Selling wholesale vehicles would not obliterate shop-made products, for a market still existed for custom work. While shop-built versions of the standard patterns could in no way compete with those from large factories, carefully conceived custom vehicles could meet some needs better than any others.[22]

The small vehicle shop should seek a special design of proven appeal to a known body of potential customers. It no longer sufficed for a carriage builder to be merely a competent craftsman, for "mechanical ingenuity and skill, refined taste, durability of product, and business integrity does not necessarily command prosperity in these days," a writer warned in 1888. The successful carriage builder also had to be a business man capable of projecting his personalized product into a unique marketplace void. Longtime Cleveland carriage maker and small shop proprietor Jacob Hoffman chose a mixed coping strategy, as revealed by a short news item in the January 1887 *Hub:* "Jacob Hoffman, Cleveland, O., is making a specialty of three-spring business wagons, and is building up quite a trade. For carriages, he handles the Columbus Buggy Co's work, as he finds it more profitable than building." In other words, Hoffman's response to cutthroat price competition was to become a wholesale agent. He did not have to forsake manufacturing altogether, for the extensive local demand for commercial vehicles, still largely untapped by wholesalers and large wagonmakers like Studebaker, meant that he could still manufacture light delivery wagons for a reasonable profit. His own three-spring business wagons ideally suited the local urban market for light- and medium-duty vehicles to deliver groceries, milk, meat, building supplies, and other goods. Some shops specialized in heavier wagons to haul milk, ice, beer, and coal, as well as moving vans, bakery wagons, and more. Small city shops able to land fleet supply contracts stood a good chance of profiting handsomely, and in any given city one could find such specialists through the end of the horse-drawn vehicle era.[23]

Specialization admitted numerous variations. Some small firms built elegant heavy carriages for clients unwilling or unable to purchase them from a prestigious New York or European maker. Others concentrated on painting new and used vehicles or took in substantial amounts of general blacksmithing and horse shoeing. Repair work, sometimes called "jobbing," continued to offer opportunities. Most small shops undertook varying amounts of this already, and many larger firms were not above such mundane chores: "Tires set while you wait," announced the huge banner on the Selle Gear Company's factory building in Akron, Ohio. An increasing number of trade journal editorials encouraged financially strapped small shops to cultivate repairs as a specialty: "Look to your repairing department as your mainsail in fair weather, and your sheet anchor in emergencies." Repairing also attracted customers interested in new work. Ever-growing numbers of vehicles on the streets ensured steady business for those actively courting repair work; though some dreaded the thought of becoming "mere repair shops," others exploited the opportunity with gusto. This option had its own perils, including the reduced number of wheel repairs due to rubber tires and widespread paving. These were late developments, however, and, even then, horse-drawn vehicles continued to require other repairs, ones that many small shops continued to perform (fig. 7.3).[24]

Yet for all the admitted efficacy of any of these approaches, small wagon and carriage builders benefited even more from widely available ready-made parts. Products of Ohio rolling mills, Connecticut drop forges, and New Jersey wheel factories complemented all the above-mentioned coping strategies. Ready-made parts expedited production and contributed to the final product's durability and customer appeal. Shop owners could purchase goods from any of the more reputable specialty manufacturers, and prominent mention of Sarven wheels, Dalzell axles, or McKinnon dashes assured customers who trusted name brands as their fathers had once esteemed seasoned oak and ash. Repair shops found ready-made parts indispensable: even when purchased only half-finished and hand-fitted into place, ready-made spokes, hubs, gear wood, and the rest facilitated swift installation. Shop owners could even assemble, trim, and paint a vehicle composed primarily of such goods.

Table 7.1. Wagon and Carriage Manufacturers by Number Employed, 1900

No employees	1,129
Under 5	3,431
5–20	2,426
21–50	370
51–100	142
101–250	99
251–500	27
501–1,000	5
Over 1,000	3
Total	7,632

Source: Adapted from James K. Dawes, "Carriages and Wagons," in U.S. Bureau of the Census, *Census Reports: Twelfth Census of the United States, 1900* (Washington, D.C., 1902), 10:312.

During the 1880s the Cleveland Hardware Company, which produced a complete line of sleigh hardware, spelled it out explicitly: carriage manufacturers need only "buy the wood from the wood manufacturers, and there is then nothing to do but bolt our irons on to the wood, and the sleigh is ready for the painter."[25]

What had happened appears logical in retrospect, but it surprised many at the time: large mass producers of vehicle parts enabled small shops to turn out batches of wagons and carriages whose specifications could be easily altered to suit market conditions. In this sense, mass and bulk production spawned batch and custom work as ready-made parts permeated the small shops. When *The Carriage Monthly's* editors surveyed the industry for their 1904 fortieth anniversary issue, they characterized these firms as "custom shops." It made for a curiously composed industry: for every Studebaker Brothers Manufacturing Company or Columbus Buggy Company, there were dozens of little partnerships, family firms, and even a few one-man enterprises. In 1901 Cleveland boasted no less than eighty-one wagon and carriage manufacturers, only one of any considerable size; the overwhelming majority remained what they had always been—small shops. Like similar situations in other cities, the majority of these small concerns made strategic use of ready-

TABLE 7.2. Wagon and Carriage Manufacturers by Business Type, 1900

Individual	5,361
Firm	1,829
Corporation	442
Total	7,632

Source: Adapted from James K. Dawes, "Carriages and Wagons," in U.S. Bureau of the Census, *Census Reports: Twelfth Census of the United States, 1900* (Washington, D.C., 1902), 10:312.

made parts. Like stones in David's hand, the parts equipped small shops to do effective battle with much larger foes. They proved a tremendous leveling influence, a counterweight to the increasing size and productive disparity between the industry's large and small manufacturers (Tables 7.1 and 7.2).[26]

Did survival of the small shop mean survival of the craftsman? On the surface, such small businesses might seem to have provided ideal employment for craftsmen eager to avoid factory work's tedium, and there certainly was an element of truth to this perception. More than one skilled carriage maker found full play for his talents in small establishments, but most shops were not throwbacks to the past. While the majority rightly touted the attention to detail, finish, and customer preference qualifying them as custom carriage builders, they also projected an image as erroneous as it was commercially useful. In an effort to distance themselves from mass producers of cheap vehicles, small shops preferred the public to think of them as building wagons and carriages the old fashioned way—by hand. While hand work, especially in the finishing operations of painting and trimming, featured prominently, it hardly predominated. "Custom" and "hand made" signified the degree of product uniqueness and attention to detail more than they did a particular work method. An early-twentieth-century commentator pointed out that while he admired custom shop work, the "hand-made" appeal was chimerical: "hand work, however, is illusory; there are no hand-made carriages, and could not be in the sense in which the term hand work is ordinarily used." Those wagon or carriage makers not using machinery to fabricate components were instead purchasing them from someone who had. In this sense, mechanization had become ubiquitous.[27]

Adoption of ready-made parts shifted shop routines from stock preparation (planing and sawing) and parts fabrication (shaping spokes or sewing seat cushions) to assembly. Assembly required only basic hand tools (hammer, screwdriver, wrench) and perhaps a simple machine or two (like a flexible-shaft screwdriver), dramatically reducing a small shop's equipment requirements. As early as the 1870s, industry observers wondered if low equipment overhead might ease entry into the trade, but then they had forgotten this had actually been the case for years. In the 1850s this principle helped Jacob Huntington make his independent debut; the surprising success of his early back-room, one-at-a-time operation owed much to ready-made parts and their equipment-obviating advantages. As with mechanization, however, ready-made parts had mixed effects upon the workman. While assembling a vehicle from parts required much less effort than laboriously building one "from scratch," it also required much less skill. If the employee was a traditionally trained vehicle craftsman, the exchange of jack plane for power planer, chisels for glue and screws, necessarily wilted his abilities. If machinery and task specialization turned the specialty factory worker into a machine tender, ready-made parts turned the vehicle builder into an assembler.[28]

In reality, however, the situation defies such neat packaging. For all the dramatic change in the trade, small shops still provided many opportunities for the traditional craftsman to exercise real skill and ingenuity. While it was certainly possible to construct a vehicle entirely from purchased parts, many small shops still produced significant portions of each vehicle from basic raw materials. In building his delivery wagons, Jacob Hoffman used "the best make of wheel in the country," as well as ready-made axles, springs, paint, and varnish—yet most of the wooden structure remained the product of his own men. Widespread availability of a part did not necessarily make it the most economical choice; everything from the continued presence of skilled workmen, plentiful and affordable local timber, and the expense of some of the larger vehicle components, entered into the economic calculation that often favored in-shop production. Even when limited to wagon boxes or buggy bodies, parts fabrication mingled with factory products in the majority of small vehicle shops. Likewise with metalwork, where tiring and ironing still occupied the shop's blacksmith even as factory-produced hardware simplified his other tasks. The finishing procedures especially remained small shop specialties. Fine striping, murals, lettering, and glossy finishes required skills that no machine could replace. Though published patterns and imported transfer ornaments found widespread use, they could not entirely displace the skilled painter's steady hand and practiced eye. Similarly, trimming required craft skill even when the shop purchased ready-made seat cushions or folding tops. Much trimming took place on the vehicle itself, so skilled trimmers remained in small shops.[29]

Even mere assembly often required substantial talent. Wagon and carriage parts did not generally come with installation instructions, and the way in which one attached a fifth wheel to a platform-spring gear or hung coach doors with a hardware kit was by no means self-evident. Wheel shop men working from stocks of prepared hubs, spokes, and rims found that incorporating each into a finished wheel required more than merely fitting "tenon A" into "mortise B." Each spoke had to be driven true into the hub, rims had to be accurately drilled to accept the spokes, and felloe plates, tire bolts, hub bands, and tires all required similar care. The difference between an experienced wheel man and a common laborer was the difference between a smooth-running wheel and a crooked assemblage of components. In other words, a wheel was more than the sum of its parts, and its assembly still mandated sufficient manual talent to disqualify it as unskilled work. Assemblers remained builders of a kind, for building still required skills beyond those of machine tenders or common laborers. The same was true for repair work, which had much to recommend it to the skilled workman, as attested by the number of older craftsmen seeking outlet for their talents in this way. "Wheel makers, the good old-time wheel makers, are getting scarce," one commentator wrote in 1907, "and what is left of them have taken to repairing." Ready-made parts did somewhat rationalize the process inasmuch as it facilitated replacement of a broken part with one purchased from a vendor. Yet contemporary automobile mechanics do nothing but replace one part with another, and we would hardly consider them unskilled. The same held true for wagon and carriage mechanics.[30]

Skilled craftsmen had still more options, some of them supervisory. The man watching over the work of others did so best when intimately familiar with all of the steps taking place on his shop floor. Experienced craftsmen were sought out as department foremen in large factories and as supervisors or foremen of sorts in small firms. Small shop foremen often worked alongside the men, a practice precluded at the factory level by supervision of many more workers. Whether assisting in planning the manufacture of some new part or vehicle, smoothing out production bottlenecks, training new workers, or inspecting completed work, the foreman drew continually from his reservoir of talent and experience. His position demanded a great deal of skill, bequeathed higher pay and prestige, and so remained eagerly sought by many of the upper-tier wagon and carriage tradesmen. While this role may have superficially resembled the hallowed tradition of the master craftsman teaching the apprentice or guiding the jour, the reality was that the supervisor instructed workers who might never rise above their station. Charles Eckhart echoed a common concern when he soliloquized at the 1910 CBNA convention: "I don't know what we are going to do in ten or twenty years, when the foremen and superintendents die off, and we have no one to take their places." He saw, as did so many others, that collapse of the trade apprenticeship decades before had eliminated the source of the most skilled vehicle craftsmen, and he worried about its implications for the future.[31]

Union Brothers

Though himself a product of a trade apprenticeship and years of practical experience, Eckhart posed his question decades after becoming head of his own large factory. What of the opinions of working, "practical" carriage makers? Like so many workers everywhere, those in the wagon and carriage industry left scant record of their thoughts about their jobs. Some voiced concerns and opinions through the trade press, commentary that informs much of this chapter. Reactions also came through benevolent or mutual aid associations and trade unions, which played a part in the lives of many nineteenth-century vehicle workmen. These provide a collective dimension to any understanding of industrialized carriage production, and the many debates over organized labor, working hours, and more tell us about the changing world of the wagon and carriage maker.

The single most influential trade organization was the Carriage Builders National Association, a body ostensibly open anyone building horse-drawn vehicles in his own firm. Most charter members had been trained as traditional craftsmen, but when they gathered in 1872 to form the association, they had long since become proprietors of firms whose employees might number in the hundreds. While rank-and-file membership included many smaller owner-operators, the organization leaned toward the interests of its larger members. Its ban on discussing wages and labor matters reduced its appeal to some employees, but by this time the more organization-minded carriage mechanics—a term which came to describe active wagon and carriage builders in another's employ—

had already turned to their own forums. When the first carriage industry trade union arose in the 1860s, it had as examples numerous unions in other trades. Perhaps the earliest and most powerful was the International Typographical Union founded in 1850, joined four years later by the National Trade Association of Hat Finishers, then the Iron Molder's Union of North America (1859), the Brotherhood of Locomotive Engineers (1863), the Cigarmakers' National Union (1864), and the Bricklayers and Masons' International Union (1865). Founded by and for skilled craftsmen, trade unions helped alleviate the effects of alarmingly altered working conditions. Although they never attracted a majority of working people, trade unions by midcentury were already profoundly shaping the debate over industrialization.[32]

The first significant wave of carriage-trade unionization came with the economic backwash of the Civil War. Some vehicle makers profited during the conflict, particularly wagonmakers like Studebaker, Milburn, Espenchied, and the others securing federal ambulance and supply wagon contracts. However, war also depressed demand for horse-drawn pleasure vehicles. As far back as 1812 carriage builders like James Brewster had commented on the public's willingness to defer purchase of luxury items such as carriages during times of trouble, but the troubles of 1861–65 outweighed those of 1812–15, just as the trade had become much larger and more complex. War's tumult and the devastated southern economy had especially dire consequences for those who had come to rely primarily on sales from this quarter.

Whatever their accustomed markets, many 1860s carriage makers found themselves with shops and factories whose production capacities now dangerously exceeded a diminished demand. One obvious remedy lay in scaling back production, but for marginal firms this meant no production at all. Some shops simply paid off their workers and closed their doors, while those with overextended credit succumbed to buyouts or liquidation. The ones managing to claw their way out did so by ruthless downsizing; the resulting wage cuts and layoffs were at least familiar if unwelcome features of a trade which continued to experience a seasonal business cycle. Most firms made every attempt to retain valuable skilled hands during slow periods, whether by shifting them to different tasks, placing them on reduced hours, or temporarily reducing their wages. The Civil War stretched these measures to the breaking point, causing unprecedented hardships for sufficient numbers of tradesmen to spark the first significant wave of carriage-trade unionization.

Despite our understanding of the general economic context, the details of the establishment of the first American carriage trade union remain obscure. This murkiness owes something to the organization's brief existence but also to later reticence of key participants, many of whom went on to become prominent and well-respected trade figures. The essential facts appear to have been these: prompted by low wartime wages and unemployment, a group of eight or ten East Coast carriage workmen met in New York in 1864 or early 1865 to consider forming a carriage mechanics' trade union. The group included William Harding, Edward Brown, James Conway, Wilmer Richardson, Mark Reeves, and several others, nearly all employed (or previously employed) in urban shops. Brown and Reeves plied their trade in Philadelphia, Richardson in Baltimore, and Conway in Albany. Though traditional carriage craftsmen, few owned their own shops, and most appear to have become resigned to the permanent wage-earner status increasingly occupied by latter-day jours unable or unwilling to start their own businesses. English-born William Harding had more tenuous trade ties, but his role as organizer and union president reveals his familiarity with labor organization. These men, along with a few others whose names have not survived, formed the Coach-Makers' International Union (CMIU) in hopes that collective action would ameliorate their distressing circumstances.[33]

They followed the well-established trade union format: a voluntary, peer-approved membership of skilled craftsmen, each active through a local chapter, and all under the aegis of the umbrella organization. We lack even a complete list of the central union's offices and officers, but William Harding became its longest-running president and Isaac D. Ware its best-known secretary. Recruitment was the highest priority from the beginning, and assisted by the efforts of the more active members, the union president spent most of his time organizing new locals. Establishment of a local required a minimum number of members merely to fill the offices, for each had its own president, vice president, secretaries (recording, corresponding, and financial), and treasurer, as well as a deputy president to the central union. A prospective inductee had first to be recommended by a member in good standing, followed by an interview with an investigating committee to confirm his status as a skilled member of the carriage-making trade. The local's vote clinched the matter and made the candidate a card-carrying, dues-paying member. New York and Philadelphia were the first to establish locals, followed by Baltimore, Albany, Washington, and Wilmington, which formed the half dozen in existence at the time of the union's second convention in 1865. Collectively they boasted some three hundred members, a figure soon to rise dramatically. Besides holding regular meetings twice monthly, the locals were to monitor area trade conditions, welcome traveling members, recruit, elect representatives to attend the annual conventions held in the late summer or early fall, and report to union headquarters via the official union journal.[34]

Much of what we know about the CMIU's brief life comes from *The Coach-Makers' International Journal.* Besides regular columns by editor Isaac Ware and union president Harding, a locals directory, and secretarial correspondence, the content also included extended editorial commentary on general labor reform and episodic reports of the activities of other American trade unions. Conventional touches such as the "International" in the title, address of fellow members as "brother," and sometimes heavy-handed reference to collective endeavor outside the workplace testified to its unabashed trade union advocacy. This soon raised the ire of the new journal's only real rival, *The New York Coach-Maker's Magazine.* Editor Ezra Stratton harbored an implacable distrust of unions, and the columns of both publications soon crackled with the vituperative language of thoroughly committed nineteenth-century journalism.[35]

The CMIU's primary income consisted of dues; at sixty cents per quarter, most found membership affordable. What did they receive in return? G. J. Carpenter, a member of the Bridgeport local, felt the benefits were numerous. Besides the union's advocacy of the eight-hour law and cooperative ventures outside work, membership provided opportunity to meet with like-minded individuals. Carpenter thought the union's push for a renewed apprenticeship system promised to protect skilled labor and uphold prices, but he concluded that the most important benefit was the collective ability to demand a living wage. J. A. Macreading, a Rhode Island member, agreed: "Coach-makers are men that are looked upon by *some* as inferior to other beings, and are to be bossed and kicked around like dogs." The reason? "Because the trade never knew the good of a Union; and now that we are united, we will show such men that we have rights, and mean to stand up for them." Members like Macreading called not for capitalism's overthrow but vigorous defense of those subject to its harsher side. Working hours and wages were staple concerns; of the hundreds flocking to the CMIU during 1865–66, many had experienced drastic wage cuts and layoffs. Yet even prior to this period, carriage tradesmen had been rocked by forces seemingly poised to destroy their way of life. However optimistic their hopes may have been, CMIU members looked to collective action as a means not only of weathering present hard times but also of coping with the long-term forces that promised a difficult future.[36]

No single categorization adequately encompasses every member's motivation. Some likely joined for social reasons or for the excitement of local and national meetings, while others may simply have signed on because they thought they should or because their friends already had. The reverse was certainly true: J. B. Hubbard, corresponding secretary of the Worcester, Massachusetts, local, scornfully noted that some potential members held back simply because an old enemy had already joined. Carpenter reported difficulty in generating much enthusiasm among central Ohio jours, and union president Harding faced similar recruitment difficulties all over the country. Walking into a Cleveland shop, he encountered a carriage body maker whose dismissive attitude shone through as clearly as his ethnicity: "you mush see *de boss; dats no use to me*," Harding recalled hearing. "Such are two-thirds of the replies you get," he added in disgust. Despite the lukewarm reception and foul weather, however, Harding managed to organize a Cleveland local. Indeed, the CMIU grew at such an impressive rate that some feared careless recruitment: "Unworthy members are like rotten timbers in a coach body, they weaken the whole structure" warned one local secretary. While the CMIU remained predictably cagey with its figures, the best estimates reveal that membership had increased from around four hundred by the end of 1865 to just under two thousand (in about forty-five locals) by the spring of 1867.[37]

Much of this increase stemmed from William Harding's assiduous recruitment; by the early fall of 1866 he had already traveled through New England and the Midwest, and he now began a second tour of both regions and beyond. Along the way he added to the membership and number of locals and hawked *Journal* subscriptions and advertising space. Traveling primarily by rail, supplemented by stage coaches and union members' buggies, he canvassed the country for weeks at a time. Harding keenly felt his responsibility in taking on a position "with about one thousand seven hundred bosses to look after me." He wrote this in an October journal column, where he promised to give a regular account of his travels and travails. Early in his journey he called on Ezra Stratton, whom he had previously met. The visit proved cordial enough, taking place before the New Yorker had commenced venting his wrath in print. Harding felt gratified in his subsequent return to New England when greeted by many familiar faces from his earlier tour; he managed to establish a new local (no. 25) in Belchertown, Massachusetts, but could not do so in Westboro, where the local tradesmen "had not yet learned the great truths of Union." He pushed on, stopping long enough in upstate New York to found local no. 26 in Schenectady before moving on though Ohio (with a side trip up to Detroit) where he proceeded south. He stopped in Columbus to visit with John B. Peek, local no. 8's loquacious corresponding secretary, wryly noting the exceptionally muddy avenues: "Most other cities have some kind of steam machinery for street cleaning, but here they have got beyond all that, and employ oxen instead."[38]

Late fall found Harding in Cincinnati experiencing stiff resistance from factory owners. He spent nearly two hours debating John W. Gosling, proprietor of one of the city's oldest carriage firms. When Gosling denounced the CMIU as a consortium of demagogues, Harding shot back that Gosling was the one defaming the union via postal circulars. In spite of the heated exchange, Harding nevertheless felt he made more progress here than with another proprietor, Mr. Miller—possibly George C. Miller, another Cincinnati old-timer. Miller demanded to know the union's real purpose, but when Harding indicated mutual protection and aid to the sick and needy, Miller pointed to the existence of hospitals and institutions. Harding soon gave up and continued across the Ohio River. Through the generosity of several successive locals he made it as far south as Nashville and Memphis and west to St. Louis before proceeding back to New England by the beginning of 1867. He proudly informed *Journal* readers that the founding of no. 43 in Lowell now meant that Massachusetts, New York, and Ohio each had seven locals, reflective of the role all three played in the 1860s trade. By now Harding and other missionary-minded union members had formed locals in the majority of the large carriage-making cities. With the exception of no. 56 (Chicago), all those organized after no. 39 (Pittsburgh) were in small towns and cities.[39]

Meanwhile, the union had already engaged in its first official arbitration, involving the Wilmington carriage firm McLear and Kendall. Henry C. McLear would eventually become president of the CBNA and its longest-term secretary. Sometime during the summer of 1866 the partners sublet their painting to a nonunion man who promptly fired three union painters. Their replacements were not only nonunion but unskilled, some merely boys. Ware pointed out the dangers of subcontracting to the *Journal*'s readers, while Harding organized a committee to meet with McLear. To their surprise, the carriage maker received them cordially. He insisted on his right to run his business as he saw fit, and the committee agreed—except in cases where it injured the jours' interests. McLear explained that his recent buyout of a competitor necessitated taking on many of the other firm's apprentices, a situation he felt contributed to the problem. Yet he confidently predicted this would subside once the boys completed their terms and moved on. In the end, McLear agreed to remove the subcontractor and reinstate the men, adding that in the event of future trouble he hoped they would "come to me like men and talk the matter over." Ware drew two conclusions: unions had sufficient power to correct wrongs, and not all employers opposed unionization. Harding shared Ware's opinion and felt that the episode also refuted the logic that unionization meant strikes. Other incidents failed to respond to diplomacy, however. That same year, O. J. Edwards, a member of the Springfield, Ohio, local, returned from the CMIU convention to find that his boss had discovered his union activities and replaced him. Edwards now commuted nineteen miles to a job in Xenia, though he insisted he willingly did so for the union's sake. Over in Warren County, New Jersey, several former factory employees left in protest over illegally long hours; their employer pointedly ignored a local ordinance setting a ten-hour limit.[40]

By the winter of 1866–67 wage and work-hour disputes precipitated a serious crisis in New York. The year 1865 had evidently been relatively prosperous for the city's carriage trade, but the following spring saw its workmen refraining from applying for a raise in fear of a depressed market. When some of the large firms put their employees on reduced hours in November 1866, some workmen felt their springtime sacrifice had been ignored. A January 1867 wage reduction sharpened the complaints. Neither the cold weather cessation of Saturday work nor the wage decrease was unusual, but coming as they did in the wake of a relatively prosperous year and the workers' subsequent abstention from applying for a raise, 1867 began looking stormy. Several additional factors added to the tension: slow sales during the previous fall left many repositories full of old stock, and a cold, wet spring delayed the normal influx of spring orders. By the middle of March, many workmen were wondering if their employers would ever resume the old wage scale, and Harding helped preside over what the *New York Herald* referred to as a "mass meeting of coach-makers" airing their grievances. Besides Harding's reminder of the CMIU's health, the meeting evidently produced the decision to request that wages be restored to pre-January levels, with an April 1 deadline. It ended up precipitating a strike, though in comparison to those in other trades, it proved a rather mild affair.[41]

According to a disgusted Ezra Stratton, few firms put up much resistance, and most simply granted the same increase they would have given later anyway. William Evans, corresponding secretary of local no. 1, said later that only one shop actually experienced a walkout. J. C. Parker & Company belonged to New Yorker John C. Parker, who had become Benjamin Brewster's partner in 1860. Parker, likely with Brewster's approval, refused to reinstate the old wages, and his men walked out of the Twenty-Fifth Street shop. By the second week of the standoff, Evans and the other members were becoming edgy as Parker started hiring nonunion men, and the protest might have collapsed had Ware and Harding not arrived on the scene with emergency funds. Meanwhile, Parker, who suffered from Bright's disease, found himself in seriously declining health. Stratton had known Parker for years, and he excoriated the CMIU for sending a delegation to "persecute and annoy" the old man on what turned out to be his death bed.[42]

Ware countered that the men merely wanted a reasonable settlement, one that Stratton's "inflammatory accounts" had only made more difficult to negotiate. Parker granted his workers their former wages during the third week of the strike, just before his death on April 26. This marked the end of the affair for the city's carriage trade, and by May any hard feelings were being forgotten under the press of springtime trade. Evans praised Parker's union hands for their "manful" resolve and discipline and felt this first test of the union found it "established on a firm basis, which it would be no light undertaking to overthrow." Harding agreed, though reiterating his conviction about the "barbarous" nature of strikes and hoping they would not again become necessary. While noting the high feeling on both sides, the union president remarked on the lack of bitterness, singling out Benjamin Brewster as particularly magnanimous. Now sole owner of Parker's business, Brewster "proscribed no one . . . but at once threw the shop open to all," telling the men that the past is forgotten and that they all should work harmoniously for the future. "These are noble sentiments," Harding concluded, and he hoped that Brewster, the other proprietors, and New York City's carriage mechanics would all prosper thereby.[43]

The Coach-Makers' International Union certainly seemed prosperous. "A few sharp conflicts between labor and capital, which ended disastrously to the latter," Ware informed his readers, "opened the eyes of many, who were previously blind to the interests of the former." Not only had the New York local won an important contest, but Harding's canvassing had also raised the total number of locals to nearly fifty. For all the optimism floating through the springtime air, however, significant problems remained. Some had to do with the locals themselves; ranging from listless or nonexistent correspondence and indecipherable handwriting ("it is not agreeable to any person to have his name *haggled* up," an exasperated Ware wrote) to late dues and missing paperwork, such disputes sapped the organization's energy and resources. There was also the matter of membership, whether resistance among non-members or apathy among those with union cards. Just prior to the 1867 strike, New York's corresponding secretary admitted that he was embarrassed by his local's nonpayment of dues, lack of enthusiasm, inability to recruit, and poorly attended meetings. He attributed its very survival to a dozen or so stalwarts who refused to give up.[44]

Others, however, did just that. One disenchanted Rochester member resigned during the winter of 1866–67 for a variety of reasons, among them what he felt to be lack of common interest among the trade's four branches. As he saw it, how could a carriage blacksmith really understand the trials of a carriage woodworker? His former union brothers dismissed this as nitpicking but admitted his other concerns carried far more weight—like the union's inability to support strikers and their families for any substantial length of time. Indeed, the CMIU's most abiding problems remained financial. Late and unpaid dues, compounded by slowing recruitment, kept the union in the red, and the pages of the *Journal* buzzed with complaints and suggestions. Economics, plus the many other problems common to unions, eventually proved the CMIU's undoing. Gradual recovery of the American carriage trade by the late 1860s further eroded its support, as new work continued to pour into the shops and members had less time to devote to collective endeavor.[45]

By the time the delegates convened in Cincinnati in August 1867 for their fourth annual convention, optimism had become hard to sustain. Though Harding was reelected and the usual rash of committees made resolutions, so little of substance transpired that John Peek's *Journal* report devoted most of its space to scenic descriptions. As winter came on, Bridgeport and New York wage disputes ended in mutually agreeable settlements made possible in part by a recovering market, and Harding found himself unable to push the number of locals any higher than sixty. Many of these had already started drying up at an alarming rate, their correspondence trickling off until the last local directory appeared in the *Journal*'s February 1868 issue. By then, William Harding had accepted a position as a traveling varnish salesmen with a New York firm, though he vowed to continue working on the union's behalf as time permitted. The following month Isaac Ware announced his purchase of the *Coach-Makers' International Journal* and his resignation as union secretary. The publication had suffered from dwindling union funding, and both parties seemed pleased by the transaction. The ex-secretary assured his readers that his resignation reflected only the time constraints of his full-time role as editor and publisher, and like Harding he promised to further the union cause. By the time the CMIU held its fifth and evidently final convention in Troy later that year, Ware declined to attend in view of his editorial workload. Much to Stratton's dismay (Ware called him "the Wise Man of Gotham" and editor of the "Monopoly Magazine"), the former union organ had become a general-interest trade journal. Ware severed the last symbolic tie in 1873, when he changed its name to *The Carriage Monthly*.[46]

The Bonds of Benevolence

The Coach Maker's International Union died with the late 1860s economic upswing, but later hard times and continuing industrialization brought about a resurgence of union activity in the 1880s and 1890s. Meanwhile, wagon and carriage workers had recourse to another important organizational venue, which in some ways proved more enduring. Though the specific details varied from group to group, mutual aid or mutual benefit societies combined the lodge's social and fraternal activities with insurance paid out to members and their dependents in the event of illness, accident, or death. Early examples such as the International Order of Odd Fellows in the 1810s and the Ancient Order of Foresters and the Ancient Order of Hibernians in the 1830s were followed by the Knights of Pythias, the Ancient Order of United Workmen, and the Irish Catholic Benevolent Union in the 1860s, then by an explosion of imitators during the 1870s and 1880s. Some had a primarily ethnic composition (like the Hibernians), others had religious requirements (Knights of Pythias), and many had confusingly similar names (depending on how one counts, there were three different Orders of Foresters and two each of the Odd Fellows, Woodmen, and United American Mechanics). While some had short life spans, many more featured robust membership rolls and inspired smaller imitations in many trades and industries, wagon and carriage manufacture included. Whether established by workmen, their employers, or both, these organizations provided real benefits to workers, and their popularity derived partially from their meeting a distinct need. Vehicle workers in eastern cities led the movement; the German Coach-Makers' Relief Association, organized in New York City in 1862, was one of the earliest. Another New York group had even more ambitious plans to assist a specific group of wage earners.[47]

James Brewster had devoted a great deal of thought to these matters during the 1820s and 1830s, and he continued to ponder the working man's condition even in his retirement. In an 1861 letter the carriage maker explained his conviction of the primacy of a "community of interest" among employees, one created and sustained by granting them a financial stake in the company. His ideas resonated particularly in son Henry's business partner John Britton. Britton shared Brewster's reform tendencies, and he quickly became Brewster & Company's point man on labor issues of the day. As the two men discussed the matter, Britton began thinking this concept might moderate labor-capital tensions. The idea of a stake in the business made sense, especially when more and more of the men who worked at the bench, anvil, and body trestle could expect to remain wage earners their entire lives.

Both Brewster and Britton were familiar with previous attempts by carriage workers to solve the problem cooperatively. The Union Carriage Company began in the late 1840s through the efforts of a group of striking Bridgeport carriage workmen, an effort that likely would not have left the planning stage were it not for Phineas T. Barnum. Barnum, like Brewster, had branched off into real estate speculation and in 1851 began developing a large tract of land across the river from Bridgeport. After platting streets and marking off lots, he began selling them at cost, with the stipulation that the purchaser build a home, store, or factory within a year's time. Meanwhile, an 1849 dispute in Bridgeport's carriage shops had prompted a group of striking carriage craftsmen to form a joint stock venture. Happily for them, their efforts coincided with Barnum's East Bridgeport project; soon they rented a building on the showman's easy terms and were in business by 1850.[48]

They struggled for survival from the very start. Operated as it was by a group of experienced workmen, the difficulty likely lay outside the purely technical. Perhaps the company lacked suitable markets, which may account for its proposed Broadway repository. Though the craftsmen went as far as offering to hire Britton away from Brewster (he turned them down), the plans remained unfulfilled. Whether lack of business acumen, internal squabbling, inadequate capitalization, or any of 101 other perils, the Union Company never became more than a shaky experiment. In 1852 James Brewster bought out the ailing concern and the remaining cooperative carriage makers became Brewster employees. The concept refused to die, however. During the 1860s the CMIU touted cooperatives as a general tonic, with Ware and Stratton trading editorial blows on the subject. Ever the astute trade historian, Stratton recalled the Bridgeport cooperative's failure, as well as that of a similar Rahway, New Jersey, contemporary as evidence of the scheme's folly, adding that human nature itself ran counter to the idea. Besides insisting that these examples were not true cooperatives but joint stock companies, Ware argued that changing conditions warranted a second try. The debate remained theoretical, and the CMIU collapsed before any significant "second generation" carriage cooperatives appeared.[49]

The cooperative idea failed to take root elsewhere in American manufacturing, and whatever other flaws it contained, its biggest problem remained distressingly human. Cooperatives required cooperation, precisely the virtue in shortest supply among groups of workmen with strongly antiauthoritarian leanings. Stratton, for all his grating condescension, correctly identified this in his sparring with Ware. The real merit of the larger ideal—that laborers did best when having a tangible stake in the enterprise—ended up informing a scheme still part of many industries today. We know it as profit sharing, and it lay at the heart of the plan that took shape in John W. Britton's mind during the 1860s. About the time the Coach-Makers' International Union took its fatal downward plunge, Britton began arranging the details of a profit-sharing plan he hoped would avert serious labor trouble and benefit Brewster & Company's several hundred workmen. Meeting initially with a small group of senior tradesmen in 1868, Britton chaired a formal committee that hammered out the details during the winter of 1868–69. Deliberations even included correspondence with leading English labor reformers John Bright, John Ludlow, Thomas Hughes, and John Stuart Mill. All gave helpful advice, but from them Britton divined what he regarded as a fatally English flaw: an excessive sense of privilege. Bright felt strongly that employees' shares should only come out after the employers had taken theirs; Britton criticized this peculiarly English notion that capital had "some special right to be paid before anybody else." His plan would operate differently.[50]

In 1869, Britton and the committee issued their "Proposal for an Industrial Partnership" to guide the drafting of a formal constitution later that year. That document determined the structure of the Brewster & Company Industrial Association, a combination mutual aid society and profit-sharing plan. The association recognized "the importance of harmonious action in business operations," averring that "Labor and Capital are not necessarily antagonistic" but "can be made identical in many respects." How? By establishing a mechanism whereby "Labor shall receive, in addition to ordinary wages, a portion of the profits accruing to Capital." It worked like this: at the end of each fiscal year (July), the company's employees would receive a percentage of the firm's profits, an amount above and beyond regular wages—which Britton proposed would equal those paid in similar firms. This payment would come from the net profits of the factory and repository combined, each man's share proportional to his wages. The profit itself depended on sales, but "having a business constantly increasing in volume and quality, we may reasonably expect that the share will be much larger," wrote Britton, "especially as we shall then have . . . the hearty co-operation of every man in the shop." Clearly, every man in the shop should view himself as a stakeholder.[51]

The association existed for Brewster & Company's workers "directly employed in mechanical labor" and excluded office and sales staff (who received salaries and commissions instead), as well as "boys, porters, apprentices, cartmen, and persons under instruction in any of the Departments." Officers had to be at least twenty-one years old, skilled workmen, and employed at Brewster & Company for a minimum of six months. The proposal certainly presumed a great deal of "hearty cooperation," for it required active participation in many ways. Each department would elect three men to a board of control charged with monitoring work and assisting the foreman. These boards would in turn elect a board of governors to manage the association, their actions subject to review by a president, one of the firm's principals. Whoever filled that role would not direct meetings or control decisions but merely exercise the power of review before any resolution became final. This amounted to a surprising concession of power for the times, as did the provision that a two-thirds board vote could override even the president's wishes. Of course Britton could conceivably dissolve the entire association, but then so could the men through their board, again with a two-thirds vote.[52]

This remarkably advanced plan stemmed from Britton's bedrock faith in the essential character and dignity of the American worker. "The mechanical classes, properly treated," he liked to say, "are as honorable in their actions, as a body, as any class of our citizens, no matter what their social condition may be." Rejecting English capitalists' disdain for wage earners, his plan took special pains to achieve a democratic balance. By voluntarily granting unheard-of privileges to Brewster & Company's employees, he had every confidence that his men would benefit, even as their actions validated his convictions. The plan's publication showed just how revolutionary it seemed to others. Letters began piling up on Britton's desk warning him of the dangers of "spoiling" working people, of making labor difficult to manage, and that his proposals "endangered capital." Some included the firm's own customers, but Britton pressed on regardless.

The arrangement appeared to succeed until early summer 1872, when a renewed wave of agitation for the eight-hour day swept through the city's trades. The ensuing rallies, ultimatums, and strikes brought unrest to Brewster & Company's workers and unhappily coincided with Britton's confinement with illness and his partners' absence from town. The board of governors had the power to mandate an eight-hour day, but their knowledge of Britton's skepticism of short hours and their own fears of injuring the firm's prosperity prompted them to stick to ten hours daily. Yet they were powerless to prevent disenchanted workmen from forming another committee that delivered Britton an ultimatum: eight hours or a strike. Britton refused to act, on grounds the men violated the association's rules in bypassing the board, and he warned that a strike would spell the plan's termination. Nevertheless, within twenty-four hours the majority of the workers joined what had become a city-wide trade strike. A few stayed at their posts, but the walkout involved somewhere between three-quarters and four-fifths of the company's three hundred-plus workers. It also spelled the end for the Brewster & Company Industrial Association.[53]

The strike dragged on for two weeks before Britton, finally well enough to be back on his feet, called the men together to state the firm's position with characteristic bluntness: since they had ignored the arbitration process, the striking hands could only return to work at the old daily hours and at the former wages. Brewster & Company would grant no concessions and the men would forfeit the dividend of approximately $11,000, which the firm would use to help cover its losses. This proclamation broke the strike, and while the arrangement granted the strikers their jobs, the profit-sharing plan became the primary casualty. Surprisingly, however, Britton remained adamant that both the plan itself and the theories behind it had been validated. Though clearly disappointed, he blamed the events on what he saw as the trade unions' "coercive power." He pointed with evident pride to the officers' loyalty to the association's ideals and procedures, and for him this had been enough to verify his convictions. Summarizing the matter in later years, Britton sounded as idealistic as ever: "I believed then, and still believe, that men belonging to the mechanical classes who have power and privileges given them by their employers voluntarily, and not under threats of strikes, would not take advantage of the power so placed in their hands." While what he regarded as extenuating circumstances cut short a noble experiment at Brewster & Company, Britton remained confident that similar plans would one day bring lasting harmony between labor and capital.[54]

Not that Britton or Brewster & Company entirely abandoned labor reform; a much less revolutionary but far more enduring arrangement replacing the profit-sharing plan. Founded with $1,000 of company money, the Brewster & Company Benevolent Association provided employees with financial assistance in the event of illness or death. The firm matched the men's regularly paid dues and built up a considerable fund in an arrangement as long-lasting as it was widely imitated. The city soon boasted the New-York Guild of Carriage Makers in 1874, joined five years later by the Milburn Wagon Company Employees' Mutual Aid Association. The 1880s saw formation of the Washington Carriage Makers' Association (Washington, Pennsylvania, 1880), J. B. Brewster & Co. Mutual Benefit Association (New York, 1881), Donigan & Nielson Employees Association (Brooklyn, 1882), the Carriage Makers' Guild of Brooklyn (1882), the Carriage and Wagon Makers' Benevolent Association of Quincy, Illinois (1885), and the Washington Guild of Carriage Makers (Washington, D.C., 1885). These accompanied a second wave of unionization that spawned the Coach-Body Makers' Association (New York, 1881), the Carriage and Wagon Makers' Union (Washington, D.C, 1884), the German Carriage and Wagon Makers' Union (New York, 1885), the Carriage Workmen's Association of New Haven, Connecticut (1885), the Carriage and Wagon Makers' Union of New-York (1885), and the Carriage Makers' Central Labor Union (New York, 1886). Some became official chapters of the Knights of Labor while others remained independent.[55]

Though many of these unions had widely fluctuating membership and short life-spans, they participated in wage and working-hour strikes in the major carriage-making centers. In most cases the disputes were brief and resulted in various compromises; some elected to join a new umbrella organization that seemed to hold great promise for all working men. Founded in 1869 by Philadelphia garment workers, the Knights of Labor evolved from a local trade union into a body whose goal was to unite all workers beneath its banner. A successful strike against financier Jay Gould bolstered the organization, and membership peaked around 1886. But another strike against Gould failed, and the rising tide of sentiment against anarchism, socialism, and labor protest hastened its decline. While some local and national wagon and carriage unions continued to affiliate with the Knights, it had a diminishing influence in the industry. The Knights organized a national assembly for wagon and carriage workers sometime after 1891, but the effort only lasted about a year. By then a renewed attempt at national organization of the wagon and carriage industry was taking place in another union. Disaffected Knights of Labor members met in Columbus, Ohio, in 1886 and united under the leadership of Samuel Gompers to form the American Federation of Labor (AFL). Rather than pursuing the all-inclusive industrial union model, the AFL built itself around local craft unions; this proved a more durable arrangement, and by 1890 the AFL had superseded the Knights on the national level.[56]

In August 1891 a handful of carriage workers from six different towns met in Pittsburgh to form the Carriage and Wagon Workers International Union of North America (CWW), affiliating themselves with the American Federation of Labor. The CWW held annual conventions beginning that year, but limited participation by the local affiliates, financial instability, and administrative incompetence plagued the new organization, which had only 874 members in 1892. Financial distress the next year hurt the CWW so that in 1894 only eight of the thirty-four locals participated in the convention. The years 1895 and 1896 were similar, and the next convention did not occur until 1903. Local unions meanwhile continued to spring up and occasionally struck over issues of hours and pay, with varying degrees of success. During the 1896–1903 hiatus the CWW engaged in internal reform designed to stabilize and strengthen the organization, but cantankerous local unions continued to sap the national's strength. Membership in the CWW rose to over a thousand after 1900, peaking at 5,500 in 1904. Despite such gains, the union was never able to achieve true stability or significant influence. It seemed almost a rerun of the CMIU some thirty years earlier; lack of funds remained a persistent problem, with insufficient reserves a particularly limiting factor. The union could not afford the full-time professional leadership it required, and the steep membership decline beginning in 1904 only exacerbated the situation.[57]

The biggest problem lay in the same external pressures affecting all labor unions. Besides employer opposition, it had also to contend with the business cycle. In both 1893 and 1903 a depressed business climate caused serious damage to the CWW and like-minded organizations in other industries. Meanwhile, opposition coalesced in the form of the National Association of Manufacturers, which took a more stridently antiunion stand than did moderate trade associations like the CBNA. Additionally, the CWW simply had limited application to the industry. It experienced the most success in organizing the small shops because, as one scholar has suggested, the skilled workers there were simply easier to organize than the semi- and unskilled laborers of the largest wagon and carriage factories. The AFL officially excluded unskilled workers in any case. The CWW's relatively greater success after the turn of the century was marred by jurisdictional quarrels with craft unions, particularly the painters, blacksmiths, and upholsterers groups.

This conflict turned the CWW from its original goal of being the overseer of all wagon and carriage labor to become an umbrella organization for the separate locals in each of the industry's branches of woodworking, blacksmithing, painting, and trimming. Even this reorganization failed to stabilize the national, and by 1910 CWW membership had dropped drastically. Only the inclusion of the new automobile workers revived the organization, but by then it was too late for the wagon and carriage industry.[58]

Unionization must be taken into account when considering workers' response to industrialization, yet persistent overemphasis on union activity has hindered historians' attempts to interpret a complicated situation. What ostensibly stands as an explanatory model for the historical past ends up a bitter turf war between capitalists and Marxists, Marxists and neo-Marxists, and the original goal of understanding the worker's world falls from view. Unionization ultimately involved a very small percentage of labor in most trades and industries, and this alone should serve as a caveat. While the limited surviving evidence on wagon and carriage unionization admits of some generalizations, these struggles serve as only part of the story of individual resistance or accommodation to industrialization. Institutional struggles for better wages and shorter hours are significant, but they were not the sole or even the predominant form of response for most tradesmen. What, then, can we conclude about the effects of industrialization on the wagon and carriage worker?

The answer really depends on the place of employment. Industrialization divided wagon and carriage workers into the fundamental categories of vehicle and parts makers, and the combination of special-purpose machinery and rationalization ensured that the average accessory-factory hand resembled the archetypical industrial laborer. An integral part of the larger wagon and carriage industry, they nonetheless had more in common with general industrial semi- and unskilled workers than they did with traditional vehicle craftsmen. Those who built the vehicles faced a somewhat broader range of opportunities. Workers punching in at one of the huge, highly rationalized vehicle factories might perform tasks similar to those of the specialty factory hand, while employees of smaller factories and shops exercised a wider range of skills. Ready-made parts played a vital role in even the smallest firms, but while the small shop remained, so did the craftsman. As parts, rationalization, and mechanization combined in various ways, they ultimately failed to eradicate skill or the man who called those skills his own. If the skilled craftsman's window of opportunity narrowed, it remained open nevertheless, and through to the very end of the industry. Whether content to work at reduced capacity in a factory, ply more of his trade in a small shop, or perhaps seek promotion to foreman in either place, the skilled vehicle craftsman was still able to make his way in the world. It was a world greatly changed from that of his predecessors, and like any change, it involved compromise. If his work now involved fewer and different skills, it also required less sheer physical exertion. Although it involved machinery, dust, and noise, it occupied fewer hours each day. It was a tradeoff, but one that many were forced, or perhaps even willing, to accept.

If this seems a fundamentally ambiguous answer, so too is industrialization's legacy in horse-drawn vehicle manufacture. Industrialized production techniques put buggies and broughams within reach of a broader spectrum of people, at the same time demanding a working rationale inimical to traditional craftsmanship. Conversely, had craft work remained the standard production format, the democratization of the horse-drawn vehicle would never have taken place. So often in the history of American manufacturing the interests of the worker and consumer parted ways on the road to industrialization, and this remains a fundamental conundrum of the contemporary material world. Although this observation may seem to beg the question, the debate over industrialization in any trade ultimately brushes up against far larger issues that remain very much contested. We can no more totally settle the matter of industrialization and the wagon and carriage worker than we can completely lay to rest the tensions inherent in a production system that has granted us so much abundance and so much strife. Industrialization brought, and continues to bring, both positive and negative developments into the lives of many people, wagon and carriage workers included. Yet a late-nineteenth-century technological innovation provided an even greater irony. Dismissed by horse-drawn vehicle manufacturers as a passing fad, the automobile heralded a new age in transportation even as it sounded the death knell for an entire industry.

[8]

That Damned Horseless Carriage

American Carriage Makers Respond to the Automobile

The convention did indeed vote that automobile makers should be eligible to join the Association, and invited them to do so. But it was a condescending gesture made in about the same spirit as might prompt a national retail grocers' association to offer the privilege of membership to sidewalk peanut vendors.

Ben Riker, 1948

"Sherman said 'War is hell,' but he was never in the automobile business."

CBNA president Charles O. Wrenn, 1915

According to Emil Hess, it had been the bankers' fault. "Cincinnati was always a conservative town," he said over lunch at the Hotel Alms, "and the bankers here were afraid of the automobile industry." The other men agreed; had it not been for the timid financial community, the Queen City might have been the Motor City. Hess spoke with the conviction of age and vast experience; eighty years old when he uttered these words in the summer of 1957, he was a former employee and long-term owner of Sayers & Scovill, once one of the city's many large carriage-making firms. He had spent most of his adult life with "S. & S.," by that time part of a custom ambulance and hearse body-making company operated by his sons. He and the informal group of former industry members who met for luncheons like this could recall when Cincinnati had been the Detroit of the American carriage trade. To them it seemed only logical that the industry that replaced their own should have had its epicenter in southern Ohio, not southern Michigan. Carriage makers who wanted to enter the automobile industry could not obtain the necessary financing, however, and in the end the opportunity slipped from their grasp.[1]

Hess's observation, by no means original to him, seemed logical to many people, and it cropped up elsewhere over the years. It presupposes that wagon and carriage makers possessed the necessary knowledge and equipment, that carriages and automobiles were analogous, and that Detroit bankers had actually been a lot more daring than the ones in Cincinnati, Amesbury, or Columbus. Yet this tidy theory breaks down in the face of historical scrutiny. The discontinuity between the old and new industries owed as much to technology as to finance, just one of the many discoveries made as American wagon and carriage manufacturers coped with the automobile. The twilight world between the age of the buggy and the reign of the motorcar, for all its dimly lit recesses and counterintuitive twists, can tell us much about them both.

One of the most challenging aspects of conjuring up the past lies in recapturing the indeterminacy that invariably permeates the present. In this particular case, the key is to remember that the horse's replacement by a mechanical contrivance was by no means a universally shared anticipation. A few visionaries had always predicted mechanized land transport, yet individuals so inclined predicted all manner of fantastic inventions, few of which anyone took seriously. Wagon and carriage manufacturers certainly failed to take the automobile seriously, at least in the beginning. More than one of them dismissed it as a passing fad like the 1860s "bone shaker" velocipede, or the 1880s high-wheel bicycle. By the 1890s the wagon and carriage industry had survived the onslaught of railroads, street railways, and bicycles, and few carriage makers perceived the automobile differently. They had every reason for confidence. According to the 1890 census, the American wagon and carriage industry consisted of more than thirteen thousand firms, including everything from one- and two-man shops to state-of-the-art factories. Some 130,000 employees turned out products valued at more than $200 million, and this did not include the accessory industry, whose own statistics pushed the totals much higher. These were the official figures, doubtless conservative ones, when the industry entered the 1890s, and they continued to climb in the years ahead. Much of this sense of collective confidence and self-satisfaction comes down to us from the Carriage Builders National Association, since 1872 the industry's preeminent trade organization.[2]

Nothing better reflected the CBNA's robust existence than its keenly anticipated annual conventions. The earliest took place in New York, Philadelphia, New Haven, and Boston, eastern cities with sizeable carriage trades and ample conference facilities. As the industry moved westward and the association evolved, it began meeting in other parts of the country. Midwestern and southern members successfully agitated for a more active role in what they perceived as an eastern clique. Tapping the considerable influence of prominent wagon makers like Clement Studebaker, this faction helped bring about the first Chicago meeting in 1880. The southerners triumphed in 1887 with a convention in Washington, D.C., and the selection process eventually became sufficiently democratized to bring it to medium-sized venues like New Haven (1877, 1883), Cleveland (1895, 1915), and Indianapolis (1899), as well as Atlanta (1906), Richmond (1920), Milwaukee (1904), and, the westernmost, St. Louis (1884, 1898, 1913). The majority took place in the obvious cities, however; over the course of its lifetime the CBNA met ten times in Cincinnati, nine in New York, six in Chicago, four in Philadelphia, and three each in Boston and Atlantic City.[3]

The organization's formal leadership followed a similar path, with early predominance by eastern carriage makers followed by gradual incorporation of midwesterners and southerners. One of the men behind the creation of the CBNA, Maine carriage maker Charles P. Kimball, held the office of president from its inception through1874 (and by default through 1875, when no convention took place). He stepped down to make way for New Haven's Henry Killam (1876–78), followed by New York's John W. Britton (1879–81), and Wilmington's Henry C. McLear (1882–84). Though by the date of McLear's tenure the Midwest had already seized the lead in the industry, the early presidents reveal the continued prestige of the old eastern houses. Greater variety came in 1885 with a pioneering Cincinnati buggy manufacturer (Lowe Emerson), followed by the nation's best-known wagonmaker (Clement Studebaker, 1886), as well as heads of prestigious firms in Columbus (C. D. Firestone, 1888), Cincinnati (Grant A. Burrows, 1890), and New Haven (Frank H. Hooker, 1891). By the early 1890s the CBNA had already witnessed the rise of second-generation leadership: Charles F. Kimball, son of founder and first president Charles P. Kimball, became president in 1892, and had he lived to see it, John W. Britton would have been proud to see his son Channing assume the mantle in 1894.[4]

Normally held in the early fall, a typical convention consisted of two or three days of business meetings interspersed with light entertainment and culminated in a lavish banquet. Concerned members first proposed the meal in 1876 to boost morale, and it quickly became a convention ritual. The business sessions were supposedly the main focus. The election of officers consumed part of this time, as did the reports of committees charged with investigating various trade matters and the debates invariably following them. In the 1870s the subjects ranged from a proposal to standardize carriage track (unsuccessful) and another to create a standard carriage guarantee (adopted) to the establishment of prize essay and carriage drafting contests (popular). The 1880s brought the idea of a technical school (established) and proposals to revive trade apprenticeships (ineffective), as well as reports on the nation's timber supply, fire insurance, and trade statistics. By the 1890s reports on oil as a fuel source, timber tests, factory design, and electric welding testified to the association's widened scope, and reports on good roads and highways formed a few additional topics under review. All of this took place with due observation of the ban on discussing wages and shop management.[5]

There was also a unique opportunity to view the latest in wagon and carriage goods. Beginning informally at the 1880 Chicago meeting, manufacturers of horse-drawn vehicle parts and hardware displayed their wares and solicited buyers in a large hall rented for the purpose. The exhibit received official sanction two years later when the association established guidelines prohibiting the display of complete vehicles (to prevent competitive friction), apportioning the available space, and more. The majority of participants were accessory manufacturers, who formed a substantial percentage of CBNA membership; by the 1890s the event routinely included well over a hundred exhibitors. Some displays became famously elaborate, the more savvy firms courting business through abundant brochures and catalogs, free cigars, and assorted trinkets designed to keep the company name fresh in the recipient's mind. Bowler-hatted salesmen traded stories and jokes between orders, conventioneers renewed old acquaintances, and a haze of cigar smoke spread out between the signs and pennants announcing important new breakthroughs in bolts, reach irons, and whip sockets.[6]

The nineteenth-century meetings radiated optimism; despite occasional wrangling over carriage guarantees, machinery, and other thorny issues, the annual conventions generally exuded an air of satisfaction. This certainly characterized the twenty-third meeting in October 1895, the first CBNA convention held in Cleveland. While the city could not boast wagon and carriage manufacture on the scale of Cincinnati, Columbus, or even Toledo, it had become home to some of the nation's largest vehicle hardware firms. The leaders included the Eberhard Manufacturing Company, the Cleveland Hardware Company, and the I. N. Topliff Manufacturing Company, while other manufacturers important to the vehicle trade included the Sherwin-Williams Company, Lamson & Sessions, and National Screw & Tack. Employees from several of these firms staffed the local arrangement committee, and each fielded a display at the accessory exhibit in the newly built Gray's Armory. Here the floors groaned beneath the load of horse-drawn vehicle hardware and forgings, bentwood carriage parts, and vehicle bodies from manufacturers nationwide. At over 120 exhibitors, it ranked as the largest exhibit yet held.[7]

CBNA president Channing M. Britton formally opened the proceedings the following day, his remarks introducing a host of reports. The members heard the secretary and treasurer on the organization's finances, technical school, freight and transportation, good roads, membership, and credit. Not that it was all work, for most conventions had long featured a variety of entertainment. Cleveland's consisted of a boxing match, which one member recalled as something of a "thriller and shocker," as well as a trip to Euclid Avenue's Opera House and a Lake Erie boat excursion. By 1895 the banquet had become a truly lavish affair, complete with elaborate table settings, abundant food, and—before temperance sentiment took hold—equally abundant beverages. Nearly seven hundred sat down in Cleveland to dine and to listen, for the CBNA boasted illustrious speakers, regardless of their trade connections. Past conventions had featured as either honored guests or speakers a Civil War general (William T. Sherman, 1884), a U.S. Senate chaplain (Edward Everett Hale, 1885), the nation's largest bicycle manufacturer (Albert A. Pope, 1889), and the Lord Mayor of Dublin (1893). Diners in Cleveland heard speeches by Mayor R. E. McKisson and Governor William McKinley, whose platitudes doubtless pleased the overwhelmingly Republican audience. After this came a rash of toasts, and it was well past midnight when the actor William H. Crane gave the closing speech to an audience that by then must have been quite full, perhaps a bit tipsy, and certainly very tired.[8]

The overwhelming impression one gets from the 1895 convention is that of members of a prosperous American industry engaged in an extended ritual of food, fellowship, and hearty self-congratulation. While the event certainly involved all of that, deeper currents ran beneath the surface. On the first day past president and executive committee chairman C. D. Firestone reported on one recent source of trouble. Noting his previous comments on the early 1890s depression and his prediction of a rapid recovery, Firestone now admitted his disappointment in the slow rebound and warned members about overproducing in a sluggish market. At the start of the decade things had looked quite rosy. When a subscriber in 1890 suggested that *The Hub* inaugurate a "Business Troubles" column, the editor happily obliged since "infelicities in these days are comparatively few and slight." He soon had cause to regret his optimism. By early 1891 he was informing his readers that there had recently been more "chattel mortgages, judgements, and executions for small amounts than perfect health on the part of small carriage and wagon-makers would seem to warrant." It soon got worse. The panic reduced demand for horse-drawn vehicles, threw many tradesmen out of work, and put more than a few carriage works out of business. Not even the big players could entirely avoid the fallout; Clement Studebaker recalled the Panic of 1893 as one of the most trying periods of his career. While such economic fluctuations were no light trial for anyone, at least the enemy was familiar.[9]

Something far less familiar loomed by the mid 1890s, and it formed the subject of a paper read by *The Hub*'s publisher W. B. Templeton. The essay concerned the horseless carriage, and it had actually been written by editor William N. FitzGerald, yet another old-time carriage maker who had turned to journalism. Trained as a body maker, FitzGerald had worked for several eastern firms prior to his Civil War service, later gravitating toward the trade press. After long tenure with both *Harness and Carriage Journal* and the *Coach, Harness and Saddlery Journal*, he succeeded George Houghton at *The Hub*, where he soon found himself contemplating the horseless carriage. The welter of automotive experimentation in the United States and abroad had finally attracted the trade's attention, and this, along with the 1895 Paris-to-Bordeaux automobile race, prompted the subject's first official airing at a CBNA convention.[10]

FitzGerald soothingly remarked that the current craze was merely one of several spread over the century; where earlier interest in horseless vehicles centered on a few cumbersome steam-powered behemoths, the latest upsurge owed itself to the "Otto principle." While he did feel that such improved internal combustion engines promised to power heavy commercial vehicles, he had every confidence in the horse-drawn carriage's continued supremacy. Beyond the numerous technical matters, there remained man's innate love of the horse and the carriage's inherent lightness and grace. Closing with a ringing defense of the horse-drawn vehicle, FitzGerald exhorted his audience to accept the horseless vehicle's limited usefulness, while at the same time casting "to the winds all fear of the 'passing of the horse' or the rejection of the carriage." The essay drew applause at several points, and the association formally thanked Templeton for presenting such an "able and timely and conservative" address. Standing as they were at the head of a sophisticated and important industry, the members understandably looked askance at a smelly and unreliable upstart. The idea was not simply going to fade away, however. Even as the conventioneers applauded Fitzgerald's editorial, another editor exhorted readers of the new *Horseless Age* that "all signs point to the motor vehicle as the necessary sequence of methods of locomotion already established and approved." Soon even *The Hub* felt compelled to run a regular horseless carriage column—if only to keep an eye on new developments.[11]

There had been new developments long before the 1890s, for a self-propelled road vehicle was a dream dating back centuries. Though a few eighteenth-century experimenters made useful contributions to the general concept, significant advances came with the nineteenth-century application of three power sources. Steam (external combustion), electricity, and internal combustion (with gasoline the preferred fuel) were once all serious contenders. The steam engine's excessive weight remained a hindrance until significant miniaturization and the flash boiler became realities during the second half of the nineteenth century. Americans seized on steam's advantages, and by the 1890s the nation boasted two leading "steamer" manufacturers. Francis and Freelan Stanley were inventive brothers from Kingfield, Maine, who in 1896 built the first of the famed Stanley Steamers. They began quantity production two years later, and several firms produced the spunky little runabouts into the 1920s. Son of a Cleveland sewing machine manufacturer and industrialist, Rollin H. White entered the automotive field in 1898–99 with his newly invented flash boiler. He and his brothers ran the branch of the family business that became the White Motor Corporation, and by 1900 produced reliable steam-propelled automobiles. These remained in production until 1909.[12]

The steamer's early success owed to its relative mechanical simplicity and forgiving manufacturing tolerances, but thermal inefficiency and the weight of fuel and water became increasingly serious drawbacks as electric and internal combustion vehicles improved. Electricity's earliest and most successful transport application consisted of urban and interurban railed vehicles; road vehicles required reduced-size electric motors and reliable storage batteries, improvements the Americans soon developed. The nation's largest bicycle manufacturer when it established a motor vehicle division in 1895, Hartford, Connecticut's Pope Manufacturing Company commenced electric automobile production in 1897. Pope's "electric" was the first American automobile produced in significant numbers, with approximately five hundred on the road by 1898. That same year, a Cleveland ball bearing manufacturer named Walter C. Baker organized the Baker Motor Vehicle Company to produce electric runabouts, and his firm quickly became a serious competitor. Baker and the others banked heavily on the electric's quietness and simplicity, but it was heavy, expensive to purchase and maintain, and it remained saddled with a severely and often unpredictably restricted range. Charging stations never became widespread, and the restricted support infrastructure limited electrics to urban use. Ultimately the concept could not withstand the challenge of internal combustion.[13]

Like most revolutionary inventions, a practical internal combustion engine owed its existence to numerous American and European scientific and technological experiments. Perhaps the biggest technical breakthrough came with the four-cycle engine, developed in 1876 by German engineer Nicholas Otto, which others soon adapted to transportation. In 1885–86 Karl Benz used one to power a three-wheeled carriage, and during 1885–89 Gottlieb Daimler made a bicycle and a carriage both powered by Otto engines. Benz and Daimler battled over primacy, but they had both been beaten by Viennese chemist and engineer Siegfried Marcus, whose four-cycle gasoline engine drove a carriage in 1875. Benz and Daimler aggressively promoted their inventions, however, and their names became indelibly connected with the automobile's origins. Americans had also been working toward the same end, the first real success coming from Massachusetts bicycle mechanics Charles and Frank Duryea. Inspired by a written description of the Benz tricycle, the Duryea brothers built their first internal combustion automobile in 1893, commencing commercial manufacture three years later. They had many competitors, including Hiram P. Maxim, Alexander Winton, Ransom E. Olds, and Henry Ford. One contemporary estimated that by 1895 over three hundred firms and individuals were working on motorized vehicles. Much of this activity took place in New England, where the tradition of precision manufacturing, an extensive tool and machine building trade, and ready financial backing from Boston and New York all aided in the pursuit. In addition to specializing in steam and electric propulsion, New England automobile manufacturers followed the European lead of making elegant and expensive models for the nation's wealthy elite.[14]

The other primary automobile manufacturing region was the Midwest, largely because of the advantages that had already attracted so many American industries: closeness to fuel and raw materials, excellent transportation links, an established industrial infrastructure, and abundant labor. Another factor, largely unexplored by automotive historians, consisted of the established wagon and carriage industry and the extensive hardware and accessory trade that supported it. Availability of ready-made parts soon translated into the kind of manufacturing efficiency that had come to characterize wagon and carriage building. The region's poor roads and locally available gasoline oriented midwestern automobile makers toward the gasoline-fueled, internal combustion engine, with a power-to-weight ratio superior to any other propulsion source. Heavily invested in steam and electricity, New England manufacturers found it impossible to cross over to this technology on a competitive scale. New England and midwestern automobile manufacturers also pursued radically different markets: the former tended toward expensive models for urban use, the latter to utilitarian transport for the rural masses. The decision to favor the large rural market created an additional impetus to explore economies of scale, which eventually resulted in the mass production of low-priced automobiles like Henry Ford's Model T. These factors thrust the Midwest into the lead, as Ohio, Michigan, and Indiana achieved unparalleled prominence in the American automobile industry.[15]

The two most important midwestern automobile manufacturing cities were Cleveland and Detroit. Cleveland's manufacturing tradition accounts for early automotive pioneers like Alexander Winton, Walter Baker, and Rollin White. Scottish-born Winton built his first internal combustion automobile in 1896, and his Winton Motor Carriage Company became an early industry leader while his celebrated endurance runs added "automobile" to the English language. By 1900 Winton owned the largest automobile factory in the world, his company remaining a key player until the early 1920s, when the saturated luxury car market turned him to diesel engine manufacture. Walter Baker's ball bearing background led him into automobile parts supply, and construction of his own electric in 1897 saw him create the Baker Motor Vehicle Company. By the 1910s, the gasoline engine's improvement sent demand for electrics into steep decline, and Baker sold his automobile division to Stevens-Duryea in 1920 in order to make electric and gasoline lift trucks. Rollin H. White's industrial family had already produced items as diverse as sewing machines, roller skates, and bicycles when White and his brothers built their first steamer in 1900. Founding the White Company in 1906 and vigorously promoting steamers though encomiums by President William H. Taft and Buffalo Bill Cody, Rollin White nevertheless recognized internal combustion's advantages and abandoned steam by 1918 for gasoline motor truck and bus manufacture.[16]

Cleveland boasted more than eighty automobile marques during the period 1896–1932. They ranged from Stearns-Knight, Gaeth, Baker, and Owen of the late 1890s, and Peerless, Hoffman, Ottokar, and Monarch of the early 1900s, to Chandler, Jordan, Merit, and the Globe Four of the 1910s and 1920s. By 1909 automobile manufacture was the city's third largest industry, with more than thirty factories and some seven thousand workers producing nearly $21 million in new automobiles. This was impressive, but it did not last. During the period 1903–4, Michigan's twenty-two automobile manufacturers employed more than two thousand people turning out nearly ten thousand vehicles. These numbers would increase, especially in Detroit. One significant reason for Detroit's lead centered on its savvy cultivation of the low-priced auto market: though Michigan produced nearly half the U. S. automobiles in 1903–4, their sales accounted for less than a third of the total value of national production. Michigan automakers concentrated on high volume production of low-priced automobiles, an approach epitomized by Henry Ford. After several years of turning out small numbers of various gasoline automobiles, Ford introduced his Model T in 1908. Produced in enormous quantities, it would soon overshadow the few thousands made by others. Cleveland retained leadership in automotive parts manufacture, in part because of the city's large wagon and carriage hardware firms. Factories producing everything from valves to horns began overshadowing the city's remaining automobile makers after 1910.[17]

When members of the CBNA met in Cleveland in 1895, this would have seemed a preposterous prediction, for in its infancy the automobile invited scorn and dismissal. The very earliest were extraordinarily temperamental, and even improved models displayed many alarming characteristics. When touring the few available good roads, they succumbed to foaming boilers, prematurely spent batteries, or carburetors that refused to carburize. The terrain only compounded the problems, for few automobiles of any type could traverse some of the quagmires that passed for thoroughfares. "Get a horse!" had all the more sting to the early motorist's ear because it really did remain a viable solution; in many instances the horse and horse-drawn vehicle could do what the automobile could not, something passersby eagerly reminded stranded motorists.

As incremental improvements began eliminating some of the more vexing mechanical problems, however, it became difficult to dismiss the automobile as a hopeless botch. Indeed, the wagon and carriage industry began moving from dismissal to rationalization: if the automobile would not fade away, then surely it would have only limited application. Even Fitzgerald's 1895 paper reflected something of this attitude in pointing to the automobile's high cost. This restricted early versions to the role of rich men's playthings, though even this came at a cost to the carriage trade. Wealthy Americans were increasingly forsaking victorias, broughams, and traps for new motorcars, so high-end carriage makers suffered the first blow delivered by the automobile. Though the shift understandably distressed the affected firms, observers took comfort that high-grade horse-drawn vehicles accounted for a relatively small percentage of the trade; losses here hardly imperiled the entire industry.[18]

These conditions informed the wagon and carriage industry's early-twentieth-century optimism. In a closing article for the 1904 fortieth anniversary issue of *The Carriage Monthly*, one editor confidently predicted that "the carriage builder of 1944 will be a prince in the handicrafts, and his workmanship will be a source of joy to thousands." Great responsibility, like fatherhood, has a conserving effect, and members of the CBNA represented one of the nation's leading industries. Could they realistically have expected it suddenly to disappear? One imagines them, like passengers of a sinking ocean liner, resisting trading the solidly familiar for the unstable confines of a much smaller and untried vessel. This of course assumes they know the ship to be doomed, which is precisely where the analogy breaks down, where it becomes difficult for the historian to recapture that indeterminacy distinguishing the present from the past. If members of the CBNA harbored doubts about the ultimate extent of automobility, they had every reason to do so. If humankind had traveled for centuries in conveyances pulled by beasts (and did not Stratton's magnum opus provide excruciatingly detailed proof?), why would a reasonable person assume the future held anything different?[19]

At the same time, the previous hundred years provided innumerable examples of mechanical solutions to basic human problems. As the last years of the last decade of the nineteenth century slipped by, even the most stubborn wagon and carriage men had to suspect that the automobile might serve as yet another poster child for the Industrial Revolution. Still, it seemed unreasonable to think every horse-drawn vehicle would disappear. Once widely feared as deadly rivals, the locomotive and the railroad had failed to kill off the carriage; would the automobile prove any different? Though speculation ran rampant, it had at least become apparent that the automobile would remain. Not that this moved the carriage trade to proceed quickly; at the 1899 convention, CBNA members voted to extend membership to automobile manufacturers. Thirteen years later, however, some outspoken individuals were still urging the association to accept automobile makers as full-fledged equals. An intensely conservative organization, it resisted all attempts to radically alter its constituency, and not until 1910 did an automobile chassis (actually a five-ton motor truck) intrude on the sacrosanct horse-drawn vehicle goods display.[20]

Joining the Revolution

Long before this many wagon and carriage makers opted to forge their own alliances with the automobile. The larger firms made the most serious attempts, many commencing soon after the turn of the century. Auburn, Indiana's William H. McIntyre Company began experimenting with its own automobile around this time, placing it on the market in 1905. A two-cylinder, air-cooled, gasoline-powered engine drove what was essentially a motorized buggy capable of carrying two people. These "buggy-type autos," as the trade pressed dubbed them, sold well at the outset, and the larger carriage firms found it possible to produce them in quantity with only relatively slight modifications to their factories. In Ohio, the Columbus Buggy Company began turning out a similar vehicle in 1903, a "Utility Carriage" driven by a ten-horsepower, two-cylinder, air-cooled gasoline engine with a top speed of twenty-five miles per hour. The firm had benefit of the enthusiasm and mechanical abilities of a lanky seventeen-year-old-employee named Eddie Rickenbacker, who liked nothing better than to subject the little auto buggies to exhaustive road tests. By 1907 over twenty firms, mostly in the Midwest, turned out similar vehicles. Some of the manufacturers were large carriage-making concerns, while others had been organized expressly to make this product, but all attempted to combine internal combustion with carriage architecture to produce practical, low-priced vehicles. As St. Louis's Victor Automobile Manufacturing Company liked to state, their "Victor sells for less than $500, and is making friends." As the makers began encountering the format's technical limitations and the public's desire for more power and convenience, some firms dropped out while others, like McIntyre and Columbus, began designing large touring cars with liquid-cooled engines.[21]

New York's venerable carriage makers, the ones that were left anyway, also tried making their peace with the automobile. Though Benjamin Brewster had died in 1902, J. B. Brewster & Company lived on under the ownership of three former employees. Cairn Cross Downey, Henry M. Duncan, and George M. White continued to market Brewster carriages, much to Brewster & Company's irritation. Not that either firm was particularly enthusiastic about change, particularly the latter firm, which had become one of the nation's premier luxury carriage builders. This specialty left Brewster & Company vulnerable to the automobile's earliest market impact, for those able to purchase broughams and victorias also formed the first significant cohort of automobile owners. Carriages continued to roll out of the old factory, for many of the city's elite still preferred the horse's familiar cadence to the pounding of an internal combustion engine, and they enjoyed dealing with a firm long accustomed to their whims and desires. Doing so remained sufficiently lucrative to permit William Brewster the luxury of putting off his reckoning with the automobile until 1905.[22]

More accommodation of a trend than a reflection of profound interest, Brewster's venture into custom automobile-body manufacture nevertheless seemed logical. The family firm had decades of coach-body-building experience, and its skilled craftsmen would have little trouble constructing what were still primarily wooden automobile bodies. Concentration on the luxury market meshed with the firm's long-standing role as a supplier of luxury goods, and it permitted continuation of the fine craftsmanship so central to its image. Old traditions of all kinds lingered at Brewster & Company, like taking a traditional rival to court. That Henry and Benjamin were both in the grave did not mean an end to their feud; eager to wring maximum profits from a famed New York name, Benjamin's former employees continued to walk a fine line between imitation and deceit. By moving its repository close to William's factory, claiming sole succession from James Brewster, and putting out vague and evasive advertising, J. B. Brewster & Company perpetuated a custom established many years before. Once again Brewster & Company claimed, with some justification, that its rival sought to generate public confusion over the firms' respective identities, a matter the New York Supreme Court found itself pondering in 1905.[23]

Meanwhile, both manufacturers grappled with new realities; Brewster & Company continued its custom auto-body work while, in 1909, J. B. Brewster & Company became the New York agents for Reading, Pennsylvania's Acme Motor Car Company. As if in validation of the legal challenge, the trade press managed to confuse the two firms in its reportage of this last fact. A year later, Brewster & Company secured the first American agency for the French Delaunay-Belleville automobiles, the chassis brand sporting the first Brewster auto body. During the period 1905–8 Brewster & Company built a small number of custom bodies for a variety of American and foreign luxury chassis, including its first for a Rolls-Royce in 1908. This arrangement suited customers, obviously wealthy, who wished to combine an advanced and prestigious automobile chassis with a traditional and prestigious custom auto body. The Brewster-Rolls connection would develop further in the coming years, but over at J. B. Brewster & Company, the Downey-Duncan-

White triumvirate found it tougher going. After a near-record year in 1905, carriage sales had once more declined in the face of market saturation and the automobile. Despite efforts to meet that trend via its agency and promoting Acme's six-cylinder runabout, the firm could not generate sufficient revenue to compensate for the drastic loss of carriage patronage. Even cut-rate sales of refurbished work failed to help; the deathblow came with the long-awaited court decision—in Brewster & Company's favor. Swamped with debt and legally enjoined from exploiting the name that formed its one major asset, J. B. Brewster & Company declared bankruptcy for the final time in 1908.[24]

Brewster & Company had once again triumphed against its old enemy, but no legal satisfaction could guarantee its own continued solvency in a time of profound transition. By 1910 William Brewster was preparing his firm to enter the automobile industry on a much larger scale, by erecting an enormous new factory. Not only had the old 1874 building become cramped, its location had become increasingly unfavorable for manufacturing. Hotels and theaters crowded the district that since 1904 had been called Times Square, and Brewster bowed to the realities of New York's constantly evolving cityscape when he ordered erection of a six-story concrete plant in Long Island City, at the western end of Long Island. Leaving new renters to convert the old carriage factory into a restaurant, Brewster & Company vacated the quarters it had occupied for thirty-six years and moved into the new building in late October 1910. The firm also vacated its once-unmatched Fifth Avenue repository but retained a Manhattan presence by leasing the stately old Gallatin house further up Fifth Avenue, at the corner of Fifty-Third. Despite protests by the Vanderbilts, Sloanes, and Harrimans, who hoped to keep commerce below Forty-Eighth, Brewster & Company converted the old home into a lavish repository in the same way it had more than forty years earlier in a similarly exclusive neighborhood. Henry's old company had long since learned to keep its products squarely before their intended purchasers.[25]

What Brewster & Company wanted to sell the Four Hundred—or rather, what the Four Hundred wanted to buy—were automobiles, and the company's new factory reflected this new reality. A trade journal news column remarked that the firm "was among the last of the carriage concerns that changed its traditional methods under the advancing popularity of the motor vehicle." By now many wagon and carriage manufacturers had already attempted automobile manufacture, parts production, or had gone out of business trying. Brewster & Company had managed to evade change in part by continued upper-class patronage, its products' high prices, and its still considerable financial reserves. The new factory and its contents cost an estimated $1 million, and editorial speculation to the contrary ("Some day, perhaps, they will make complete automobiles"), Brewster & Company's stated purpose was to better facilitate custom automobile body manufacture. Carriage production ceased for good not long after the move, and by 1912 the firm relinquished its controversial residential lease for a less expensive space nearby. Most sales now took place at the factory anyway.[26]

In 1914 William Brewster expanded his company's commitment to the automobile by becoming a Rolls-Royce sales agent. The first chassis arrived from England that July, and before the war interrupted shipments, Brewster & Company built custom bodies for forty-five more. Whether making bodies for new chassis or those already owned by Americans, the company had already established a method little different from the previous century's customer-driven custom production techniques. In conference with each client, managers filled out an extensive order form that included the weights of the occupants and chauffeur; specifications for the upholstery, seat springs, and interior woodwork; and the serious matter of color scheme. The firm reserved specific family colors for prominent customers, inevitably creating conflict as more of the visually appealing combinations were taken. Fortunately Brewster & Company had a great deal of experience in the delicate diplomacy required to solve such crises.[27]

Despite earlier indications to the contrary, William Brewster had been giving serious thought to building complete automobiles. In 1915 the firm finished a prototype and commenced production the following year. "After overhauling the leading makes, both foreign and domestic chassis, for the last ten years," a January 1916 press release announced, "we have at last decided to capitalize our experience, and now offer a complete car, built in our Long Island City factory." The accompanying photograph showed a low, long-wheelbase (125 inches) automobile, sporting a modified brougham body, called a town brougham or town car. Heavily based on the Rolls-Royce chassis, the "Brewster" featured a four-cylinder, liquid-cooled gasoline engine on the "Knight principle," chosen for its simplicity and power adequate "for town and suburban use, without being high-powered and heavy." The automobile's smaller size (Brewster called it a medium-sized car) and lighter weight made it more economical, "a much less expensive car to keep in tires and gasoline" than heavier foreign makes. Drawing on its decades of coachbuilding experience, the company had designed the chassis specifically for this particular body (fig. 8.1).[28]

Though the firm outsourced components such as the Zenith carburetor, Stewart Warner fuel pump, Bosch ignition, Timken roller bearings and Valentine paints and varnish, the automobile was indeed a product of the Long Island plant. Assembly and frame and body fabrication took place at the new factory. The results cost a great deal of money ($8,300 to $8,500, depending on options), and the firm continued to target the familiar upper-class market. It also continued its commitment to aftermarket custom bodies for individuals, whether those wanting a new body for an older automobile or purchasers of new "rolling chassis" desiring a Brewster body built to order. William Brewster proved equal to the task of carrying the firm's design ethic into the age of internal combustion, and he pioneered a number of automobile body innovations designed to improve driver efficiency and passenger comfort. Brewster & Company continued to work with Rolls-Royce, but in fielding his own motor car William Brewster had attained a real measure of independent automotive achievement.[29]

By this time a New York automobile manufacturer had become something of an anomaly, for the American automobile industry had become thoroughly midwestern. Many firms had already entered the industry, but by the 1910s a few leaders had emerged, one with an enviable wagon- and carriage-making pedigree. Clement, J. M., and Peter Studebaker realized their company would eventually have to confront automotive developments. Actual entry of the Studebaker Brothers Manufacturing Company into the automobile business, however, owed much to an outsider. Son of a Baptist minister, Frederick S. Fish opted for the law instead of the cloth, and after passing the bar in 1876 began a successful corporate law practice. His increasingly high profile in Newark and New York complemented his growing involvement in state-level politics, and he became president of the New Jersey Senate in 1887, the year he married J. M. Studebaker's daughter Grace. His legal acumen and marriage into the Studebaker clan contributed to his desirability, and in 1891 the company retained him as general counsel and a director. Fish was convinced that Studebaker should get an early start in automobiles, and he found a sympathetic ear in his father-in-law. J. M. had an adventurous streak, and he shared at least a degree of his son-in-law's interest in self-propelled vehicles. Perhaps more important, he had sufficient faith in the younger man to allow him the time and resources to engage in serious experimentation.[30]

Peter Studebaker's sudden 1897 death made Fish chairman of the firm's executive committee, giving him even more control over the automobile project. At a meeting that spring, Fish gave a verbal report on the horseless vehicle prototype, an electric runabout, which would soon be complete. Clement Studebaker, though hardly dismissive of the entire concept, still viewed the automobile with skepticism. In his opinion, 1890s automobiles were still in an experimental stage, and he feared overcommitting company resources. The importance of protecting the Studebaker reputation naturally concerned Clement, who had spent decades cultivating public trust, and he knew the pitfalls of blindly following trends. "The competition is going to be sharp," he predicted in 1899, and there remained a great many technological challenges to be overcome. But he added, "we are keeping in close touch with the movement and are watching developments with interest." Watching implied idleness, but Studebaker had already converted part of its enormous South Bend works to auto body construction. In 1899 the firm accepted a contract with New York's Electric Vehicle Company to build a hundred bodies for electric cabs, also pursing similar arrangements with other automakers. By this time Clement had grown sufficiently comfortable with the idea to announce the firm's intent to manufacture its own automobile. He still harbored misgivings, however. "The first thing to be accomplished is to get a machine that is safe and reliable under all conditions," he said in 1900, for "those now in use are demonstrating daily that they are not." He wanted to keep moving ahead slowly, predicting three more years of experimentation before production could commence.[31]

In accounts of these years Clement Studebaker frequently looks like a hopelessly antiquated holdover, and in a sense he was. Like so many of his fellow wagon and carriage makers, Clement knew wood and iron, wagons and carriages, and he used this knowledge to build up a manufacturing enterprise of stellar proportions. Always one of the most conservative of the brothers, Clement realized that caution and duty had guided his personal and professional conduct his entire life, and he naturally erred on the side of restraint when it came to the automobile. If he seemed anachronistic, and he had every reason to be, he still defies such easy dismissal. Clement, his brothers, and their company had not succeeded through temperamental Luddism. The Studebaker Brothers Manufacturing Company pioneered the use of everything from drop forges to electric welding in the mass fabrication of horse-drawn vehicles. Such achievements seldom flow from hidebound reactionaries, and neither did Clement's attitude toward the automobile. As early as the fall of 1899, Studebaker's board of directors gave Fish permission to create an automobile company, a resolution doubtless made with Clement's permission. Clement willingly ceded Fish the resources to make a Studebaker automobile a reality; all he asked in return was a moderate pace along the way.[32]

To Fish and the other younger managers, this pace may have seemed glacial. Indeed, had it not been for Clement's death in 1901, the firm might have followed his drawn-out timetable. His passing appears to have removed the remaining restraints, and with J. M.'s blessing, Fish accelerated his plans. In 1902 the company unveiled its first effort, a chain-drive electric runabout. It resembled little more than a bicycle-wheel road wagon spruced up with a few accessories. Indeed, one early publicity release suggested Studebaker's long horse-drawn vehicle experience allowed it "to solve, peculiarly well, the problems of suitable body dimensions, correct weight distribution and easy spring suspension, regardless of the load carried." The vehicle allegedly embodying this received wisdom consisted of a piano-box body sporting an upholstered spindle seat, dash, fenders, and a pair of coach lamps. This conventional arrangement perched atop an unconventional gear consisting of a tubular metal frame riding on front and rear elliptic springs. The gear rolled on wire-spoked, thirty-inch wheels equipped with double-tubed pneumatic tires ("a complete tire repairing outfit is furnished with each vehicle"), and the operator turned the front pair with a left-hand mounted vertical tiller. A foot lever activated rear-axle drum brakes, and an auxiliary drum rode on the end of the motor's armature shaft.[33]

The whole ensemble took its motion from a low-mounted, melon-sized electric motor linked by chain to the rear axle sprocket. By virtue of a "controller" mounted beneath the seat the operator could choose four speeds, the maximum thirteen miles per hour. The motor took its power from a pair of twelve-cell batteries ("of the latest and most durable type") housed in the body's rear, and the company estimated that a charge would suffice for about forty miles of average streets and grades. The purchaser could recharge the batteries from any 110 volt direct current outlet ("It is well known that such direct-current circuits are available and convenient in practically every city and town having an electric-lighting system"), as well as those as low as seventy volts. Its light appearance masked an astonishing weight of 1,300 pounds, nearly 600 of which rested in the lead-acid batteries. Still, this was the state of the art for electric automobiles in 1902, and the firm had gone to great lengths to turn out a product worthy of the Studebaker name.[34]

Not that Studebaker produced the entire product, for the motor and the majority of the drive train came from a variety of outside firms. Like most early automobile pioneers and the horse-drawn vehicle manufacturers before them, the South Bend giant understood the utility of ready-made parts. Studebaker's claims for its newest product came from a rigorous testing program. As the three thousand or so Studebaker factory hands worked at their accustomed tasks, they perhaps noticed the odd horseless carriage rumbling overhead on specially constructed wooden tracks. A 1902 publicity release announced that in the course of this testing "a chain has never been broken, and the same is true of the frame and all parts of the driving mechanism." The late Clement Studebaker's concern for reliability remained very much a priority. The company clearly intended its offering to appeal to the widest possible market segment, noting its suitability for pleasure and business (trap and phaeton bodies were options) and that it had deliberately kept down the cost of purchase and maintenance. Missourian F. W. Blees purchased the first on February 12, 1902, but the firm scored a major publicity triumph when it sold the second to Thomas A. Edison himself.[35]

Edison loved his Studebaker electric, but it was not his first automobile. The inventor counted Walter Baker among his close friends, and in 1899 he purchased the Baker Motor Vehicle Company's very first product, a sprightly three-quarter horsepower electric buggy. Not surprisingly, the apostle of electricity predicted a tremendous future for electrically propelled vehicles, and his opinion carried weight. So too did the electric itself, whose own power source formed its heaviest component and single greatest drawback. Edison responded with characteristic flair when telling a colleague "I don't think Nature would be so unkind as to withhold the secret of a *good* storage battery if a real earnest hunt for it is made. I'm going to hunt." By the time he became a Studebaker owner, Edison had already embarked on a campaign to improve the lead-acid storage battery, a task that would occupy him for several years. In 1907 J. M. Studebaker announced that Edison, "a friend of the family," was perfecting an electric battery that would place the electric far ahead of its steam- and gasoline-propelled competition. While his experiments eventually yielded significant improvements, even the Wizard of Menlo Park could not completely overcome the battery's excessive weight and frustratingly brief discharge cycle.[36]

Perhaps Edison's endorsement helped move the eighteen other Studebaker electrics sold by the end of 1902, but these were modest sales for a firm accustomed to five-digit annual production figures. While J. M. knew intimately the lengthy process and delayed rewards of building up a new business branch, it surely made him nervous. The low sales figures reflected not only limited production but limited demand for what remained an expensive novelty. Studebaker continued to make refinements, published new specifications, and engaged in the public endurance trials which had become a rite of passage. Besides pursuing foreign sales via its extensive branch house and sales agency network, the firm gradually added a Victoria as well as a truck and light delivery wagon body. The commercial bodies were designed to appeal to urban merchants and, despite their continued reliance on horse-drawn units, some gave the electric a try. Still, the core of electric vehicle sales remained primarily urban, upper-class, and female. Particularly when the alternative involved a hot boiler or vigorous hand-cranking, women preferred the electric's quiet convenience. "Electrics are used by ladies," J. M. stated in 1907, "and it is not advisable they should have a speed over 18 mph." Whether this reflected the ladies' own wishes or J. M.'s opinion of their driving abilities, he clearly recognized his market. This accounted for the elegant styles like the Victoria, whose gracefully sweeping body really did resemble that of the carriage of the same name. *The Carriage Monthly* ran a photograph of one with J. M. Studebaker at the tiller, his gray-bearded demeanor lending credibility to his company's newest departure.[37]

This photograph, featured in a 1908 issue, belied the fact that Studebaker's automotive approach had already become two-pronged. When Fish began pushing for a Studebaker automobile no one could reliably predict which propulsion system would become standard, or even the most popular. Studebaker was not the only firm to grasp this uncertainty, but unlike so many of its competitors, it had the resources to pursue manufacturing on both tracks. Equally important, the firm's leadership wisely enlisted outside help. Acutely aware of his lack of precision metalworking experience, Fish entered into a cooperative arrangement with an established manufacturer. At some point he became acquainted with Elyria, Ohio's Garford Manufacturing Company. Arthur L. Garford had garnered years of business and banking experience when he became interested in bicycles in the 1890s, and this led to his 1901 establishment of Elyria's Federal Manufacturing Company. A large auto parts producer with plants in several major cities, this became the Garford Manufacturing Company in 1905. During 1903–4 Fish engaged in serious negotiations with Garford, who was already producing automobile chassis for several other firms. In 1904 the two men reached an agreement whereby Studebaker would buy into Garford's firm, which would produce chassis (frame and drive train, including engines) for Studebaker, then complete them with its own bodies. To better accommodate this branch of the business, and perhaps to shield the much larger wagonmaking core, Fish created the Studebaker Automobile Company in 1904, the same year the first Studebaker-Garford came on the market. A sixteen-horsepower, two-cylinder gasoline automobile with an open body, it sold in July to a customer in South Bend.[38]

During 1904–11 Studebaker received shipments of Garford chassis and motors and added its own bodies to produce passenger cars selling as either Studebaker-Garfords or Studebakers. The firms constantly revised their joint product, starting out with two-cylinder models, adding a four-cylinder in 1905, and switching to the latter exclusively by 1906. These models E-20, F-28, and G-30, ranged from $2,600 to $4,000, typical for the product but still far outside Fish's intended price range. They sold moderately well (President Theodore Roosevelt purchased one in 1907); the best available figures indicate just under 2,500 marketed by 1911. Yet Fish hoped for something far greater. In 1906 a Detroit automobile manufacturer named Henry Ford articulated what many industry leaders were beginning to grasp; the greatest possible market existed for a low-priced gasoline automobile "powerful enough for American roads and capable of carrying its passengers anywhere that a horse-drawn vehicle will go without the driver being afraid of ruining his car." Fish wanted to harness Studebaker's wagon mass production success to automobile production, and he continued to explore any route promising to take him there.[39]

His ambition led to yet another joint venture, this time with a brand-new Detroit firm called the Everitt-Metzger-Flanders Company. E-M-F, as it quickly became known, took its name from founders Barney F. Everitt, William Metzger, and Walter E. Flanders (rather than "Every Mechanical Failure" or another quip supplied by general store philosophers). Everitt was a Detroit body and top maker and Metzger a leading Cadillac sales manager, but Flanders shone as a skilled and highly innovative manufacturer. A Vermont machinist steeped in New England armory practice, Flanders spent 1906–8 at Ford, where he passed on the importance of uniformity, interchangeable parts, efficient process sequencing, and several additional concepts that became central to Fordist mass production. Fish may or may not have been aware of all this, but when Flanders began forming his own automobile company in 1908, Fish surely sensed great promise.[40]

Fish's interest in E-M-F derived in part from knowing the newcomer intended to produce a medium-priced gasoline automobile, a goal exactly matching his own. Incorporated in August, 1908, E-M-F had actually come into being via acquisition and combination of several previously established firms, mostly from Detroit. In the subsequent negotiations, the firms' respective strengths proved mutually attractive: E-M-F possessed precision manufacturing ability while Studebaker offered unparalleled financial backing and an extensive dealer network. The talks culminated in a 1908 agreement whereby E-M-F would produce the chassis and Studebaker the bodies, but sharp disputes erupted over the automobile's name. Both firms sought to become full-fledged automakers in their own right. E-M-F's recent successes only raised the stakes: from the time it commenced production in 1908 until the end of 1910, its two plants and four thousand employees turned out more than 23,000 automobiles. These were serious production figures, and they only whetted Fish's appetite. Legal sparring soon gave way beneath the weight of Studebaker's advantageous stock holdings, and Fish finally persuaded Flanders to cut a deal. At the end of 1910, E-M-F became property of the Studebaker Automobile Company, and Flanders became Studebaker's vice president and a director. Everitt and Metzger opted to return to Detroit. Flanders would remain with Studebaker for only a short time, but while there he made important contributions to the firm's automotive success. Sales of the E-M-F Model 30, which continued after the merger, helped Studebaker break into volume auto production for the first time.[41]

By 1910 Studebaker had reached a historic turning point. Fish and other like-minded managers had worked tirelessly for nearly a decade to maneuver the company into the best possible position to enter the automobile industry. The E-M-F purchase capped off Fish's longstanding policy of strategic acquisition, which would continue in the years ahead. It had also created an unwieldy conglomeration of large and small companies grouped around the venerable Studebaker Brothers Manufacturing Company. Fish's corporate law and finance experience likely suggested the next step, a major reorganization that finalized the transition from nineteenth-century wagonmaker to twentieth-century automobile manufacturer. In 1910 Fish put the finishing touches on the paperwork that combined the Studebaker Brothers Manufacturing Company, the Studebaker Automobile Company, and E-M-F into the Studebaker Corporation. This promised to supercharge production of what appeared to be a winning product capable of successfully competing in the American automobile market.[42]

The reorganization mandated divestiture of marginal interests such as Garford and other remnants of forays into the higher price brackets, including terminating the Studebaker electric. While still in demand for city use, the market for these had never been large, and the batteries remained a vexing weak point. Continuing improvement of gasoline automobiles and the advent of a practical electric starter (all Studebakers came so equipped by 1913), removed most of consumers' remaining objections, sealing the fate of competing systems. The final Studebaker electrics came out in 1912, capping off a ten-year, 1,841 unit production run. Gasoline automobile production had already far outstripped those figures, with 2,481 built by this time, exclusive of E-M-F's output. Liquidation of these various assets proved costly, but not so dear as continuing to back a technology with no future. Fish wanted Studebaker to produce an affordable, gasoline-powered automobile with easily replaced parts, an automobile that would occupy a favored place in consumers' sheds, barns, and garages. The sometimes tortuous maneuvers of the last several years already showed signs of paying off: by 1913 the best available production figures indicated that Detroit's Ford Motor Company led all others in American automobile production. In second place, and significantly ahead of any runners up, was the Studebaker Corporation.[43]

If Frederick Fish provided the necessary driving force and managerial acumen, and E-M-F the manufacturing ability, funding formed an equally important reason for Studebaker's successful transition. When considering this period in the firm's history, it is easy to forget that Studebaker remained primarily a horse-drawn vehicle manufacturer all the while. In 1904, the year it introduced its first gasoline automobile, Studebaker's production totaled some 100,000 vehicles, the overwhelming majority of them horse-drawn. In 1910, on the eve of a major reorganization, the firm still produced approximately that many wagons and carriages, in addition to several thousand automobiles. Obviously horse-drawn vehicle production had ceased to increase as formerly, but it remained a company mainstay and provided ample funds for Fish's automotive game plan. Studebaker rode the wagon, as it were, until the very last minute. When the company finally ceased all horse-drawn vehicle production in 1920 by selling its wagonmaking operation to a Kentucky wagon firm, it was essentially liquidating a production branch with no future. Yet Studebaker's extensive and still-expanding branch house and dealer network made a natural outlet for Studebaker automobiles. At a time when the automobile market was flooded with a confusing variety of products by dozens of obscure small firms, Studebaker's long years of creative advertising, aggressive market extension, and assiduous sales network creation gave it a market presence the envy of its automotive competition. The firm lacked precision manufacturing experience, but it also had sufficient funds to purchase that expertise and the factories that best applied them. Studebaker's successful rebirth as an automaker owed less to any ability instilled by wagon building than it did to the enormous corporate structure this humble product had allowed it to erect.[44]

William Brewster perhaps harbored similar ambitions, but his company's finances proved unequal to the task. Production of "The Brewster" commenced in 1916, and by the end of 1917 the firm had expanded available body styles to include town, converted, and glass-quarter broughams; town, falling front, and touring landaulets; three kinds of runabouts; two kinds of limousines; and an open phaeton. On the surface this sounded like the old days: high-class vehicle production in all its varietal glory. Yet none of these several body styles accounted for large production figures, and they all rode on the same unchanging chassis. By the early 1920s the combination of static automotive technology and small output of an expensive product contributed to the firm's mounting indebtedness, which by 1925 threatened its very existence. The last Brewster automobile left the plant that year, and the story might have ended there were it not for another company's intervention. Headquartered in Springfield, Massachusetts, the American division of Rolls-Royce (established 1919) was very interested in acquiring the venerable carriage firm, and in 1926 Brewster & Company became the body division of Rolls-Royce America, Incorporated. Yet the increasingly streamlined body-fabrication process, including innovations such as a production line, conflicted with William Brewster's ideas about quality, and he resigned in 1927. Less than ten years later Rolls-Royce America would itself be gone, and with it the last tangible link to the venerable Brewster & Company.[45]

Carriages Equal Automobiles, and Other False Analogies

What, exactly, accounted for the failure of so many other wagon and carriage makers in the automobile industry? One explanation, which seems so logical in retrospect, was adoption of the "wrong" technology. While highly specialized electric vehicles ended up having important industrial uses, the electric passenger automobile turned out to be a technological dead end, and firms manufacturing them found it extremely difficult to switch to internal combustion. Although technological obsolescence certainly played a part in the demise of firms like McIntyre, DeMars, Broc, and dozens of others, overemphasis on "right" and "wrong" technologies masks a more fundamental explanation. Most wagon and carriage makers failed as auto manufacturers because they lacked the necessary metalworking expertise. This at first seems unlikely, for one might logically assume that carriage makers would make good automakers; indeed, this was a widespread assumption among wagon and carriage makers themselves. One carriage firm insisted that automobiles were rightly the carriage makers' domain because their power source "does not alter the fact of its being a carriage whether the motor is gasoline, steam or electric." "It seemed to be the natural trend of manufacture and business," another stated, "that the men who made horse-drawn vehicles should evolve directly into the manufacture and selling of the natural successor of the horse-drawn vehicle, the automobile." The logic of this inference rests squarely on the false assumption of an underlying similarity between the two products. Unlike horse-drawn vehicles, automobiles were primarily metal objects whose successful manufacture required precision metalworking. A good deal of metal went into a typical horse-drawn vehicle, but the object itself remained fundamentally wooden. Blacksmiths, woodworkers, painters, and trimmers produced horse-drawn vehicles; engineers and machinists built automobiles.[46]

Even the evolution of automotive nomenclature suggests as much. To whatever degree the 1890s horseless carriage resembled its name, reality emphasized the horseless portion of the appellation. As a vehicle requiring no animate power—one that was *automotive*—it really was more than simply a cart without a horse. The progenitors of these early hybrids soon discovered that wagon and carriage technology was unequal to these new demands. Willowy patent wheels shivered and split from increased speed and torque; nails and screws lost their grip to vibration; springs and axles sagged and broke; and the entire assemblage of once-adequate parts failed repeatedly under familiar forces accelerated to levels no horse ever generated. In some instances the remedy lay in making the part a little more robust, a solution taken in many "buggy type" automobiles. In more cases, however, it required wholesale recasting of the vehicle's architecture. Carriages and automobiles each had wheels, gears (or chassis, as the French called them), and bodies, but they increasingly had little else in common beyond general function and name.

How then to account for Studebaker? While that firm had adopted some advanced metal manufacturing technology in drop forging and electric resistance welding, it nevertheless could hardly be considered a machine shop. Studebaker's successful transition flowed from wise acquisition of firms with the knowledge and manufacturing experience the parent company lacked. Realizing the wisdom of such a plan was not the same as the ability to implement it, yet Studebaker nevertheless found itself financially capable of acquiring the best of those firms and individuals even before it became apparent which propulsion system held the greater promise. While many smaller firms had to make a choice and stick with it, Studebaker could afford to bankroll investments in electrics and gasoline automobiles until one pulled ahead of the other. Sufficient horse-drawn vehicle demand continued to exist to reward the company with the profits necessary to write off repeated losses and explore fresh options. By the time Studebaker reorganized, it had already overcome many of the hurdles that had or would soon defeat most other wagon and carriage manufacturers striving for the same goal. A fundamental technological disparity lay at the heart of the matter, but Studebaker had the rare asset of enormous financial resources that permitted it to overcome even that.

If Studebaker was an anomaly, and the wagon and carriage industry proved incapable of wholesale rebirth, there was another transportation industry that did act as something of a midwife to the automobile. By virtue of both the materials and their manipulation, bicycle manufacturing made enormous contributions to practical automotive design. Tubular steel, chain-and-sprocket drive components, metal rims, steel spokes, and pneumatic tires combined with electric welding, gear cutting, and drop forging became as important to the automobile as they were to the bicycle. It was certainly no accident that new transportation developments came from the tinkering of scattered bicycle makers. Whether the Wright Brothers or Colonel Albert A. Pope, bicycle makers and their technology were at the right place at the right time. Henry Duncan, one of J. B. Brewster & Company's final owners, certainly thought so. He felt that the bicycle industry had been "ready-made" for early and effective entry into automobile manufacturing. "The metal construction of the bicycle brought it more closely in touch with the all metal automobile," and this did not require a great structural change to the factory or its manufacturing methods. "In the case of the carriage," he said, "it was a transition from an all wood construction to an all metal one," and to him this had made all the difference in the world.[47]

The same logic explains the accessory industry's much greater success in accommodating the automobile. Many wagon and carriage parts manufacturers were already mass-producing huge quantities of uniform metal goods; drop forges could shape automotive steering knuckles as well as wagon box braces; punch presses could stamp auto body panels as easily as bow socket blanks; factories and workmen accustomed to precision, high-volume metalwork could take on motor vehicle parts as well as they had those intended for horse-drawn vehicles. The tolerances might be more exacting, but then many of them had already encountered the daunting demands of uniformity in their creation of "bolt on" wagon and carriage parts. So when early automobile manufacturers sought to outsource some of the components required to build a motor vehicle, they had only to contact any of the hundreds of well-established wagon and carriage hardware firms to obtain what they required. Like astute manufacturers everywhere, the accessory industry quickly accommodated these new customers by turning more and more of their plants over to auto parts production (fig. 8.2).

This trend had become evident by the end of the nineteenth century, but it soon became almost universal. The venerable Dalzell Axle Company offered its "Motor Vehicle Guiding Axle" along with its familiar carriage axles, while Henry Timken applied his versatile tapered roller bearings to nearly anything that rolled. Virtually all the big spring manufacturers produced automotive springs and suspension components, likewise wheel makers who found that wooden spokes did not necessarily disqualify a wheel from automotive use. Hardware firms like Eberhard and Cleveland Hardware turned their formidable array of equipment to the automotive items that rapidly became their primary products. Towing hooks, slam latches, door hinges, spare tire hangers, and all the other impedimenta of a new industry soon mandated separate catalogs, while those devoted to wagon and carriage goods gradually disappeared. Even producers of more ancillary products like carriage lamps and sunshades turned to the automobile. New Haven's C. Cowles & Company added automotive lighting to its line, and Norwalk, Ohio's Sprague Umbrella Company began turning out folding tops for the Model T Ford. Soon this great gout of auto parts spilled over into the catalogs of hardware retailers like Cray Brothers in Cleveland; Pittsburgh's Love, Thompson Company; and many others. Even the Chicago general merchandise mail order houses joined in; though both Sears & Roebuck and Montgomery Ward continued carrying wagon and carriage items into the early twentieth century, they devoted diminishing space to a line clearly on the wane.[48]

Internal combustion's dominance drove the majority of wagon and carriage makers to abandon their attempts to manufacture complete automobiles. The early horseless carriages fell from favor in the face of increasingly powerful and reliable motorcars whose very design, materials, and construction had already outpaced transitional "buggy type automobiles." If most carriage makers found themselves outside the ranks of automobile manufacturers, however, many found a niche as custom body builders. Though automotive running gears mandated metal, body construction still relied heavily on wood. Taking advantage of their long-standing wooden fabrication abilities, many horse-drawn vehicle builders began producing auto bodies. This was certainly a more tenable proposition, and many early automobile manufacturers contracted body work to horse-drawn vehicle makers.[49]

Some of the latter specialized in exquisitely crafted "coach-built" bodies mated to expensive American and foreign chassis. Such custom coach work remained viable through the 1920s and a little beyond, a period one automotive historian has dubbed the "Olympian Age" of the luxury automobile, even if it meant body-making firms assumed subservient roles to a handful of exclusive automobile manufacturers. Even here, however, more of the upscale automobile companies eventually chose vertical integration, gradually eroding the independent body maker's opportunities. One alternative lay in catering to automotive mass producers like Ford Motor Company or General Motors, precisely the route taken by one group of former carriage makers. Sandusky, Ohio's Frederick J. Fisher came from a large carriage-making family whose trade pedigree reached back to Germany. In 1902 Frederick secured a drafting job with Detroit's C. R. Wilson Carriage Works. Five years later the young man had worked his way up to superintendent and had been joined by brothers Charles, Alfred, Lawrence, William, and Edward. Wilson had since added auto bodies to its output, but in 1908 Frederick and Charles formed their own firm specifically for auto body manufacture. Fisher Body Company began as a custom auto body shop, but the brothers wisely sought larger orders, such as the 150 bodies they built for Cadillac in 1910. After numerous expansions they combined their various branches into the Fisher Body Corporation in 1916. With an annual production capacity of around 370,000 units, it was the nation's largest auto body company.[50]

Mirroring their dedication to traditional wooden craftsmanship and to classic horse-drawn vehicle manufacture, the brothers chose an elaborate Napoleonic coach as the company trademark. Conceived from the beginning as a firm dedicated to meeting the needs of automobile mass producers, Fisher Body embraced a streamlined, standardized production format differing sharply from the custom and batch practices of firms like Brewster & Company. Fisher developed close ties with automotive giant General Motors, which bought a significant share in Fisher immediately after the World War and contracted for nearly all the latter's output. Brewster family lore has it that General Motors initially approached the old New York firm, but William Brewster, fearing a drastic deviation from his firm's traditional quality, turned them down. Though he may not have realized the full consequences of his decision at the time, doing so tied his company to a lucrative but diminishing market, virtually guaranteeing its eventual demise. By contrast, Fisher Body expanded beyond all expectation, and while it eventually became a division of General Motors (1926), the brothers remained intimately involved with the company, enormously successful as auto body manufacturers, and particularly loyal to each another.[51]

Fisher Body survived not just by its willingness to strike a reasonable balance between quantity and quality but because it also followed the automobile industry away from wooden body construction. The shift from open touring cars to the closed, hard-topped bodies that eventually became standard only accelerated this trend. Fisher's 1910 Cadillac contract had been for closed bodies, and this had been what prompted formation of the Fisher Closed Body Company in 1912, one of the several firms which became Fisher Body Corporation. Though the earliest closed bodies still contained a great deal of wood, particularly in their framing, improvements in metal stamping combined with speed of manufacture finally eliminated wood entirely. All-metal closed designs were watertight, simpler to construct, cheaper to produce, and often functionally superior to the composite bodies of even the finest body shops. Whereas a few of the larger firms like Fisher made the wood-to-metal transition, most lacked the capital and expertise to acquire and operate expensive and sophisticated stamping machinery. Provided they did not find some other wooden commodity to manufacture, such firms could anticipate little more than declining orders and dim futures. By the 1930s the independent auto body maker was largely a thing of the past.[52]

Wood held on longer in motor trucks, where its insulating and impact-absorption qualities gave it an advantage over other materials. Perhaps it had something to do with the familiar raw material, or maybe the more utilitarian nature of their product, but wagonmakers appeared more willing than carriage makers to adapt to internal combustion. Many became heavily involved not only in motor truck body fabrication, but also in the new semi-trailer field, itself a direct offshoot of wagon manufacture. Wood framing and panels remained common in semi-trailers for decades, and some former wagon firms like Freuhauf became nationally recognized trailer manufacturers. Custom commercial body work opened whole new vistas for those willing to pursue it. Commercial vehicles were often available as "rolling chassis" for purchasers requiring customized bodies. Heavy automobiles lent themselves to hearse and ambulance use, and at least one former carriage maker, Cincinnati's Sayers & Scovill, thrived by putting such bodies on

TABLE 8.1. Rise of Automobile Industry, 1904–1929

	Number of Firms	Number Employed	Wages ($)	Cost of Materials ($)	Value of Products ($)
1904	121	10,239	6,178,950	11,658,138	26,645,064
1909	265	51,294	33,180,474	107,731,446	193,823,108
1914	300	79,307	66,934,359	292,597,565	503,230,137
1919	315	210,559	312,165,870	1,578,651,574	2,387,903,287
1921	385	143,658	221,973,586	1,107,062,085	1,671,386,976
1923	351	241,356	406,730,278	2,147,463,352	3,163,327,874
1925	297	197,728	341,210,401	2,108,191,812	3,198,122,633
1927	264	187,910	321,664,093	1,889,426,249	2,848,442,843
1929	244	226,116	366,579,233	2,401,511,763	3,722,793,274

Source: Adapted from U.S. Bureau of the Census, "Motor Vehicles," *Census Reports: Fifteenth Census of the United States, 1930* (Washington, D.C., 1933), 2:1223.

heavy limousine chassis. Light and heavy motor trucks had even more commercial applications, including ambulance and police "paddy wagon" duty, fire fighting, passenger and school bus service, and even "house cars," recreational vehicle forerunners that by the 1920s flowed from such unexpected quarters as the vehicle department of St. Louis brewer Anheuser-Busch. Many towns and cities had custom truck body shops like Cleveland's Schaefer Body, which began as a wagon shop in 1880 and lasted as a truck and trailer body maker through the 1970s. Some continue today, like Richmond, Indiana's Wayne Works, a leading manufacture of the ubiquitous yellow school bus.[53]

Many of these trends were well under way when the Carriage Builders National Association held its forty-third annual convention in 1915. Twenty years had passed since it first met in Cleveland, and a great deal had happened in the meantime. A year before the 1895 convention Cleveland's population stood at approximately 368,000, but by 1915 it had almost doubled. In 1895 auto making had been a minor cipher in the statistics of American manufacturing; by 1915 it had become a leading industry. By the end of that year, American auto makers' production of just under a million automobiles had nearly doubled the previous year's output (Table 8.1).[54]

There was some continuity between 1895 and 1915; many of the old local arrangements committee members were still on duty, and the convention once again headquartered in the Hollenden, still one of Cleveland's finest hotels. The business meetings were at the capacious Central Armory, where the exhibit area, now reduced to around fifty firms, remained the most popular gathering place. While the usual items graced the booths, automobile products sat alongside horse-drawn vehicle hardware, many of them made by firms once wholly dedicated to wagon and carriage parts. It was in the Central Armory that association president Charles O. Wrenn officially opened the convention on Tuesday, September 21. A carriage maker from Norfolk, Virginia, Wrenn came from a family of horse-drawn vehicle manufacturers who supplemented their trade with an automobile dealership. A man of considerable size (one wonders of his reaction to editorial commentary insisting "everybody loves a fat man"), Wrenn's bulk at least added gravitas to his proclamations. Acknowledging the preceding year's exceedingly poor business, he expressed pleasure that so many members had weathered the storm. He certainly did not gloss things over, relaying the obvious conclusion that the automobile had killed the high-class carriage trade. His own foray into auto sales had been motivated in part by this very development. While helpful, the dealership had been a major headache causing him "trouble, trouble all the time." Still, he continued to keep the faith: "I contend that the class of persons buying the medium-priced buggy is not a prospective automobile buyer, and will not be for many years to come," he told the assembly. Wrenn urged perseverance and closed with a ringing validation of the CBNA as the best affiliation for the remaining horse-drawn vehicles makers.[55]

The following two days featured the usual variety, including talks on business trends and the state of the American horse industry, as well as reports on vehicle warranties, standardization, and membership. In some ways the news was not all bad: the executive committee reported that the association was in sound financial shape, the technical school flourishing, and the federal Bureau of Animal Industry's G. Arthur Bell reported on dramatically increased horse production. The automobile was thriving beyond all expectations, however, and even the most stubborn horse-and-buggy advocate could not deny the obvious. Business was far from good, and a search for solutions consumed the greater part of the meetings. Many of the conventioneers placed their hopes on the American farmer. Bell emphasized the horse's continued agricultural service, which in turn continued to provide a market for wagons and buggies. Charles Adams told his audience that he was still clinging to a maxim relayed to him in the 1890s: when the farmers had good crops it meant good business for the wagon and carriage industry. He cited the current boom in farm production as evidence that things would soon change for the better. It was not that simple, and many of the conventioneers knew it. Even though most realized farmers still required wagons and carriages, improved rural roads and lower priced automobiles meant that the automobile would make inroads even here. Adams admitted as much when he closed by predicting that carriage makers would eventually "do some other things besides making carriages. I think they will make other things, and think they will be doing business at the old stand for many years to come." From a strictly moneymaking point of view this was perhaps reassuring, but wagon and carriage makers who loved their craft found little comfort in Adam's prophecy.[56]

Horse-drawn vehicles simply did not sell like they used to. Statistics committee chairman O. B. Bannister summed up the problem and his view of the solution when he stated that the buggy was now "a poor seller and calls for greater effort—but the facts are, as we see them, is that less effort is being made to sell than heretofore." In other words, once-guaranteed horse-drawn vehicle sales were still possible, but too many dealers exerted insufficient effort. One southern member complained that carriage makers "instead of getting out and hustling for business . . . have simply laid down and allowed the auto industry to jump all over them." The solution, he felt sure, lay in vastly intensified salesmanship, and many of his fellow conventioneers agreed. Another relayed a recent conversation with one of his traveling representatives regarding fewer salesmen afield: as they saw it, a dealer who had not been constantly "drummed for business" tended to suspect diminished demand and became reluctant to make any purchases at all. He added that such situations called for a more active and aggressive salesman, "a man that is able to go to the dealer and convince him that he is losing out by not keeping buggies on his floor." Over and over again the various speakers returned to this conviction of a latent demand calling for only a little more persistent and perceptive salesmanship.[57]

Such dark rumination gave way on day three to elections, expressions of thanks to their host city, and a few other internal matters before business adjourned and the banquet could begin. Though the meal was certainly a fine affair, compared to 1895 it must have seemed almost austere. Diminished attendance made it possible to hold the event in the Hollenden's own dining facilities, and it no longer featured a long series of speakers. There were, in fact, no speeches at all. This was also the first officially temperate banquet in the CBNA's long history, and heavy-handed jests about water formed a recurring theme in many conversations. If the conventioneers could not drink, however, they could at least dance, for many members now brought their wives to this hitherto predominantly male event. Spousal participation reflected not only loosening gender boundaries, but also the CBNA's increasingly social focus.[58]

If 1895 had been a time of hearty self-congratulation, 1915 was one of strict economy with uneasy overtones. As the attendees packed their bags and trunks, bought their last-minute souvenirs, and caught trains, or drove automobiles homeward, perhaps they reflected on some of the many factors contributing to this pall, whether the trials of coping with the automobile or the inexorable decline of horse-drawn vehicle sales. The heady optimism of the early-twentieth-century conventions had materially eroded. Back at the 1911 meeting, the members formally aired the subject of the future of their trade, but bleak speculation had not entirely obscured Atlantic City's charms, a venue so popular that the association voted overwhelmingly to meet there the following year, and after a break for appearances' sake in St. Louis, returned once more in 1914. These gatherings had all been well attended, but the 1915 Cleveland meeting turned out to be the last truly large CBNA convention. Just four years later the entire event, including the exhibition, could be held within the confines of a single hotel. Even the trade journal reportage began looking threadbare. Attenuated columns under headings like "Carriage Builders Optimistic Over Future" and "Use of Horse to Be Urged at Session" conveyed more, at least in retrospect, than their writers perhaps realized.[59]

Along with an emphasis of the efficacy of sharp salesmanship, speakers at the 1915 convention had also expressed an almost touching faith in advertising. American business increasingly stressed brand identification and national advertising, and some CBNA members looked to these to arrest declining sales. In 1914 the association had established a committee to promote the industry through newspaper and magazine publicity. At Cleveland it proudly reported it had gathered over a thousand newspaper clippings that, added with misplaced enthusiasm, "if pasted end to end would make a strip 7,400 inches long." Though pleased at its accomplishments, the committee sought more funding to ensure that "people will never have a chance to forget that the carriage is still on top of the earth." For decades a sympathetic and eminently helpful ally, the trade press actively assisted. Not long after the 1915 meeting the journals ran reproductions of the committee's newest effort: a series of four-color posters designed to "give favorable publicity to the horse and all his accessories." The first, titled "The Pleasure of the Road," showed a young swain and his lady out for a buggy ride. "Happiness Supreme," perhaps an unintentional sequel, portrayed a family outing in a fringed-top surrey. The posters were available in quantity at low cost, and as the accompanying article noted, "a few hundred put up in public places in the dealer's territory will turn the attention of passersby to the sterling worth of the horse-drawn vehicle." All people had to do was look at them.[60]

The CBNA had no monopoly on this faith in promotion. In 1914 a group of prominent eastern horse enthusiasts formed the National Association of Allied Horse Interests (NAAHI). Headed by New York City horse-racing giant August Belmont, the organization pledged itself to "boost the horse in a legitimate manner whenever and wherever it can be done" to prove the passing of the horse was only a "motor myth." Another group, the Horse Association of America (HAA), also actively promoted what had once been an unquestioned part of everyday life. One of its many efforts was a leaflet from the late 1920s, intended to persuade the reader on purely economic grounds. "Horse Supreme on the Milk Route" attempted to show that horse-drawn wagons remained ideal for delivery use. They had a point. Horses remained widely used into the 1920s for such purposes for entirely practical reasons. Large dairies already owned considerable stables and fleets of delivery vehicles, and delivery men enjoyed the convenience of horses that would move slowly down the streets on their own. No motor truck could boast as much, and the Horse Association of America wanted this fact to extend beyond the ranks of knowing milkmen (fig. 8.3).[61]

Regardless of the specific content of such efforts and the degree to which they accorded with reality, their very existence richly reflects the tenor of the age. Taken as a whole they indicate not some quirk peculiar to wagon and carriage makers but a widespread and particularly American belief in reason and persuasion in the guise of vigorous salesmanship and clever advertisement. It was as if one might redeem an entire faltering industry by virtue of honest, unremitting toil. In an age when the notion of progress was a widely shared faith, such sentiments indeed seemed eminently reasonable. Wagon and carriage makers still believed in their product, and many of them honestly felt that the buying public would do the same if properly approached. Yet the buying public had been buying automobiles whose speed rendered four-color promotional posters little more than polychrome blurs. Despite the best efforts of the CBNA, NAAHI, HAA, and the others, it had become distressingly apparent by the 1910s that horse-drawn vehicles were an endangered species. High-grade automobiles and cabs continued to erode the urban markets they had first stormed during the century's first decade, and now the Model T and its many imitators spread steadily out over the rural hinterland. Even urban delivery had started to succumb. In 1914 a large New England brewer confessed that he would "willingly consign his trucks to unmentionable regions if others would do the same." Others were not doing the same, however, and the brewer would soon be followed by the baker, the dairyman, and others in discarding the wagon. Members of the horse-drawn vehicle industry likely could not help but feel they were witnessing a veritable revolution in how Americans moved from place to place.[62]

Even the trade press reflected this transformation. In an effort to straddle two industries *The Carriage Monthly* changed its name to *The Vehicle Monthly* in 1916, but by 1921 bowed to the inevitable when it became *The Motor Vehicle Monthly*. *The Hub* did not even bother with the evolutionary approach, simply becoming *The Automobile Manufacturer* in 1919. After the 1910s, *The Spokesman* (consolidated with *Harness World* in 1924) carried the largest amount of horse and horse-drawn vehicle news. Based in Cincinnati and close to the horse-loving South, it circulated among those who remained involved with all things equine. But titles such as "City Horse Is Holding His Own against the Motor Car," "What Would Happen in This Country If Horses Were Eliminated (Famine would ensue, multitudes would perish and scale of living would be like China's, noted authority states)," and "Buggy and Wagon Industry Goes On Despite Increase of Autos" reflected an increasingly unrealistic tone.

The authors of these pieces were simply telling their readers what they thought they wanted to hear, for the industry that was "going on" was rapidly diminishing. In 1914 nearly three thousand firms nationwide produced horse-drawn vehicles and the materials for their manufacture, but by the end of the 1910s the number had dropped to about 1,500. At the start of the 1920s it dipped below a thousand, and 1923 saw it fall to less than five hundred. Though the remaining firms often experienced momentary demand spikes from loss of competitors, confidence in these surges was misplaced. While in 1923 wagon and carriage manufacturers and the accessory industry reported a production increase of over 100 percent above economically depressed 1921, their 236,091 vehicles paled in comparison to figures for the automobile industry. In 1923 the United States turned out over 3 million automobiles (Table 8.2).[63]

Just a few years before, the automobile had seemed more a minor annoyance than a harbinger of doom, and the early mating of internal combustion to wheeled road transport appeared to warrant the dismissal it elicited from most wagon and carriage makers. As the automobile be-

TABLE 8.2. Decline of Wagon and Carriage Industry, 1904–1929

	Number of Firms	Number Employed	Wages ($)	Cost of Materials ($)	Value of Products ($)
1904	4,982	61,306	31,174,188	61,709,541	126,510,756
1909	4,894	53,204	29,942,370	64,605,299	126,992,390
1914	4,622	41,627	26,697,287	52,602,002	107,639,309
1919	2,297	18,464	19,664,837	49,333,552	93,142,763
1921	837	8,025	9,903,327	16,883,611	33,113,372
1923[a]	396	8,109	9,318,763	18,682,274	37,372,825
1925	152	4,833	5,791,729	12,859,793	24,934,814
1927	117	3,387	4,211,610	10,648,271	19,422,235
1929	88	2,873	3,417,963	8,458,294	16,449,879

Source: Adapted from U.S. Bureau of the Census, "Carriages, Wagons, Sleighs, and Sleds," *Census Reports: Fifteenth Census of the United States, 1930* (Washington, D.C., 1933), 2:1202.
[a]Figures for 1923 and later exclude repair shops, which in 1921 numbered a little over 300 of the 837 firms.

gan entrenching itself in American life, however, the trade was forced to respond. Many pursued the seemingly logical transition to automobile manufacture, but the two transportation technologies were separated by a technological gulf that most could not bridge. While a surprisingly large numbers of firms turned out "horseless carriages," few could ultimately keep pace with changing technology or economically produce a true automobile in saleable quantities. Exceptions such as Studebaker owed more to size, wealth, canny acquisitions, and a long-established marketing structure than to any technological similarity between the two industries. The demands of precision metalworking had more in common with bicycle making than it did with wagon and carriage building, which also accounts for the great success of wagon and carriage parts manufacturers in coping with the automobile.

Wagon and carriage makers experienced transitory success in auto body manufacture, but eventually even this failed in the face of vertical integration and the all-metal body. Ultimately only a few firms like Fisher breached the barriers between batch and mass production. As wagons and carriages assumed the role of quaint reminders of a horsy past, the once-mighty industry that built them dispersed into the making of automobile parts and accessories, took up the manufacture of entirely unrelated items, or simply paid their employees and closed their doors forever. The transition from carriages to automobiles was a messy, indistinct overlap of several decades, but to so many who lived through it, it seemed to happen so quickly. Then again, was this not in keeping with the automobile's very nature? In a sense, the whole encounter resembled another repeated on countless early-twentieth-century American back roads: appearing suddenly from behind, the roaring motor car overtakes the unwitting pedestrian, only to leave him in rueful wonder, blinking in a cloud of dust.

Epilogue

Hail and Farewell

Good weather greeted the men who gathered at Cincinnati's Grand Hotel at noon on September 15, 1926. The sunshine was particularly welcome given their plans for a large outdoor picnic. Colonel Tom Cody, owner of an extensive horse farm in nearby Kentucky, would be hosting the gathering. Only one of several ironies that day was the procession of automobiles lined up in front of the hotel to take them there. For these men, over a hundred in number, were all members of the Carriage Builders National Association, and this meeting, officially their fifty-fourth, would also be their last. It was no business convention. Besides the vast quantities of southern-cooked food, the main attraction was the opportunity for the few active and many retired wagon and carriage men to renew old friendships, find out who was still in business, who had changed occupations, retired, or, increasingly, died. Like old men everywhere, they reminisced. The majority of them could recall a time when wagon, carriage, and buggy were the kings of the road and when making them constituted a prestigious national industry. Times had changed, however. While new horse-drawn vehicles continued to be sold each year, they came from a drastically reduced number of firms. Occasional demand surges had less to do with an equine revival than it did with the elimination of competitors. Cincinnati, so long a leader in the American carriage trade, boasted only a single carriage factory in 1926.[1]

The trade everywhere had been declining for years. Even as CBNA conventioneers had strolled the boardwalks of Atlantic City in 1911 and 1912, during what might be described as the trade's Indian summer, the downward spiral was already apparent. As much as a decade before those meetings, the association had noted the dip in sales as the purchasers of expensive carriages eschewed broughams and opera buses for the new brass-bedecked automobiles. By 1914, when the CBNA met once more to romp in the Jersey surf and feed french-fried potatoes to the seagulls, Henry Ford had been making his Model T for six years—to the acclaim of former rockaway and buggy purchasers. That same year the U.S. Census Bureau estimated that a few more than five thousand wagon- and carriage-making firms remained; by 1919 the number had dropped to 2,666, and it continued to plummet. Three years later, an announcement for the upcoming CBNA convention admitted that despite its best efforts, the association could best be described as "moribund." Its technical school had lasted until 1918, succeeded by a correspondence course in auto body design, taught by Andrew Johnson, the school's longstanding principal.

The conventions barely hung on; the 1922 meeting featured the final accessory exhibit: thirty-four lonely exhibitors, down from a high of nearly 150 just thirty years earlier. The constrained financial condition of the surviving manufacturers prompted the convention's cancellation in 1924, and the association might have ceased then had it not been for the momentum of tradition and nostalgia. The 1925 convention was the last to feature any official business at all, and the press had already become accustomed to refer to the events as reunions. Those who gathered in Cincinnati in 1926 had no illusions; there were no business sessions because, as nearly all had finally admitted, there was no business. They met for social purposes, and their gaze was, as it had been for some time, fixed squarely on the past.[2]

It had certainly been a past shot through with momentous change. Entering the nineteenth century as a traditional handcraft, the wagon and carriage trade had grown into a sizeable industry, one whose heyday formed the topic of countless conversations that day in Kentucky. It had been a tumultuous transition, one beginning long before most of them were born. The forces of rationalization and mechanization spawned a productive revolution that split the trade into a pair of distinct but highly interdependent branches: wagon and carriage factories turned out horse-drawn vehicles while accessory factories produced their parts and accessories. Proprietors of each pursued mechanization to increase production, but machinery conferred other benefits, among them greater uniformity and even parts interchangeability. Special-purpose machinery contributed enormously to this achievement even as it allowed the accessory trade to turn out unprecedented numbers of wagon and carriage goods. Unlike so many other trades where the factory swept away the small shop, however, the industrialization of wagon and carriage manufacture actually gave the old shop a new lease on life. The interchangeable parts flowing from the large accessory factories permitted minimally capitalized small firms to successfully assemble and market wagons and carriages in competition with large mass producers.

Just as the factory failed to wipe out the shop, so too did industrialization fail to eradicate the skilled worker. Factories, machinery, and ready-made parts all diminished the traditional craftsman's role, but not all machines were inimical to skill, and while unskilled labor gained favor, there remained many tasks for those with craft training. From department foremen and factory supervisors to repairmen and custom builders, the tradesman remained in the trade even though it had now become an industry. Industrialization altered work and the workplace, but what finally brought the entire enterprise to a halt was a fundamental shift in transportation technology. The first sputtering horseless carriages marked the beginning of a series of technological developments that made the horse-drawn vehicle obsolete and turned "horse power" into an allegorical measurement of mechanical force. The change was not instantaneous, nor was it without setbacks and reversals, but it moved ahead steadily and in the end its results were sure: the automobile replaced the carriage, the motor truck replaced the wagon, the automobile industry replaced the wagon and carriage industry, and the automobile worker replaced the wagon and carriage craftsman. Perhaps some of this ran through the picnickers' minds as they listened to the familiar old songs sung by the Carriage Makers' Quartet. But as the group's mellow harmonizing gave way to a jazz band's syncopated tunes, even the most resolutely nostalgic were reminded that they were already over halfway through the roaring twenties. None needed to be told that the long reign of the horse and the horse-drawn vehicle was over for good.

Notes

Abbreviations

BLCB	Brewster lawsuit copybook
GCRL	Gerstenberg Carriage Reference Library, The Long Island Museum of American Art, History, & Carriages, Stony Brook, N.Y.
NHCHS	New Haven Colony Historical Society, New Haven, Conn.
NMAH	Division of Transportation, National Museum of American History, Smithsonian Institution, Washington, D.C.

Introduction. Remembering a Forgotten Industry

1. *Authorized Visitors' Guide to the Centennial Exhibition and Philadelphia, May 10th to November 10th, 1876* (Philadelphia: J. B. Lippincott & Co., 1876), 13, 16, 21, map of grounds; United States Centennial Commission, *Grounds and Buildings of the Centennial Exhibition, Philadelphia, 1876* (Philadelphia: J. B. Lippincott & Co., 1878), 30, 53–54, 97, 143–44; United States Centennial Commission, *Official Catalogue of the U.S. Centennial Exhibition, 1876, Complete in One Volume* (Philadelphia: John R. Nagle & Co., 1876), part 1, "catalog for annex to Main Building," 295–301; Ezra M. Stratton, *The World on Wheels; or, Carriages, with their Historical Associations from the Earliest to the Present Time* (New York, 1878; reprint, New York: Benjamin Blom, 1972), 475; "Report of the Centennial Exhibition," *The Hub* 18 (September 1876): 215.

2. *Authorized Visitors' Guide*, 7–8, 13–23, map of grounds.

3. Ibid., 13–15; *Official Catalogue*, 295–97; John F. Kasson, *Civilizing the Machine: Technology and Republican Values in America, 1776–1900* (New York: Grossman, 1976; reprint, New York: Penguin, 1977), 162–64; Alan I. Marcus and Howard P. Segal, *Technology in America: A Brief History* (New York: Harcourt Brace Jovanovich, 1989), 136–37.

4. Ben Riker, *Pony Wagon Town* (New York: Bobbs-Merrill, 1948), 24; Cleveland City Directory, 1891.

5. *G. & D. Cook & Co.'s Illustrated Catalogue of Carriages and Special Business Advertiser* (New Haven, 1860; reprint, New York: Dover Publications, 1970), 225; "Clement Studebaker," *The Hub* 21 (January 1880): 434.

6. Brooke Hindle and Steven Lubar, *Engines of Change: The American In-*

dustrial Revolution, 1790–1860 (Washington, D.C.: Smithsonian Institution Press, 1986), 213–15; Diana Appelbaum, "In Search of Ten-Footers," *Yankee* 59 (October 1995): 144.

7. Philip Scranton, *Proprietary Capitalism: The Textile Manufacture at Philadelphia, 1800–1885* (Cambridge: Cambridge University Press, 1983), xi, 3; Philip Scranton, "Diversity in Diversity: Flexible Production and American Industrialization, 1880–1930," *Business History Review* 65 (spring 1991): 28–31.

8. David A. Hounshell, *From the American System to Mass Production, 1800–1932: The Development of Manufacturing Technology in the United States* (Baltimore: Johns Hopkins University Press, 1984), 8–11, 122; Scranton, "Diversity in Diversity," 30–31; Philip Scranton, *Endless Novelty: Specialty Production and American Industrialization, 1865–1925* (Princeton: Princeton University Press, 1997), 10–11.

9. Scranton, *Endless Novelty,* 135; Joanne Abel Goldman, "From Carriage Shop to Carriage Factory: The Effect of Industrialization on the Process of Manufacturing Carriages," in *Nineteenth Century American Carriages: Their Manufacture, Decoration and Use* (Stony Brook, N.Y.: The Museums at Stony Brook, 1987), 9–33.

10. Advertisement for the Easy Wagon Gear Company, *The Hub* 29 (April 1887): 56; Don H. Berkebile, ed., *Carriage Terminology: An Historical Dictionary* (Washington, D.C.: Smithsonian Institution Press, 1978), s.vv. "Malleable Casting," "Malleable Iron"; advertisement for the Eberhard Manufacturing Company, *The Hub* 29 (March 1888): 886; *Industries of Cleveland: A Resume of the Mercantile and Manufacturing Progress of the Forest City* (Cleveland: Elstner, 1888), 104.

11. Isaac D. Ware, ed., *The Coach-Makers' Illustrated Hand-Book,* 2d ed. (Philadelphia, 1875; reprint, Mendham, N.J.: Astragal Press, 1995), 363; "Subdivision of Labor in Carriage-Building," *The Hub* 17 (July 1875): 122.

12. W. R. Crozier, "Patent Wheels and Hub-Mortising," *The Hub* 14 (October 1872): 159.

Chapter 1. Rich Men's Vehicles at Poor Men's Prices

1. Sears, Roebuck & Co., *Consumer's Guide,* catalog no. 104 (Chicago, 1897; reprint, New York: Chelsea House, 1968), 708; Carl Bridenbaugh, *The Colonial Craftsman* (New York: New York University Press, 1950; reprint, New York: Dover Publications, 1990), 91. Chapter epigraphs: Benjamin Franklin, *Autobiography* (New York: Vintage, 1990), 131; Francis T. Underhill, *Driving for Pleasure; or, The Harness Stable and Its Appointments* (New York: D. Appleton, 1897; reprint New York: Dover Publications, 1989), 133–34.

2. Seymour Dunbar, *A History of Travel in America* (New York: Tudor Publishing, 1937), 14–16; George W. W. Houghton, "The Coaches of Colonial New-York," (paper read at the New-York Historical Society, 4 March 1890; published by the Hub Publishing Company, New York, 1890), 5; Ezra M. Stratton, *The World on Wheels; or, Carriages, with their Historical Associations from the*

Earliest to the Present Time (New York, 1878; reprint, New York: Benjamin Blom, 1972), 398; "The Old-Time Baltimore Dray," *The Carriage Monthly* 51 (December 1915): 46; Don H. Berkebile, ed., *Carriage Terminology: An Historical Dictionary* (Washington, D.C.: Smithsonian Institution Press, 1978), s.vv. "Cart," "Dray," "Truck."

3. "The Rise and Development of the Carriage Building Industry in America," *The Carriage Monthly* 40 (April 1904): 97; Berkebile, *Carriage Terminology*, s.vv. "Conestoga Wagon," "Farm Wagon," "Truck"; Albert E. Lownes, "Farm Implements of the 18th Century" (part 5), *The Chronicle* 1 (July 1937): 4; chap. 8 of Mark L. Gardner's *Wagons for the Santa Fe Trade: Wheeled Vehicles and Their Makers, 1822–1880* (Albuquerque: University of New Mexico Press, 2000) provides numerous examples of migrants' wheeled vehicles.

4. Berkebile, *Carriage Terminology*, s.vv. "Albany Cutter," "Bob-sled," "Booby-hut," "Cutter," "Portland Cutter," "Sled, Sledge, or Sleigh"; Museums at Stony Brook, *The Carriage Collection* (Stony Brook, N.Y., 1986), 92–93; Stratton, *World on Wheels*, 412–13, 422–24; Albert E. Lownes, "Farm Implements of the 18th Century" (part 4), *The Chronicle* 1 (May 1937): 4; Houghton, "Coaches," 5; C. Max Hilton provides a detailed account of winter timber hauling in *Rough Pulpwood Operating in Northwestern Maine, 1935–1940* (*The Maine Bulletin* 45 [August 1942]; University of Maine Studies, 2d ser., no. 57 [Orono: University of Maine Press, 1942]), chaps. 2 and 4, while Charles Asa Post's *Those Were the Days: When Hearts Were Kind and Sports Were Simple* (Cleveland: Caxton, 1935) contains many accounts of nineteenth-century city sleigh races.

5. Stratton, *World on Wheels*, 398–99, 410–11, 418–19; Houghton, "Coaches," 19; Oliver Wendell Holmes, "The Deacon's Masterpiece, or, The Wonderful 'One-Hoss Shay,'" in *The Literature of America*, ed. Arthur H. Quinn et al. (New York: Charles Scribner's Sons, 1929), 1:372–73; Berkebile, *Carriage Terminology*, s.vv. "Carriage," "Chair," "Chaise," "Gig," "Phaeton."

6. Stratton, *World on Wheels*, 410–11, 418–19; Houghton, "Coaches," 6–7, 16; Berkebile, *Carriage Terminology*, s.vv. "Carriage," "Chariot," "Coach."

7. Bridenbaugh, *Colonial Craftsman*, 89; Stratton, *World on Wheels*, 399, 402, 407–8; Alexander Hamilton, "Report on Manufactures, December 5, 1791," in *The Reports of Alexander Hamilton*, ed. Jacob E. Cooke (New York: Harper and Row, 1964), 155–56; Richard E. Powell Jr., "Coachmaking in Philadelphia: George and William Hunter's Factory of the Early Federal Period," *Winterthur Portfolio* 28 (Winter 1993): 248–49.

8. Houghton, "Coaches," 13–16; Stratton, *World on Wheels*, 409–15, Bridenbaugh, *Colonial Craftsman*, 90–91; 4–5; Powell, "Coachmaking in Philadelphia," 250–51.

9. Don H. Berkebile, ed., *Horse-Drawn Commercial Vehicles: 255 Illustrations of Nineteenth-Century Stagecoaches, Delivery Wagons, Fire Engines, Etc.* (New York: Dover Publications, 1989), v, 64–71; William Louis Gannon, "Carriage, Coach, and Wagon: The Design and Decoration of American Horse-Drawn Vehicles," (Ph.D. diss., Iowa State University, 1960), 44–45; Berkebile, *Car-*

riage Terminology, s.vv. "Concord Coach," "Concord Wagon," "Stagecoach," "Troy Coach."

10. Berkebile, *Carriage Terminology*, s.vv. "Cab," "Omnibus"; Stratton, *World on Wheels*, 432, 437–40.

11. "Address of Hon. C. P. Kimball at the Convention," *The Hub* 14 (December 1872): 219; "Rise and Development of the Carriage Building Industry," 100; "Carriage Building in Amesbury," *The Carriage Monthly* 40 (April 1904): 116; untitled historical note, *The Carriage Monthly* 40 (April 1904): 176; Victor S. Clark, *History of Manufactures in the United States* (Washington, D.C.: Carnegie Institution, 1929; reprint, New York: Peter Smith, 1949), 1:476; U.S. Bureau of the Census, "Manufactures in the States and Territories," *Census Reports: The Seventh Census of the United States, 1850* (Washington, D.C., 1850), 36, 121; U.S. Bureau of the Census, "Totals of Manufactures [by state]," *Census Reports: The Eighth Census of the United States, 1860* (Washington, D.C., 1865).

12. Berkebile, *Carriage Terminology*, s.vv. "Concord Coach," "Concord Wagon"; "Manufactures in the States and Territories," *Seventh Census, 1850*, and "Totals of Manufactures," *Eighth Census, 1860;* U.S. Bureau of the Census, "General Statistics of Manufactures," *Census Reports: The Ninth Census of the United States, 1870* (Washington, D.C., 1872), vol. 3, Tables 8B and 8C; "Rise and Development of the Carriage Building Industry," 97–98, 100; U.S. Bureau of the Census, "Manufactures of 20 Principal Cities," *Census Reports: The Tenth Census of the United States, 1880* (Washington, D.C., 1882), 2:1–32; Horace Greely et al., *The Great Industries of the United States* (Hartford: J. B. Burr & Hyde, 1873), 805; "Story of Early Carriage Building in Baltimore," *The Carriage Monthly* 40 (April 1904): 144; George T. Reed, "Reminiscences of the Carriage Trade Half a Century Ago," *The Carriage Monthly* 47 (April 1911): 38–40.

13. "Rise and Development of the Carriage Building Industry," 100; "A Microcosmic History of the Carriage Industry of the United States," *The Hub* 39 (October 1897): 427; "James Brewster, the Father of American Carriage Making," *Varnish* 4 (August 1891): 352; "Harness and Carriage Items, Etc.," *Harness and Carriage Journal* 17 (January 1874): 396; Stratton, *World on Wheels*, 459; combined total number of wagon and carriage firms for Massachusetts and Connecticut from "Totals of Manufactures," *Eighth Census, 1860*, and "General Statistics of Manufactures," *Ninth Census, 1870*.

14. "Rise and Progress of Carriage Building in Cincinnati," *The Carriage Monthly* 40 (April 1904): 107–8; "Microcosmic History," 427; "Rise and Development of the Carriage Building Industry," 100–101; "Some Early Wagon Builders," *The Carriage Monthly* 40 (April 1904): 161.

15. *Webster's International Dictionary of the English Language* (Springfield, Mass.: G. & C. Merriam, 1909), s.v. "buggy." Theories abound on this term's origins, including speculation that it derived from the Hindu *bagghi*, a light traveling cart; see, for example, "The Word 'Buggy,'" *The Hub* 56 (February

1915): 34. The *Oxford English Dictionary* (2d ed., 1989), s.v. "buggy," discounts this, though it admits that the etymology remains unknown.

16. Stratton, *World on Wheels*, 421–22; Berkebile, *Carriage Terminology*, s.vv. "Buggy," "Concord Wagon," "Pleasure Wagon," "Road Wagon," "Side-Bar Wagon"; Berkebile suggests that "buggy" first appeared in print in William Felton's 1796 *Treatise on Carriages*, while the *Oxford English Dictionary* 2d ed., finds it in use as early as 1773.

17. Henry W. Meyer, *Memories of the Buggy Days* (Cincinnati, 1965), 5–6; Berkebile, *Carriage Terminology*, s.v. "Buggy"; Museums at Stony Brook, *Carriage Collection*, 123; Clark, *History of Manufactures*, 2:128; Merri McIntyre Ferrell, "Before the Cart: The Relationship Between Horses and Carriages," in *Nineteenth Century American Carriages: Their Manufacture, Decoration and Use* (Stony Brook, N.Y.: The Museums at Stony Brook, 1987), 160–61.

18. George Sturt, *The Wheelwright's Shop* (Cambridge: Cambridge University Press, 1923), chap. 9; R. A. Salaman, *Dictionary of Woodworking Tools, c. 1700–1970, and Tools of Allied Trades* (London: George Allen & Unwin, 1975; reprint, Newtown, Conn.: Taunton Press, 1990), 503–19; "O. B. Bannister on Hickory," *The Carriage Monthly* 45 (April 1909): 17.

19. Museums at Stony Brook, *Carriage Collection*, 35; Merri McIntyre Ferrell, "A Harmony of Parts: The Aesthetics of Carriages in Nineteenth-Century America," in *Nineteenth Century American Carriages*, 40; Stratton, *World on Wheels*, 464–65; Ralph Straus, *Carriages and Coaches: Their History and Their Evolution* (London: Martin Secker, 1912), 274–75.

20. Hamilton S. Wicks, "The Manufacture of Pleasure Carriages," *Scientific American* 40 (8 February 1879): 79; A. P. Daire, "Popular Carriages at Popular Prices" (part 2), *The Hub* 30 (January 1889): 752; Berkebile, *American Carriages*, vi; Sears, Roebuck & Co., *Consumer's Guide*, catalog no. 111 (Chicago, 1902; reprint, New York: Bounty Books, 1969), 363.

21. "Carriage Building in Amesbury," *The Carriage Monthly* 40 (April 1904): 116–17; *G. & D. Cook & Co.'s Illustrated Catalogue of Carriages and Special Business Advertiser* (New Haven, 1860; reprint, New York: Dover Publications, 1970), 225; *The World's Carriage Building Center, Cincinnati, Ohio, U.S.A.* (Cincinnati: Robert Clarke, 1893), 10–12; U.S. Bureau of the Census, "General Statistics for Each Specified Industry," *Census Reports: The Tenth Census of the United States, 1880* (Washington, D.C., 1883), 2:26–27.

22. James K. Dawes, "Carriages and Wagons," in U.S. Bureau of the Census, *Census Reports: The Twelfth Census of the United States, 1900* (Washington, D.C., 1902), vol. 10, part 4:296; Nelson Courtlandt Brown, *Forest Products: Their Manufacture and Use* (New York: John Wiley & Sons, 1919), 2–3; "Statistics of the Carriage Industry," *The Carriage Monthly* 40 (April 1904): 177; Clark, *History of Manufactures*, 2:128.

23. "Manufactures of 20 Principal Cities," *Tenth Census, 1880*, 2:1–32: *Statistics of Manufactures;* Hiram Percy Maxim, *Horseless Carriage Days* (New York: Harper, 1936), 105–6.

24. U.S. Patent Office, *Annual Report of the Commissioner of Patents* (Washington, D.C., 1864, 1869); "Col. C. W. Saladee," *The Carriage Monthly* 30 (October 1894): 216; "Col. C. W. Saladee," *The Carriage Monthly* 40 (April 1904): 106; Charles B. Sherron, "The Story of 'The Hub,'" *The Hub* 50 (October 1908): 233; "Cleveland Shops," *The Coach-Maker's Illustrated Monthly Magazine* 1 (February 1855): 10; "Ready-Made Wheels," *The Coach-Maker's Illustrated Monthly Magazine* 1 (October 1855): 107–8; "The French Rule—Introduction," *The Coach-Maker's Illustrated Monthly Magazine* 1 (February 1855): 24–25.

25. Sherron, "Story of 'The Hub,'" 233; J.L.H. Mosier, "Recalling Early History," *The Hub* 50 (October 1908): 242; "John W. Britton," *The Carriage Monthly* 28 (September 1886): 161–63; Berkebile, *Carriage Terminology*, 9; "Modern Apprenticeships," *The New York Coach-Maker's Magazine*, 14 (August 1861): 52–53; "Col. C. W. Saladee," 216; New York City Directory, 1861.

26. Sherron, "Story of 'The Hub,'" 233; "To Readers and Correspondents," *The New York Coach-Maker's Magazine* 2 (June 1859): 13 and 2 (September 1859): 73; "A View from an Editorial Stand-Point," *The New York Coach-Maker's Magazine* 2 (June 1859): 13–14; Ezra M. Stratton, "Preface to Volume Two," *The New York Coach-Maker's Magazine* 2 (April 1860): n.p. [in bound version only]; "Purity of Motives," *The Coach-Makers' International Journal* 4 (December 1868): 34–35; "The Wise Man of Gotham," *The Coach-Makers' International Journal* 4 (February 1869): 66; "Chronicles" (part 1), *The Coach-Makers' International Journal* 4 (February 1869): 72; "Chronicles" (part 3) *The Coach-Makers' International Journal* 4 (May 1869): 122; "The Culmination of Genius," *The New York Coach-Maker's Magazine* 2 (February 1860): 175; "The Scene of a Late Disaster in Crossing the Alleghenies," *The New York Coach-Maker's Magazine* 2 (October 1859): 98.

27. "The Last But One," *The New York Coach-Maker's Magazine* 2 (April 1860): 215; Mosier, "Recalling Early History," 242; Sherron, "Story of 'The Hub,'" 233; "Culmination of Genius," 175; "To Readers and Correspondents," *The New York Coach-Maker's Magazine* 2 (April 1860): 214; "Col. C. W. Saladee," 216, presents a great deal of unverified information, which should be treated with caution. Sherron knew both protagonists, and his generally factually accurate account favor's Saladee while Mosier, long-time Brewster employee and friend of Britton, favored Stratton.

28. Sherron, "Story of 'The Hub,'" 233; New York City Directory, 1861–66, 1881; Mosier, "Recalling Early History," 242, relates John W. Britton's financial and administrative support to Stratton's publication and states that Stratton closed his shop due to an expired lease. Berkebile, *Carriage Terminology*, erroneously lists Stratton as leaving the trade in 1858, but in "Our Next Volume," *The New York Coach-Maker's Magazine* 8 (May 1867): 184, Stratton indicates it was late 1866 or early 1867, by which time the city directory listed him as a coal dealer.

29. Saladee's long list of patents and dwelling places can be traced through

the *Annual Report of the Commissioner of Patents* (Washington, D.C.: U.S. Government Printing Office, 1856–92).

30. Sherron, "Story of 'The Hub,'" 233–34; Mosier, "Recalling Early History," 242; "Obituary: George W. W. Houghton," *The Carriage Monthly* 27 (April 1891): 56, incorrectly states that Houghton began under Stratton with the *The New York Coach-Maker's Magazine;* Holmes's reference to the Boston state house comes from his *Autocrat of the Breakfast Table* (1858); Berkebile, *Carriage Terminology,* 10, 487; New York City Directory, 1872–73.

31. Berkebile, *Carriage Terminology,* 10; Sherron, "Story of 'The Hub,'" 233–34; *Appleton's Cyclopedia of American Biography* (New York: D. Appleton, 1888), s.v. "George Washington Wright Houghton"; index to vol. 14, *The Hub* (1872); Houghton published two works of poetry, *The Legend of St. Olaf's Kirk* (Boston: Houghton, Mifflin, 1881) and *Niagara, and Other Poems* (Boston: Houghton, Mifflin, 1882), and he also edited Chauncey Thomas's novel *The Crystal Button* (Boston: Houghton, Mifflin, 1891).

32. "The History of 'The Carriage Monthly,'" *The Carriage Monthly* 40 (April 1904): 200Y, 200CC (this incorrectly puts the name change in vol. 5; it was vol. 9); "From a Veteran Accessory Manufacturer," *The Carriage Monthly* 40 (April 1904): 174; Berkebile, *Dictionary,* 487; "I. D. Ware as a Carriage Builder," *The Carriage Monthly* 40 (April 1904): 200P; Charles B. Sherron, "Early Days of 'The Carriage Monthly,'" *The Carriage Monthly* 40 (April 1904): 200R, 200S; Philadelphia City Directory, 1865–69; Peter C. Marzio, *The Democratic Art: Pictures for a 19th Century America* (Boston: David R. Godine; Fort Worth: Amon Carter Museum of Western Art, 1979), xi–xii, 4, 8–9, 42, 53–54.

33. "History of Working Drafts in The Carriage Monthly," *The Carriage Monthly* 40 (April 1904): 200M; Anson K. Cross, *Mechanical Drawing: A Manual for Teachers and Students* (Boston: Ginn, 1895), 28–32; "History of 'The Carriage Monthly,'" 200Y, 200Z, 200CC; "'The Carriage Monthly' the Originator of Perspective Carriage Drawing," *The Carriage Monthly* 40 (April 1904): 200G, 200H.

34. Berkebile, *Carriage Terminology,* 10, 487.

35. "The Carriage Builders' National Association," *The Carriage Monthly* 40 (April 1904): 94; "Origin and History of the Carriage Builders' National Association," *The Carriage Monthly* 40 (April 1904): 102; "Address of Hon. C. P. Kimball at the Convention," *The Hub* 14 (December 1872): 219–20; John M. Davis, "First Ten Years of the C.B.N.A.," *The Hub* 50 (October 1908): 247.

36. "Origin and History," 102; "Prominent Carriage Builders Prior to 1870," *The Carriage Monthly* 40 (April 1904): 200CC; "John W. Britton" [obituary], *The Carriage Monthly* 28 (September 1886): 162; "The Carriage-Builders' Convention. Positive Appointment of the First Meeting," *The Hub* 14 (September 1872): 145; C.H.E. Redding, "A Fifty Year Old Association," *The Automotive Manufacturer* 64 (September 1922): 22.

37. "Obituary: George W. W. Houghton," 56; "History of 'The Carriage

Monthly,'" 200Z; Redding, "Fifty Year Old Association," 22; "Carriage-Builders' Convention," *The Hub* 14 (December 1872): 205; "Carriage Builders' National Association," 94; "Positive Appointment," 145–46.

38. "Positive Appointment," 203–4; "Address of Hon. C. P. Kimball," 218–20; "Carriage-Builders' Convention," 203–6.

39. "History of the Carriage Builders' National Association, 1872–1913," *The Carriage Monthly* 50 (October 1914): 42–43; Davis, "First Ten Years," 248–49; "Carriage-Builders' Convention," 205; Redding, "Fifty Year Old Association," 22.

40. "Origin and History," 102–3; "History of the Carriage Builders' National Association," 42–43; Andrew F. Johnson, "History of the Technical School for Carriage Draftsmen and Mechanics," *The Carriage Monthly* 40 (April 1904): 134–35; "A Quarter of a Century Ago," *The Carriage Monthly* 40 (April 1904): 144.

41. Johnson, "History of the Technical School," 134–36.

42. John K. Brown, "Design Plans, Working Drawings, National Styles: Engineering Practice in Great Britain and the United States, 1775–1945," *Technology and Culture* 41 (April 2000): 200, 207, 217.

43. "Carriage Builders' Convention," 204; "Carriage-Makers' Societies," *The Hub* 28 (September 1886): 369–71; "Associations Connected with the Vehicle Industry and Their Officers," *The Carriage Monthly* 50 (February 1915): 58–59. The claim that carriage makers had founded the first national trade association surfaced in the carriage trade press sometime during the 1890s and has been widely repeated in secondary works (particularly Ben Riker's *Pony Wagon Town*). In reality several other industries established earlier national associations, among them the American Iron and Steel Institute, the National Association of Cotton Manufacturers, the National Association of Wool Manufacturers, the United States Brewers Association, and the Writing Paper Manufacturers Association—all before 1865. Jay Judkins, *National Associations of the United States* (Washington, D.C.: U.S. Department of Commerce, 1949), viii.

44. Dawes, "Carriages and Wagons," 291–322.

Chapter 2. Knights of the Draw Knife

1. Ezra M. Stratton, "Autobiography of Caleb Snug, of Snugtown, Carriage-Maker" chap. 2: *The New York Coach-Maker's Magazine* 2 (August 1859): 44, chap. 3 [mistakenly numbered chapter 16]: *The New York Coach-Maker's Magazine* 2 (October 1859): 81–82; Colonel James H. Sprague, "When the Old-School Wagon Maker Was in Flower," *The Carriage Monthly* 38 (April 1902): 21–22. Chapter epigraphs: George Sturt, *The Wheelwright's Shop* (Cambridge: Cambridge University Press, 1923), 55; Stratton, "Autobiography" chap. 3, 82.

2. Carl Bridenbaugh, *The Colonial Craftsman* (New York: New York Uni-

versity Press, 1950; reprint, New York: Dover Publications, 1990), 1–2; Ian M. G. Quimby, ed., *The Craftsman in Early America* (New York: W. W. Norton, for the Henry Francis du Pont Winterthur Museum, 1984), 12–15; L. F. Salzman, *English Industries of the Middle Ages* (Oxford: Clarendon Press, 1923): 340–51; W. J. Rorabaugh, *The Craft Apprentice: From Franklin to the Machine Age in America* (Oxford: Oxford University Press, 1986), 4, 8.

3. David S. Landes, *The Unbound Prometheus: Technological Change and Industrial Development in Western Europe from 1750 to the Present* (Cambridge: Cambridge University Press, 1969), 43–44; Rorabaugh, *Craft Apprentice*, 4–5.

4. Rorabaugh, *Craft Apprentice*, 3–5; Ian M. G. Quimby, "Apprenticeship in Colonial Philadelphia," (M.A. thesis, University of Delaware, 1963; reprint, New York: Garland, 1985), 32–33, 40.

5. Stratton, "Autobiography" chap. 1: *The New York Coach-Maker's Magazine* 2 (July 1859): 23; James Brewster, "Autobiography of James Brewster," 1856, chaps. 1 and 2, typescript copy, NHCHS.

6. Stratton, "Autobiography" chap. 1, 23; Brewster, "Autobiography," chap. 2; "Biography of James Brewster," *The New York Coach-Maker's Magazine* 1 (June 1858): 1; Stratton speculates in the latter piece that both James Goold and Jason Clapp also apprenticed with Chapman, an error repeated in many later sources.

7. Edward S. Cooke Jr., *Making Furniture in Preindustrial America: The Social Economy of Newtown and Woodbury, Connecticut* (Baltimore: Johns Hopkins University Press, 1996), 34; Rorabaugh, *Craft Apprentice*, 135–36; Stratton, "Autobiography," chap. 4: *The New York Coach-Maker's Magazine* 2 (November 1859): 102–3.

8. "Apprenticeship in the Good Old Days," *The Hub* 43 (March 1908): 376; Quimby, "Apprenticeship in Colonial Philadelphia," 29.

9. Quimby, "Apprenticeship in Colonial Philadelphia," xiii, 29, 32–35; Charles F. Hummel, "The Business of Woodworking," in *Tools and Technologies: America's Wooden Age* ed. Paul B. Kebabian and William C. Lipke (Burlington: Robert Hull Fleming Museum and the University of Vermont, 1979), 51; "Apprenticeship in the Good Old Days," 376.

10. Charles Eckhart, "How Carriages Were Built in Early Days," *The Spokesman* 26 (September 1910): 367; "Education in Mechanism," *The Coach-Makers' International Journal* 2 (June 1867): 145.

11. Charles F. Hummel, *With Hammer in Hand: The Dominy Craftsmen of East Hampton, New York* (Charlottesville: University Press of Virginia, 1968): 238.

12. Stratton, "Autobiography," chap. 3, 81.

13. Cooke, *Making Furniture in Preindustrial America*, 34–35; Stratton, "Autobiography," chap. 2, 43–44, and chap. 3, 81–82; Quimby, "Apprenticeship in Colonial Philadelphia," 50–51, 76.

14. Cooke, *Making Furniture in Preindustrial America*, 34.

15. Stratton, "Autobiography," chap. 5: *The New York Coach-Maker's Magazine* 2 (December 1859): 121, and chap. 7: *The New York Coach-Maker's Magazine* 2 (February 1860): 162.

16. Ibid., chap. 2, 42; chap. 3, 82; chap. 8: *The New York Coach-Maker's Magazine* 2 (February 1860): 163–64; Sprague, "Old-School Wagon Maker," 21.

17. Stratton, "Autobiography," chap. 4, 103; Brewster, "Autobiography," chap. 7.

18. Stratton, "Autobiography," chap. 2, 44; Brewster, "Autobiography," chap. 2.

19. Rorabaugh, *Craft Apprentice*, 68–69.

20. Brewster, "Autobiography," chap. 3; Stratton, "Autobiography," chap. 9: *The New York Coach-Maker's Magazine* 3 (June 1860): 5.

21. For examples of "freedom dues" in various trades in Philadelphia, see Quimby, "Apprenticeship in Colonial Philadelphia," appendix C, and 52–53; "Apprenticeship in the Good Old Days," 376; Stratton, "Autobiography," chap. 2, 44 and chap. 9, 5–6; and Brewster, "Autobiography," chap. 3.

22. Stratton, "Autobiography," chap. 9, 5–6; Brewster, "Autobiography," chaps. 4 and 5; "Reports from Subordinate Unions," *The Coach-Makers' International Journal* 2 (June 1867): 154.

23. Stratton, "Autobiography," chap. 5, 121.

24. Ibid., 121–22; chap. 6, 142; chap. 8: *The New York Coach-Maker's Magazine* 2 (May 1860): 222.

25. Brewster, "Autobiography," chap. 5.

26. Ibid., chap. 3; M. T. Richardson, ed., *Practical Carriage Building* (New York, 1892; reprint, Mendham, N.J.: Astragal Press, n.d.), 1:10–11.

27. Stratton, "Autobiography," chap. 2, 81; Eckhart, "How Carriages Were Built," 367.

28. Cooke, *Making Furniture in Preindustrial America*, 14–17 and notes; Victor S. Clark, *History of Manufacturing in the United States* (Washington, D.C.: Carnegie Institution, 1929; reprint, New York: Peter Smith, 1949), 1:476; "How Carriage Makers Are Made Out West," *The New York Coach-Maker's Magazine* 2 (September 1859): 62.

29. Stratton, "Autobiography," chap. 10: *The New York Coach-Maker's Magazine* 3 (November 1860): 109–10; Brewster, "Autobiography," chap. 5.

30. "How Carriage Makers Are Made," 61.

31. Sprague, " Old-School Wagon Maker," 21; Richardson, *Practical Carriage Building*, 1:11–25.

32. Ben Riker, *Pony Wagon Town* (New York: Bobbs-Merrill, 1948): 47; Stratton, "Autobiography," chap. 2, 44; Richardson, *Practical Carriage Building*, 1:23; examples of incremental additions, as well as separate structures, can be found in Paul A. Kube, "A Study of the Gruber Wagon Works at Mount Pleasant, Pennsylvania," (M.Ed. Thesis, Millersville State College, 1968), 179–80.

33. Isaac D. Ware, ed., *The Coach-Makers' Illustrated Hand-Book*, 2d ed. (Philadelphia, 1875; reprint, Mendham, N.J.: Astragal Press, 1995), 344; Sprague,

"Old-School Wagon Maker," 21; J. G. Holmstrom, *Modern Blacksmithing, Rational Horse Shoeing, and Wagon Making* (Chicago: Frederick J. Drake, 1904): 33–34, relates this common anvil-anchoring practice; Richardson, *Practical Carriage Building,* 1:11–14.

34. Stratton, "Autobiography," chap. 2, 44; Eckhart, "How Carriages Were Built," 367; "Reflections by an Old-Time Wood Worker," *The Hub* 36 (February 1895): 806; Ware, *Coach-Makers' Illustrated Hand-Book,* 344–45.

35. Richardson, *Practical Carriage Building,* 1:17, 19, 21; William Byron, "Art of Carriage Painting," *The Coach Painter* 1 (February 1880): 3–4.

36. In 1865 Ezra Stratton began serial publication of such a dictionary, and after his journal became part of *The Hub,* George Houghton and several others expanded on Stratton's work in a serial installment later published separately. William N. Fitzgerald began serially publishing a carriage dictionary in 1871 in his *Carriage and Harness Journal,* and there were a few others compiled at various times by the trade press.

37. "The New York Coach-Maker's Monthly Magazine" and "The New York Coach-Maker's Magazine Charts," *The New York Coach-Maker's Magazine* 9 (August 1867): unpaginated advertising following p. 48; Isaac D. Ware, editorial prospectus, *The Coach- Makers' International Journal* 3 (September 1868): 266; "Office Chart," *The Coach-Makers' International Journal* 4 (December 1868): 35.

38. For the English system during the 1840s, see "A Day at a Coach-Factory," *The Penny Magazine* 10 (1841): 501–8; for a discussion of this method versus the French system, see "Working Drafts for Body-Makers," *The Carriage Monthly* 27 (March 1891): 382, 384; for use of wooden patterns, see "Reflections by an Old-Time Wood Worker," 806; "The 'French Rule,'" *The Carriage Monthly* 42 (February 1907): 300.

39. Sprague, "Old-School Wagon Maker," 21; Eckhart, "How Carriages Were Built," 367.

40. Sprague, "Old-School Wagon Maker," 21.

41. Ibid.; Eckhart, "How Carriages Were Built," 367; "Lumber Dry-Kilns," *The Hub* 35 (May 1893): 166; air-dried lumber remained common in the trade throughout its existence, its proponents claiming—with some justification—that wood so cured retained far more strength than kiln-dried lumber.

42. Stratton, "Autobiography," chap. 3, 82; Sprague, "Old-School Wagon Maker," 21; "A Day at a Coach-Factory," 503; Eckhart, "How Carriages Were Built," 367.

43. Eckhart, "How Carriages Were Built," 209; Henry C. Mercer, *Ancient Carpenters' Tools,* 5th ed. (Doylestown, Pa.: Bucks County Historical Society, 1975), 222; Peter Haddon Smith, "The Industrial Archeology of the Wood Wheel Industry in America," (Ph.D. diss., George Washington University, 1971), 9, 11–12; Sprague, "Old-School Wagon Maker," 21.

44. R. A. Salaman, *Dictionary of Woodworking Tools, c. 1700–1970, and Tools of Allied Trades* (London: George Allen & Unwin, 1975; reprint, Newtown,

Conn.: Taunton Press, 1990), s.v. “Wheelwright’s Equipment, Tyring Tools”; Sprague, “Old-School Wagon Maker,” 21–22; “An Improvement for Bending Tires,” *The Carriage Monthly* 11 (April 1875): n.p.; Smith, “Industrial Archeology,” 14–17.

45. “Reflections by an Old-Time Wood Worker,” 806; Sprague, “Old School Wagon Maker,” 21; Eckhart, “How Carriages Were Built,” 367; Stratton, “Autobiography,” chap. 3, 81–82, chap. 6: *The New York Coach-Maker’s Magazine* 2 (January 1860): 141–42, and chap. 7, 163; while most nineteenth-century American works on carriage building omit detailed treatment of woodworking tools—they were probably perceived as too familiar to warrant comment—Richardson, *Practical Carriage Building*, vol. 1, chap. 1, contains some useful information. A more complete treatment of the hand tools used in American as well as English carriage shops can be found in John Philipson, *The Art and Craft of Coach Building* (London: George Bell and Sons, 1897), chap. 6 and plates.

46. Eckhart, “How Carriages Were Built,” 209; Sprague, “Old-School Wagon Maker,” 21; “A Day at a Coach-Factory,” 503–7; U.S. Bureau of Labor, *Thirteenth Annual Report of the Commissioner of Labor, 1898: Hand and Machine Labor* (Washington, D.C., 1899), 2:714–15.

47. Sprague, “Old-School Wagon Maker,” 21–22; Eckhart, “How Carriages Were Built,” 209; Kube, “Gruber Wagon Works,” 185–86; *Thirteenth Annual Report*, 2:714–17; “A Day at a Coach-Factory, 504–5; Riker, *Pony Wagon Town*, 114–15, 133–34.

48. Eckhart, “How Carriages Were Built,” 209; Sprague, “Old-School Wagon Maker, 22; Stratton, “Autobiography,” chap. 4, 101.

49. Sprague, “Old-School Wagon Maker,” 22; Don H. Berkebile, ed., *Carriage Terminology: An Historical Dictionary* (Washington, D.C.: Smithsonian Institution Press, 1978), s.vv. “Rough Stuff,” “White Lead;” Riker, *Pony Wagon Town*, 146–60; while he describes the process as he knew it late in the nineteenth century, Riker’s detailed coverage of the painting process applies to far earlier periods given the relatively static technology of nineteenth-century carriage painting; likewise for Mayton Clarence Hillick’s *Carriage and Wagon Painting* (Chicago: Press of the Western Painter, 1898; reprint, Mendham, N.J.: Astragal Press, 1997), 17–24.

50. Sprague, “Old-School Wagon Maker,” 22; Eckhart, “How Carriages Were Built,” 367; Riker, *Pony Wagon Town*, 146–60; Hillick, *Carriage and Wagon Painting*, chap. 6.

51. Eckhart, “How Carriages Were Built,” 367; Riker, *Pony Wagon Town*, 22–23; *Thirteenth Annual Report*, 2:716–17.

52. Eckhart, “How Carriages Were Built,” 367; Riker, *Pony Wagon Town*, 47–48.

53. Stratton, “Autobiography,” chap. 2, 44, chap. 6, 141, and chap. 9, 5; Brewster, “Autobiography,” chap. 3; Eckhart, “How Carriages Were Built,” 367.

54. Bruce Laurie, *Artisans into Workers: Labor in Nineteenth-Century America* (New York: Noonday Press, 1989): 64, 79; Rorabaugh, *Craft Apprentice*, 89; Stratton, "Autobiography," chap. 2, 44, chap. 6, 141, and chap. 9, 5; Brewster, "Autobiography," chap. 3. Like many other aspects of the system, craft shop work hours remain ripe for additional research, particularly in their still poorly understood relation to the changing seasons and the political calendar.

55. George W. W. Houghton, "Are the Times Ripe for Reduced Working Hours?" *The Hub* 30 (October 1888): 499–500; Sprague, "Old-School Wagon Maker," 21; Eckhart, "How Carriages Were Built," 367; Richard E. Powell Jr., "Coachmaking in Philadelphia: George and William Hunter's Factory of the Early Federal Period," *Winterthur Portfolio* 28 (winter 1993): 261.

56. Stratton, "Autobiography," chap. 10, 109–10; Brewster, "Autobiography," chaps. 2–3.

57. Stratton, "Autobiography," chap. 8, 221; Brewster, "Autobiography," chap. 12.

58. Brewster, "Autobiography," chap. 14.

59. Powell, "Coachmaking in Philadelphia," 261; Stratton, "Autobiography," chap. 10, 109; "Reflections by an Old-Time Wood Worker," 806; U.S. Bureau of the Census, "Carriage and Wagon Works," *Census Reports: Tenth Census of the United States, 1880* (Washington, D.C., 1886), 20:410–11, 414; this report is built around average wage samples recorded in carriage firms in various parts of the country from 1849 to 1880.

60. Stratton, "Autobiography," chap. 8, 222; Ware, *Coach-Makers' Illustrated Hand-Book*, 351–54; "Carriage and Wagon Works," *Tenth Census, 1880*, 20:410; James K. Dawes, "Carriages and Wagons," in U.S. Bureau of the Census, *Census Reports: The Twelfth Census of the United States, 1900* (Washington, D.C., 1902), 10:308, reveals the seasonal cycle still evident in 1900.

61. Brewster, "Autobiography," chap. 9; Ware, *Coach-Makers' Illustrated Hand-Book*, 352, 354; "Commercial Aspect of Carriage-Making," *The New York Coach-Maker's Magazine* 2 (September 1859): 75; "Carriage and Wagon Works," *Tenth Census, 1880*, 20:410.

62. Riker, *Pony Wagon Town*, 174–76, 179; Richardson, *Practical Carriage Building*, 1:15, 23; Ware, *Coach-Makers' Illustrated Hand-Book*, 343.

63. Richardson, *Practical Carriage Building*, 1:17–18; Stratton, "Autobiography," chap. 2, 44, chap. 5, 121, chap. 8, 221–22.

64. Brewster, "Autobiography," chap. 12; advertisement for name plates, *The New York Coach-Maker's Magazine* 9 (October 1867): unpaginated advertising following p. 80.

65. Eckhart, "How Carriages Were Built," 367; Ezra M. Stratton, "Cheap Carriages," *The New York Coach-Maker's Magazine* 8 (December 1866): 106.

66. Saladee resided in at least a dozen different places before and after the Civil War, but he managed to get his name into the Columbus, Ohio (1856–59), Pittsburgh (1874–75), and Cleveland (1887–90) city directories, and probably others; advertisement for Jacob Lowman, Cleveland City Directory,

1845–46; Paddock advertisement quoted in Bridenbaugh, *Colonial Craftsman*, 90; sampling of James Brewster's early advertising from *The Dana Collection Scrapbook*, vol. 127:13, in the NHCHS library.

67. Stratton, "Autobiography," chap. 10, 109–10.

68. Brewster, "Autobiography," chaps. 6–9, 11.

Chapter 3. From Shop to Factory

1. *G. & D. Cook & Co.'s Illustrated Catalogue of Carriages and Special Business Advertiser* (New Haven, 1860; reprint, New York: Dover Publications, 1970), frontispiece, 1–5, 225; "The Largest Carriage Mart in the World," *Harness and Carriage Journal* 14 (August 1870): 30–31. Chapter epigraphs: "Does Machinery Pay?" *The Hub* 17 (April 1875): 6; "A Microcosmic History of the Carriage Industry of the United States," *The Hub* 39 (October 1897): 420.

2. David S. Landes, *The Unbound Prometheus: Technological Change and Industrial Development in Western Europe from 1750 to the Present* (Cambridge: Cambridge University Press, 1969), 1–12; Brooke Hindle and Steven Lubar, *Engines of Change: The American Industrial Revolution, 1790–1860* (Washington, D.C.: Smithsonian Institution Press, 1986), 9–11, 21; Maxine Berg, *The Age of Manufactures: Industry, Innovation, and Work in Britain, 1700–1820* (Oxford: Oxford University Press, 1986), preface, 316–17; Robert B. Gordon and Patrick M. Malone, *The Texture of Industry: An Archaeological View of the Industrialization of North America* (Oxford: Oxford University Press, 1994), 231; the oft-repeated phrase about satanic textile mills comes from William Blake's *Milton* (1804–8), see Bartlett, *Familiar Quotations*, 13th ed., s.v. "William Blake."

3. Bruce Laurie, *Artisans into Workers: Labor in Nineteenth-Century America* (New York: Noonday Press, 1989), 16; Gordon and Malone, *Texture of Industry*, 231; Susan E. Hirsch, *Roots of the American Working Class: The Industrialization of Crafts in Newark, 1800–1860*. (Philadelphia: University of Pennsylvania Press, 1978), xix, 21; Walter Licht, *Industrializing America: The Nineteenth Century* (Baltimore: Johns Hopkins University Press, 1995), 40, 44–45.

4. Andrew Ure, *The Philosophy of Manufactures* (London: Charles Knight, 1835), 13, quoted in Gordon and Malone, *Texture of Industry*, 298; Jean Gimpel, *The Medieval Machine: The Industrial Revolution of the Middle Ages* (New York: Penguin, 1977), 3, 5, 67, is the chief expositor of the medieval factory, and Maxine Berg, *Age of Manufacturers*, 40–41, similarly argues for the existence of "machineless" factories in early-eighteenth-century Britain in cloth, paper, and iron manufacturing. "Manufactory" sometimes crops up in nineteenth-century sources and sees sporadic use among some contemporary historians to describe smaller, less intensively organized and mechanized production formats, as in Gordon and Malone, *Texture of Industry*, 298, and Sean Wilentz, *Chants Democratic: New York City and the Rise of the American Working Class, 1788–1850* (Oxford: Oxford University Press, 1984), 31 and throughout. Still, inasmuch as the English textile model arguably provided the inspiration and some of the technology for the establishment of factories elsewhere, I find it argues against ap-

plication of the term to what might be better characterized as large shops, as with Richard E. Powell, "Coachmaking in Philadelphia: George and William Hunter's Factory of the Early Federal Period," *Winterthur Portfolio* 28 (winter 1993).

5. "A Pioneer Carriage Builder," *The Carriage Monthly* 40 (April 1904): 129; "Carriage Building in Amesbury," *The Carriage Monthly* 40 (April 1904): 116; "Origin of Amesbury Industry: Founder of Vehicle Business and His Early Struggles," *Newburyport Herald*, January 1903, n.p., from Jacob R. Huntington files, Amesbury Public Library, Amesbury, Mass.

6. "Jacob R. Huntington Expires," *The Hub* 50 (October 1908): 244; Margaret S. Rice, "Brief History of the Carriage Industry in Amesbury, Massachusetts," n.d., 4, typescript in the collections of the Amesbury Public Library; "Carriage Building in Amesbury," 116; "Origin of Amesbury Industry."

7. The extant sources differ on Huntington's early output, with some placing his second batch of thirteen vehicles in 1853 and others listing this number for the following year. "Origin of Amesbury Industry"; Rice, "Brief History," 4–5; "Jacob R. Huntington Expires," 244; "Pioneer Carriage Builder,"129; Cincinnati City Directory, 1859.

8. Rice, "Brief History," 4–5; "Carriage Building in Amesbury," 116–17; "Origin of Amesbury Industry"; "Pioneer Carriage Builder," 129.

9. "Pioneer Carriage Builder," 129.

10. Ibid.; "The Workings and Interior Views of a Large Carriage Factory," *The Hub* 51 (May 1909): 53–56; Ezra Stratton, "Systemized Carriage-Making," *The New York Coach-Maker's Magazine* 8 (October 1866): 74–75; Lindy Biggs, *The Rational Factory: Architecture, Technology, and Work in America's Age of Mass Production* (Baltimore: Johns Hopkins University Press, 1996), 2–6.

11. U.S. Bureau of Labor, *Thirteenth Annual Report of the Commissioner of Labor, 1898: Hand and Machine Labor* (Washington, D.C., 1899), 1:11, 16; *G. & D. Cook*, 225–26.

12. Stratton, "Systemized Carriage-Making," 74.

13. Scholars have not always used the terms "uniformity" and "interchangeability" in the most precise fashion, but the general consensus seems to be that uniformity represents an increasing degree of sameness and interchangeability, the ideal culmination of that process. "The Duplication of Parts," *The Carriage Monthly* 41 (September 1905): 173; Hindle and Lubar, *Engines of Change*, 218; Henry W. Meyer, *Memories of the Buggy Days* (Cincinnati, 1965), 15; *The World's Carriage Building Center: Cincinnati, Ohio, U.S.A.* (Cincinnati: Robert Clarke, 1893), 15.

14. "Pioneer Carriage Builder," 129; Hindle and Lubar, *Engines of Change*, 218–35; David A. Hounshell, *From the American System to Mass Production, 1800–1932: The Development of Manufacturing Technology in the United States* (Baltimore: Johns Hopkins University Press, 1984), 3–4, 25–28; Chauncey Jerome, *History of the American Clock Business* (New Haven: F. C. Dayton, 1860), 136.

15. "Rise and Progress of Carriage Building in Cincinnati," *The Carriage Monthly* 40 (April 1904): 107–8; Steven J. Ross, *Workers on the Edge: Work, Leisure, and Politics in Industrializing Cincinnati, 1788–1890* (New York: Columbia University Press, 1985), 106–7.

16. "Carriage Building in Amesbury," 116–17; Charles F. Raun, "Use of Steam-Power in Carriage-Building," *The Hub* 14 (July 1872): 97; "Steam Power for Carriage-Building," *The Hub* 13 (January 1872): 199; "Labor-Saving Appliances in Carriage and Wagon Shops," *The Carriage Monthly* 43 (November 1907): 237; "Microcosmic History," 420; Polly Anne Earl, "Craftsmen and Machines: The Nineteenth-Century Furniture Industry," in *Technological Innovation and the Decorative Arts*, ed. Ian M. G. Quimby and Polly Anne Earl (Charlottesville: University Press of Virginia, 1974), 307–29; Gary Paul Lehmann, "Foot Power," *The Chronicle* 47 (December 1994): 115–18; Bob Horner and Steve Johnson, "Foot-Powered Woodworking Machinery of W. F. & John Barnes Company," *The Chronicle* 48 (December 1995): 75–88; Gordon and Malone, *Texture of Industry*, 353–54.; T. K. Derry and Trevor I. Williams, *A Short History of Technology, from the Earliest Times to a.d. 1900* (Oxford: Oxford University Press, 1961),74, 130–31; Ivan H. Crowell, "Horse Power," *The Chronicle* 18 (December 1965): 49–51, 56; John Didsbury, "Horse Power," *The Chronicle* 21 (June 1968): 20–21, 31; Paul A. Kube, "A Study of the Gruber Wagon Works at Mt. Pleasant, Pennsylvania," (M.Ed. thesis, Millersville State College, 1968), 37–38; "Dog Labor in Workshops," *The Hub* 31 (June 1889): 191.

17. Louis C. Hunter, *A History of Industrial Power in the United States, 1780–1930*, vol. 1: *Waterpower in the Century of the Steam Engine* (Charlottesville: University Press of Virginia, for the Hagley Museum and Library, 1979), 103, 159, and vol. 2: *Steam Power* (Charlottesville: University Press of Virginia, for the Hagley Museum and Library, 1985), xix–xxi.

18. Carroll W. Pursell Jr. *Early Stationary Steam Engines in America: A Study in the Migration of a Technology* (Washington, D.C.: Smithsonian Institution Press, 1969), vi–vii, 2–6; Hunter, *Waterpower*, 343, 536, and *Steam Power*, xix–xxi, 67–68.

19. Hunter, *Steam Power*, 67–68; Sears, Roebuck & Co., *Consumer's Guide*, catalog no. 104 (Chicago, 1897; reprint, New York: Chelsea House, 1968), 104, 151.

20. English inventor Samuel Bentham invented the most common woodworking machines, including the planer, molder, dovetail cutter, rotary mortiser, as well as improvements to table saws, circular saw blades, and mortising chisels. John Richards, "Wood Working Machinery: A Treatise on its Construction and Application, with a history of its Origin and Progress," *Journal of the Franklin Institute* 59 (May 1870): 306–10; Derry and Williams, *Short History of Technology*, 351; Carolyn C. Cooper, *Shaping Invention: Thomas Blanchard's Machinery and Patent Management in Nineteenth-Century America* (New York:

Columbia University Press, 1991), chap. 2; Hounshell, *American System to Mass Production*, 147, 360, n.16; F. R. Hutton, "Report on Machine Tools and Wood-Working Machinery," in U.S. Bureau of the Census, *Census Reports: Tenth Census of the United States, 1880* (Washington, D.C., 1888), 22:ix; advertisement for Lane & Bodley, *Lumberman's Gazette* 7 (April 1876): 270; M. Powis Bale, *Woodworking Machinery: Its Rise, Progress, and Construction* (London: Crosby, Lockwood & Co.; reprint, Lakewood, Colo.: Glen Moor, 1992), 13; advertisement for the Silver Mfg. Co., *The Hub* 40 (September 1898): 468; "Labor-Saving Appliances," 237.

21. Nathan Rosenberg, *Perspectives on Technology* (Cambridge: Cambridge University Press, 1976), 37; Paul E. Rivard, *Maine Sawmills: A History* (Augusta: Maine State Museum, 1990), 4–6; M. T. Richardson, ed., *Practical Carriage Building* (New York, 1892; reprint, Mendham, N.J.: Astragal Press, n.d.), 1:10; "Microcosmic History," 420; "Labor-Saving Appliances," 237; "Does Machinery Pay?" *The Hub* 17 (April 1875): 6; advertisement for the Egan Company, *The Hub* 30 (November 1888): 651.

22. Bale, *Woodworking Machinery*, 84, 88; Rosenberg, *Perspectives on Technology*, 41; John Richards, "Wood-Working Machinery," *Journal of the Franklin Institute* 59 (May 1870), 311–12; John Richards, "Wood-Working Machinery," *Journal of the Franklin Institute* 60 (November 1870), 305, 307; "Labor-Saving Appliances," 237; "Does Machinery Pay?" 6.

23. Samuel Bentham seems to have been the originator of a practical mortising machine, but other inventors like Fay produced their own versions. "The Manufacture of Wood-Working Machinery," American Industries, no. 28 [C. B. Rogers & Co., Norwich, Conn.], *Scientific American* 42 (17 January 1880): 37, and "The Manufacture of Wood-Working Machinery," American Industries, no. 85 [J. A. Fay & Co., Cincinnati, Ohio], *Scientific American* 47 (30 December 1882): 419; The Egan Company, advertisement, *The Hub* 80 (November 1888), 651; "Does Machinery Pay?" 6.

24. Cray Bros., *Carriage and Wagon Materials* (Cleveland, 1902), 15–18, 25-27; "Labor-Saving Appliances," 237; R. A. Salaman, *Dictionary of Woodworking Tools, c. 1700–1970, and Tools of Allied Trades* (London: George Allen & Unwin, 1975; reprint, Newtown, Conn.: Taunton Press, 1990), s.v. "Drill, Press; the Beam Drill." "Stiver's Hand-Drilling Machine," *The New York Coach-Maker's Magazine* 6 (July 1864): 25–26; "Machinery and Tools in Modern Carriage and Wagon Factories," *The Carriage Monthly* 37 (March 1902): 459–60; advertisement for the Silver Mfg. Co., *The Hub* 40 (September 1898): 468; George Worthington Co., *1909 Catalog* (Cleveland, 1909), partially reprinted in Towana Spivey, ed., *A Historical Guide to Wagon Hardware & Blacksmith Supplies* (Lawton, Okla.: Museum of the Great Plains, 1979), 179; "Silver's Upright Post Drills," *The Hub* 36 (June 1894): 220; "Does Machinery Pay?"5.

25. Derry and Williams, *Short History of Technology*, 352–53; advertisement for the Standish foot-power hammer, "Trade News," *The Hub* 29 (September

1887): 415; "Does Machinery Pay?" 5; "Machinery and Tools," 453–54, 458, 461; Hutton, "Report on Machine Tools and Wood-Working Machinery," *Tenth Census, 1880*, 22:5–20.

26. "Does Machinery Pay?" 5; Ben Riker, *Pony Wagon Town* (New York: Bobbs-Merrill, 1948), 126–27; Cray Bros., *Carriage and Wagon Materials* (1902), 70; Millers Falls Co., *Catalog of Millers Falls Company* (Millers Falls, Mass., 1887; reprint, Mendham, N.J.: Astragal Press, n.d.), 34.

27. "Does Machinery Pay?" 6; "Machine Tufting of Cushions and Backs," *The Carriage Monthly* 39 (September 1903): 169; "Novelty Tufting Machine," *The Carriage Monthly* 42 (April 1906): 61; Raun, "Use of Steam Power," 97; "Stitching Machines in the Carriage Shop," *The Carriage Monthly* 40 (April 1904): 115; Singer Manufacturing Co., advertising brochure for Singer's "Class 67" sewing machine (n.p., 1904), in "Carriage Parts" subject file, NMAH; advertisement for the Elliot Dash Stitching Machine Co., *The Hub* 26 (February 1885): 794; advertisement for Valentine & Co., *The Hub* 15 (February 1874): 380; "Labor-Saving Appliances," 237; advertisement for Sherwin, Williams & Co., *The Coach Painter* 1 (March 1880): 36.

28. Betsy Hunter Bradley, *The Works: The Industrial Architecture of the United States* (New York: Oxford University Press, 1999), 1–8, 262; Hunter, *Waterpower*, 432–36.

29. Newspaper advertisement for James Goold, 7 April 1837, in Goold file, GCRL; Bradley, *The Works*, 29–31; H. F. LeMont, "A New Factory in an Old Industry," *The Carriage Monthly* 49 (October 1913): 65.

30. *G.& D. Cook & Co.*, frontispiece.

31. Bradley, *The Works*, 13, 65–66; plant descriptions in trade press articles, as well as thousands of examples from fire insurance maps, letterhead illustrations, and similar sources, all confirm these general layout patterns in wagon and carriage works.

32. "Literature on Carriage Building and Accessories," *The Carriage Monthly* 40 (April 1904): 194; "Biographies of Prominent Carriage Draftsmen," *The Carriage Monthly* 40 (April 1904): 145; "Vehicle Workers," *The Hub* 31 (February 1890): 855; Richardson, *Practical Carriage Building*, 2:v; "Is Knowledge of the French Rule Necessary?" *The Hub* 30 (April 1888): 29; Don H. Berkebile, ed., *Carriage Terminology: An Historical Dictionary* (Washington, D.C.: Smithsonian Institution Press, 1978), s.v. "French Rule," credits Zablot with being the originator of the system, but it was clearly a team effort.

33. "Biographies of Prominent Carriage Draftsmen," *The Carriage Monthly* 40 (April 1904): 145; Berkebile, *Carriage Terminology*, s.v. "French Rule"; Cyrus W. Saladee, "The French Rule—Introduction," *The Coach-Maker's Illustrated Monthly Magazine* 2 (February 1855): 24; "A Body-Maker's Testimony, *The New York Coach-Maker's Magazine* 8 (December 1866): 103; "Vehicle Workers," 855; "History of the Technical School," 134–35; "Working Drafts," 382, 384; "French Rule," 300; "Literature on Carriage Building," 194–95, says Thomas

began his series in 1867, while Houghton in "The French Rule; Introduction," *The Hub* 13 (April 1871): 5, says it began in February 1869.

34. Ezra M. Stratton, "A Whisper in the Ears of Those Concerned," *The New York Coach-Maker's Magazine* 2 (September 1859): 76; Saladee, "The French Rule—Introduction," 24; "History of Working Drafts in the Carriage Monthly," *The Carriage Monthly* 40 (April 1904): 200M; "French Rule," 300; Berkebile, *Carriage Terminology*, s.v. "French Rule."

35. "Our Prospects," *The Coach-Makers' International Journal* 2 (October 1866): 17; Biographies of Prominent Carriage Draftsmen," 146; Isaac D. Ware, ed., *The Coach-Makers' Illustrated Hand-Book*, 2d ed. (Philadelphia, 1875; reprint, Mendham, N.J.: Astragal Press, 1995), 43.

36. *Thirteenth Annual Report*, 2:718–19; J. B. Hampton, *A Pocket Manual for the Practical Mechanic; or The Carriage Maker's Guide* (Indianapolis: Frank H. Smith, 1886), 7–8, 30–32.

37. *Thirteenth Annual Report*, 2:718–19; Hampton, *Carriage Maker's Guide*, 30–32.

38. *Thirteenth Annual Report*, 2:718–19; Hampton, *Carriage Maker's Guide*, 16–18.

39. *Thirteenth Annual Report*, 2:718–19; Ware, *Coach-Makers' Illustrated Hand- Book*, 106–109.

40. *Thirteenth Annual Report*, 2:718–19; Ware, *Coach-Makers' Illustrated Hand- Book*, 106–109.

41. *Thirteenth Annual Report*, 2:718–21; Riker, *Pony Wagon Town*, 149; Ware, *Coach-Makers' Illustrated Hand-Book*, 176–77, 287, 359–60.

42. *Thirteenth Annual Report*, 2:718–21; Riker, *Pony Wagon Town*, 154; Ware, *Coach-Makers' Illustrated Hand-Book*, 198, 287.

43. *Thirteenth Annual Report*, 2:718–21; Ware, *Coach- Makers' Illustrated Hand-Book*, 295–97.

44. *Thirteenth Annual Report*, 2:718–21.

45. Riker, *Pony Wagon Town*, 149.

46. John K. Brown, "Design Plans, Working Drawings, National Styles: Engineering Practice in Great Britain and the United States, 1775–1945," *Technology and Culture* 41 (April 2000): 198–99; Ohio Carriage Manufacturing Co., *Those Good Split Hickory Buggies and How They Are Made* (Columbus, Ohio, 1913), 5, in NMAH.

47. Alliance Carriage Co., *Successful Salesman, "On the Road for the Alliance Carriage Co."* (Cincinnati: Alliance Carriage Co., 1895),7, in NMAH; Studebaker Brothers Manufacturing Co., *Catalogue and Price List, Studebaker Wagons* (South Bend, Ind., 1883), 11, 13, 78, 80, GCRL.

48. *Thirteenth Annual Report*, 2:720–21; Ohio Carriage Manufacturing Co., *Portfolio and Book of Split Hickory Vehicles for 1910* (Columbus, Ohio, 1910; reprint, Hampshire, U.K.: John Thompson, n.d.), 15; J.L.H. Mosier, "Recalling Early History," *The Hub* 50 (October 1908): 242.

49. James Brewster, "Autobiography of James Brewster," 1856, chap. 13, typescript copy, NHCHS; "Pioneer Carriage Builder," 129.

50. While providing some useful facts *The Carriage Monthly's* fortieth anniversary issue also gives a mildly hagiographical account of Jacob Huntington: "The Wholesale Carriage Industry in the United States," *The Carriage Monthly* 40 (April 1904): 95, and "Pioneer Carriage Builder," 129.

51. "Wholesale Carriage Building in the Western States," *The Carriage Monthly* 40 (April 1904): 166, 168; "Wholesale Carriage Building in the Eastern and Middle States," *The Carriage Monthly* 40 (April 1904): 184; Riker, *Pony Wagon Town*, 42, 287; Charles E. Tuttle, "The Walborn & Riker Company, Saint Paris, Ohio," *The Carriage Journal* 17 (Autumn 1979): 70. Widespread remarketing of vehicles under different names raises interesting questions about the contemporary worth of builder's plates as a means of positive identification.

52. Sears, Roebuck & Co., *Consumer's Guide*, catalog no. 104 (Chicago, 1897; reprint, New York: Chelsea House, 1968), 709; Montgomery Ward & Co., *Catalogue and Buyer's Guide*, no. 56 (Chicago, 1894–95; reprint, Northfield, Ill.: D.B.I. Books, 1977), 556–58; Tuttle, "Walborn & Riker Company," 70.

53. *G. & D. Cook & Co.;* Brewster & Baldwin, *Catalogue of Carriages* (New York, 1860), copy in GCRL.

54. Alliance Carriage Co., *Successful Salesman.*

55. *Portfolio and Book of Split Hickory Vehicles*, 4–7; *Those Good Split Hickory Buggies*, 1–4.

56. *Catalogue and Price List, Studebaker Wagons* (1883), 2.

57. *Portfolio and Book of Split Hickory Vehicles;* 1908 advertisement for the Columbus Carriage and Harness Company and 1908 advertisement for Sears & Roebuck's Solid Comfort Surreys reproduced in Charles Philip Fox, *Working Horses* (Whitewater, Wis.: Heart Prairie Press, 1990), 11, 117; advertisement for Fouts & Hunter's Cozy Cab buggy, *The Breeder's Gazette* (22 September 1909): 498; other trade names and trade marks appear in Meyer, *Memories of the Buggy Days*, 79, 80–81, 112–13, 148.

58. Meyer, *Memories of the Buggy Days*, 9–11.

59. *Portfolio and Book of Split Hickory Vehicles*, 10; Hounshell, *American System to Mass Production*, 263–64.

60. *World's Carriage Building Center*, 11–12; "Hard on Cincinnati," *The Hub* 28 (August 1886): 310; "Rise and Progress of Carriage Building in Cincinnati," *The Carriage Monthly* 40 (April 1904): 107; "Reports from Subordinate Unions," *The Coach-Makers' International Journal* 2 (September 1866): 15.

61. "Rise and Progress," 107–11; A. P. Daire, "Popular Carriages at Popular Prices" (part 1) *The Hub* 30 (December 1888): 672; "Carriage Building in Amesbury," 116–20; "Origin of Amesbury Industry."

62. "Rise and Progress," 108; Daire, "Popular Carriages" (part 1), 672–73.

63. Ezra Stratton, "How Carriage Makers Are Made out West," *The New*

York Coach-Maker's Magazine 2 (September 1858): 61–62; *Those Good Split Hickory Buggies*, 46.

64. Walborn & Riker advertisement reprinted in Tuttle, "Walborn & Riker Company," 61–62; Sears, Roebuck & Co., *Consumer's Guide* (1897), 709.

65. "The Carriage-Builders' Convention," *The Hub* 14 (December 1872): 204–205; "History of the Carriage Builders' National Association, 1872–1913," *The Carriage Monthly* 50 (October 1914): 44, 46; "The C.B.N.A. Warranty," *The Carriage Monthly* 42 (January 1907): 282; "On Warranty," *The Carriage Monthly* 43 (April 1907): 18.

66. Brewster & Baldwin, *Catalogue of Carriages* (1860), 86–91; *G. & D. Cook & Co.*, 6- 7; "Rise and Progress," 110; sales brochure for the Jacob Hoffman Wagon Company (Cleveland: J. B. Savage, [ca. 1885]) in the collection of the Western Reserve Historical Society, Cleveland, Ohio; A. P. Daire, "Popular Carriages at Popular Prices" (part 2) *The Hub* 30 (January 1889): 752.

67. "The Apotheosis of Cheapness," *The Hub* 27 (July 1885): 235; Riker, *Pony Wagon Town*, 27; Montgomery Ward & Co., *Catalogue and Buyer's Guide* (1894–95), 556–57; Sears, Roebuck & Co., *Consumer's Guide*, catalog no. 111 (Chicago, 1902; reprint, New York: Bounty Books, 1969), 363, 367.

Chapter 4. The Coming of Parts

1. "Trade News," *The Hub* 29 (October 1887): 481. Chapter epigraphs: M. T. Richardson, ed., *Practical Carriage Building* (New York, 1892; reprint, Mendham, N.J.: Astragal Press, n.d.), 1:116; "Cleveland, Ohio," *The Hub* 37 (October 1895): 469.

2. William Louis Gannon, "Carriage, Coach, and Wagon: The Design and Decoration of American Horse-Drawn Vehicles" (Ph.D. diss., Iowa State University, 1960), 183–85; Richard Hegel, *Carriages from New Haven: New Haven's Nineteenth-Century Carriage Industry* (Hamden, Conn.: Archon, 1974), 22; John L. H. Mosier, "Reminiscent: 1858–1908," *The Hub* 50 (October 1908): 241; "The Late Thomas Skelly," *The Carriage Monthly* 38 (January 1903): 331; "Supplemental Summary," *The Carriage Monthly* 40 (April 1904): 200II; "Accessory Product Manufacturers in the Eastern and Middle States," *The Carriage Monthly* 40 (April 1904): 193.

3. Don H. Berkebile, ed., *Carriage Terminology: An Historical Dictionary* (Washington, D.C.: Smithsonian Institution Press, 1978), s.v. "Coach Lace"; "Lace Making in the United States," *The Carriage Monthly* 40 (April 1904): 180–81; U.S. Bureau of the Census, "Manufactures in the States and Territories," *Census Reports: The Seventh Census of the United States, 1850* (Washington, D.C., 1850), 130–35; U.S. Bureau of the Census, "Introduction," *Census Reports: The Eighth Census of the United States, 1860* (Washington, D.C., 1865), cii.

4. U.S. Bureau of the Census, "Manufactures by States, Territories, and Districts," *Census Reports: The Third Census of the United States, 1810* (Philadelphia, 1814), 12, 43; U.S. Bureau of the Census, "Manufactures by States," *Cen-*

sus Reports: The Fourth Census of the United States, 1820 (Washington, D.C., 1823), 9; R. A. Salaman, *Dictionary of Woodworking Tools, c. 1700–1970, and Tools of Allied Trades* (London: George Allen & Unwin, 1975; reprint, Newtown, Conn.: Taunton Press, 1990), s.v. "wheelwright"; Ezra M. Stratton, *The World on Wheels; or, Carriages, with Their Historical Associations from the Earliest to the Present Time* (New York, 1878; reprint, New York: Benjamin Blom, 1972), 397–98, 402.

5. "History of Wheel Construction," *The Carriage Monthly* 40 (April 1904): 144C; "S. N. Brown, Dayton, Ohio," *The Carriage Monthly* 40 (April 1904): 139; "New Haven Wheel Co.," *The Carriage Monthly* 40 (April 1904): 180; "Supplemental Summary," 200GG; "Accessory Product Manufacturers in the Eastern and Middle States," 190, 193; "Accessory Product Manufacturers in the Western and Southern States," *The Carriage Monthly* 40 (April 1904): 200E.

6. Berkebile, *Carriage Terminology*, s.vv. "Patent Wheel," "Sarven Wheel"; "History of Wheel Construction, with Improvements in Manufacture," *The Carriage Monthly* 40 (April 1904): 144C; advertisement for the New Haven Wheel Company, *The Hub* 14 (November 1872): 194.

7. Berkebile, *Carriage Terminology*, s.v. "Warner Wheel"; "History of Wheel Construction," 144C; Howard M. DuBois, "Tests of Vehicle Wheels," *Journal of the Franklin Institute* (July 1886): 36–37; advertisement for the Royer Wheel Company, *The Hub* 28 (August 1886): 324; major patent wheel types and legal disputes may be conveniently traced in the trade journal excerpts reprinted in Don Peloubet, ed., *Wheelmaking: Wooden Wheel Design and Construction* (Mendham, N.J.: Astragal Press, 1996), 194–212.

8. Henry P. Jones, "Memoirs of a Wheel Maker," *The Hub* 50 (October 1908): 243–44.

9. "Totals of Manufactures," *Eighth Census, 1860;* U.S. Bureau of the Census, "General Statistics of Manufactures," *Census Reports: The Ninth Census of the United States, 1870* (Washington, D.C., 1872), 3:443.

10. Berkebile, *Carriage Terminology*, s.vv. "Axle," "Axle-Skein," "Collinge Axle," "Mail Axle," "Thimble-Skein."

11. "Old Time Axle and Spring Makers," *The Carriage Monthly* 40 (April 1904): 144G; Dalzell Axle Company, *How Carriage Axles Are Made* (South Egremont, Mass., [ca. 1905]), "carriage parts" subject file, NMAH.

12. "Henry Timken," *The Carriage Monthly* 40 (April 1904): 125; *History of the Timken Company* (n.p.: The Timken Company, 1990), 5–9.

13. Berkebile, *Carriage Terminology*, s.vv. "C-Spring," "Suspension"; "Accessory Product Manufacturers in the Eastern and Middle States," 188; "Old-Time Axle and Spring Makers," 144G, puts the Rowlands in the spring business as of 1842, but this appears to be an error. The 1850 census lists wagon and carriage hardware firms in Connecticut, Massachusetts, New York, and Pennsylvania, some of which almost certainly produced vehicle springs; "Manufactures in the States and Territories," *Seventh Census, 1850*, 59.

14. "The Spring and Axle Industry," *The Carriage Monthly* 40 (April 1904): 127.

15. Ibid., 125; Berkebile, *Carriage Terminology*, s.v. "Side Bar."

16. Berkebile, *Carriage Terminology*, s.vv. "Brewster Gear," "Timken Spring"; "Spring and Axle Industry," 124–25; "Changes in Forms and Suspensions of Carriages in Twenty-Five Years," *The Hub* 39 (October 1897): 432; "A Man of Ideas: The Story of Henry Timken, Carriage Builder and Founder of a World-Wide Enterprise, *The Carriage Journal* 17 (spring 1980): 185–86.

17. "General Statistics of Manufactures," *Ninth Census, 1870*, 3:476; advertisements for the Cleveland Spring Company, *The Hub* 24 (March 1883): 759 and 29 (April 1887): 56.

18. "Supplemental Summary," 200GG; "Accessory Product Manufacturers in the Eastern and Middle States," 189; "History of Carriage Bodies and Parts," *The Carriage Monthly* 40 (April 1904): 144A; advertisement for the New Haven Folding Chair Company, *The Coach-Makers' International Journal* 4 (May 1869): 121.

19. Supplemental Summary," 200JJ, 200LL; "Accessory Product Manufacturers in the Western and Southern States," 200A, 200C, 200E.

20. "Accessory Product Manufacturers in the Eastern and Middle States," 188, 191; "Accessory Product Manufacturers in the Western and Southern States," 200D; "Some Essentials in Carriage Trimming," *The Carriage Monthly* 37 (September 1901): 204, 211; "Trade News," *The Hub* 31 (June 1889): 215; "Carriage Lamps," *The Hub* 42 (February 1901): 507–9, 512.

21. Scott W. Robison, ed., *History of the City of Cleveland: Its Settlement, Rise and Progress* (Cleveland: Robison & Crockett and the Sunday World, 1887), 498–99; "Trade News," *The Hub* 15 (April 1873): 24, 13 (August 1871): 95, 13 (September 1871): 116, 17 (January 1876): 336; "Report of the Centennial Exhibition," *The Hub* 18 (August 1876): 186; "The Centennial Judges' Report," *The Hub* 18 (November 1876): 315; "Trade News," *The Hub* 19 (August 1877): 219; advertisement for Topliff & Ely, *The Hub* 20 (February 1879): 552.

22. Robison, *History*, 498–99; Berkebile, *Carriage Terminology*, s.vv. "Bow," "Bow Socket," "Top"; "Topliff's Metal Bows," *The Hub* 22 (May 1880): 73; advertisement for I. N. Topliff, *The Hub* 29 (April 1887): 52.

23. "Trade News," *The Hub* 20 (June 1878): 140; Robison, *History*, 499; "Trade News," *The Hub* 22 (April 1880): 34, 22 (May 1880): 82; Ohio, Chief State Inspector of Workshops and Factories, *Third Annual Report of the Chief State Inspector of Workshops and Factories, 1886* (Columbus: Westbote, 1887), 42; "Trade News," *The Hub* 22 (September 1880): 272, 24 (December 1882): 565, 25 (February 1884): 728; Ohio, State Inspector of Shops and Factories, *First Annual Report of the State Inspector of Shops and Factories, 1884* (Columbus: Westbote, 1885), 30; "Trade News," *The Hub* 26 (April 1884): 44, 28 (April 1886): 44, 28 (July 1886): 248, 28 (March 1887): 785, 29 (June 1887): 191, 29 (January 1888): 724, 32 (July 1890): 291, 33 (March 1892): 569; advertisement for the I. N. Topliff Manufacturing Company, *The Hub* 29 (November 1887): 577;

Peter C. Marzio, *The Democratic Art: Pictures for a 19th Century America* (Boston: David R. Godine; Fort Worth: Amon Carter Museum of Western Art, 1979), 322.

24. "Trade News," *The Hub* 29 (June 1887): 191, 29 (September 1887): 415, 32 (July 1890): 291; Robison, *History*, 499; "Trade News," *The Hub* 29 (March 1888): 875, 33 (August 1891): 272; "Representative Men of the Vehicle Industry," *The Carriage Monthly* 40 (September 1904): 362.

25. "Trade News," *The Hub* 30 (May 1888): 128, 30 (December 1888): 681, 33 (January 1892): 493, 32 (July 1890): 291; "The End of a Gigantic Litigation," *The Hub* 36 (August 1894): 370; "Trade News," *The Hub* 35 (November 1893): 688.

26. "Accessory Product Manufactures in the Eastern and Middle States," 188; "Paint and Varnish," *The Carriage Monthly* 40 (1904): 144B; Charles B. Sherron, "The Story of 'The Hub,'" *The Hub* 50 (October 1908): 234.

27. U.S. Bureau of the Census, "Totals of Manufactures," *Census Reports: The Tenth Census of the United States, 1880* (Washington, D.C., 1882), 2:8–9, 12, 17–18, 20–25, 30–31; David D. Van Tassel and John J. Grabowski, eds., *The Encyclopedia of Cleveland History* (Bloomington: Indiana University Press, 1987), s.v. "Industry"; *Industries of Cleveland: Commerce and Manufactures for the Year 1878* (Cleveland: Richard Edward, 1879), 102, 198; "Office," *The Hub* 16 (December 1874): 267; "Fire Losses by the Carriage and Accessory Trades during the Year 1888," *The Hub* 30 (March 1889): 912–14; "Trade News," *The Hub* 25 (February 1884): 728, 29 (September 1887): 415; *Industries of Cleveland: A Resume of the Mercantile and Manufacturing Progress of the Forest City* (Cleveland: Elstner, 1888) 148; "Trade News, *The Hub* 29 (December 1887): 649, 34 (March 1893): 595; "Made in Cleveland," *The Carriage Monthly* 51 (September 1915): 43; "The Late Charles J. Forbes," *The Carriage Monthly* 50 (December 1914): 56.

28. William Ganson Rose, *Cleveland: The Making of a City* (Cleveland: World, 1950), 372; Van Tassel and Grabowski, *Encyclopedia*, s.v. "Glidden Coatings & Resins Division (Imperial Chemical Industries)"; "Trade News," *The Hub* 31 (December 1889): 690; "Harmony and Good Will," *The Carriage Monthly* 43 (July 1907): 109; "New Varnish Plant Covers Seventeen Acres," *The Carriage Monthly* 43 (February 1908): 347; Rose, *Cleveland*, 372, erroneously reports the firm as buying out the Forest City Paint & Varnish Co. in 1875 (a fact repeated in Van Tassel and Grabowski) when the latter was still in business as late as 1902: "Trade News," *The Hub* 43 (March 1902): 569.

29. Van Tassel and Grabowski, *Encyclopedia*, s.v. "Sherwin-Williams Company"; Rose, *Cleveland*, 333; Luther H. Schroeder, *The Story of Sherwin-Williams* (Cleveland: Sherwin-Williams, [ca. 1954]); "Trade News," *The Hub* 18 (April 1876): 20; advertisements for Sherwin, Williams & Co., *The Coach Painter* 1 (February 1880): 16, 1 (April 1880): 56; "Trade News," *The Hub* 25 (February 1884): 727; "Obituary Notices" [Edward P. Williams], *The Carriage Monthly* 39 (June 1903): 93.

30. Van Tassel and Grabowski, *Encyclopedia*, s.v. "Sherwin-Williams Com-

pany"; Rose, *Cleveland*, 333; "Trade News," *The Hub* 28 (February 1887): 717, 26 (July 1884): 276, 31 (February 1890): 858; "Latest News in the Trade," *The Carriage Monthly* 39 (April 1903): 28; advertisement for the Sherwin-Williams Co., *The Spokesman* 24 (August 1901): 403; "The Sherwin-Williams 'Perfect Method' of Carriage Painting," *The Carriage Monthly* 37 (January 1902): 377–78; "Sherwin-Williams Co. Celebrates with Golden Jubilee," *The Hub* 58 (November 1916): 28.

31. Robert Friedel, "Crazy about Rubber," *Invention & Technology* 5 (winter 1990): 45- 8; John S. Bowman, ed., *The Cambridge Dictionary of American Biography* (Cambridge: Cambridge University Press, 1995), s.v. "Charles Goodyear."

32. Friedel, "Crazy about Rubber," 48–49; Bowman, *Cambridge Dictionary of American Biography*, s.v. "Charles Goodyear"; Berkebile, *Carriage Terminology*, s.vv. "Rubber," "Rubber Cloth"; advertisement for J. W. Munson in *The New York Coach-Maker's Magazine* 2 (May 1860): unpaginated advertising following p. 236; advertisements for Flesche & Perpente and the Odorless Rubber Company in *The Coach-Makers' International Journal* 4 (September 1869): unpaginated advertising following p. 193.

33. H. T. Firestone, "The Rubber-Tire Industry," *The Carriage Monthly* 40 (April 1904): 200EE; Berkebile, *Carriage Terminology*, s.v. "Rubber Tires."

34. Firestone, "Rubber-Tire Industry," 200EE; Berkebile, *Carriage Terminology*, s.v. "Rubber Tires"; Bowman, *Cambridge Dictionary of American Biography*, s.v. "Benjamin F. Goodrich"; Harlan Hatcher, *The Western Reserve: The Story of New Connecticut in Ohio* (Indianapolis: Bobbs-Merrill, 1949), 271–78.

35. Firestone, "Rubber-Tire Industry," 200EE; Berkebile, *Carriage Terminology*, s.v. "Rubber Tires"; Bowman, *Cambridge Dictionary of American Biography*, s.v. "Benjamin F. Goodrich"; Hatcher, *Western Reserve*, 271–78; "Accessory Product Manufacturers in the Eastern and Middle States," 188, 190, 192; "Accessory Product Manufacturers in the Western and Southern States," 200A–200F; "Supplemental Summary," 200LL; "Tires of Today," *The Carriage Monthly* 37 (June 1901): 95.

36. Van Tassel and Grabowski, *Encyclopedia*, s.v. "Industry."

37. Cleveland City Directory, 1870–71, 1876; *Industries of Cleveland* [1887], 75, 186; Van Tassel and Grabowski, *Encyclopedia*, s.v. "Lamson and Sessions Company"; Rose, *Cleveland*, 331; *Industries of Cleveland, 1878*, 92; "Trade News," *The Hub* 31 (August 1889): 364, 35 (March 1894): 1020; "Smith-Shop," *The Hub* 15 (March 1874): 387–88; "Items of Interest," *The Hub* 43 (September 1901): 267, 43 (October 1901): 320; "The Cleveland Bolt and Mfg. Co.," *The Carriage Monthly* 51 (October 1915): 82; "Items of Interest," *The Hub* 45 (August 1903): 178; advertisement for the National Screw & Tack Company, *The Hub* 44 (1902–1903): unpaginated advertising section.

38. W. David Lewis, *Iron and Steel in America* (Greenville, Del.: Hagley Museum, 1976), 9–14; Robert B. Gordon, *American Iron, 1607–1900* (Baltimore: Johns Hopkins University Press, 1996), 7–8, 14.

39. Lewis, *Iron and Steel*, 13–15, 23; Gordon, *American Iron*, 15–17.

40. "Does Machinery Pay?" *The Hub* 17 (April 1875): 5; H. A. Schwartz, *American Malleable Cast Iron* (Cleveland: Penton, 1922), 8–10; Gordon, *American Iron*, 257.

41. Schwartz, *American Malleable*, 9–13, 16, 36; Gordon, *American Iron*, 257.

42. "Wagon Building: Transition from Malleable Iron to Open Hearth Cast Steel," *The Carriage Monthly* 42 (May 1906): 35; Schwartz, *American Malleable*, 9–13, 16, 36.

43. Schwartz, *American Malleable*, 15–19; Civil Engineers' Club of Cleveland, *Visitors' Directory to the Engineering Works and Industries of Cleveland, Ohio* (n.p., 1893), 67; *Industries of Cleveland* [1888], 89, 104; "Death of the President of the Eberhard Manufacturing Co.," *The Hub* 55 (September 1913): 216.

44. "Death of the President," 216; *Industries of Cleveland* [1888], 89, 104; Schwartz, *American Malleable*, 21; Eberhard Manufacturing Co., *Saddlery Hardware, Catalogue and Price List no. 8* (Cleveland, 1887); Eberhard Manufacturing Co., *Eberhard Manufacturing Co.'s Production in Malleable Iron Casting of Carriage and Wagon Hardware . . . Catalog no. 8* (Cleveland, 1915); both catalogs in NMAH.

45. "Trade News," *The Hub* 29 (October 1887): 481; *Industries of Cleveland* [1888], 104; "Trade News," *The Hub* 32 (April 1890): 55; "Items of Interest," *The Hub* 40 (September 1898): 442; *Visitors' Directory*, 67; Ohio, Department of Inspection of Workshops, Factories, and Public Buildings, *Twenty-Fourth Annual Report of the Department of Inspection of Workshops, Factories and Public Buildings, 1907* (Columbus: F. J. Heer, 1908), 57.

46. Rose, *Cleveland*, 351–52; Elroy M. Avery, *A History of Cleveland and Its Environs* (Chicago: Lewis Publishing, 1918), 2:291–93; *Visitors' Directory*, 67–68; Eberhard evidently had two primary catalog series. Wagon and carriage hardware: no. 1 (ca.1880–81?), no. 2 (ca.1883?), no. 3 (1888), no. 4 (1893), no. 5 (1896), no. 6 (1900), no. 7 (1906), and no. 8 (1915). Saddlery and harness hardware: nos. 1–7 (ca.1880–85), no. 8 (1887), no. 13 (1909), and likely beyond. A partial list of automobile catalogs includes nos. 1–2 (ca. 1900–13?) and no. 3 (ca. 1915?). Eberhard Manufacturing Company, *Catalog no. 8* [wagon and carriage hardware], 7; "Trade News," *The Hub* 25 (April 1883): 32, 30 (May 1888): 128; "Wagon Building," 35.

47. The best extant sources indicate Brown & Curtiss's establishment as 1876 or 1878; by the latter date the name appeared in the city directory. Cleveland City Directory, 1872–78; *Industries of Cleveland, 1878*, 88; "Made in Cleveland," 45; Van Tassel and Grabowski, *Encyclopedia* s.v. "Root & McBride Co."; "Death Ends Work of Lee M'Bride," *Plain Dealer* (21 April 1909): 16.

48. Cleveland Hardware & Forging Company, *75 Years of Progress: Cleveland Hardware & Forging Company* (Cleveland, [ca.1956]); "Made in Cleveland," 45; "Cleveland, Ohio," *The Hub* 37 (October 1895): 470; advertisement for the

Cleveland Hardware Company, *The Hub* 24 (February 1883): 703; "Trade News," *The Hub* 25 (November 1883): 498.

49. "Cleveland, Ohio," 470; "Trade News," *The Hub* 29 (May 1887): 116; specific items come from two representative 1897–98 catalogs: Cleveland Hardware Company, *Wagon Hardware Catalogue* and *Carriage Hardware Catalogue* (Cleveland, 1897–98), both in company records.

50. Cleveland City Directory, 1886–88; "Trade News," *The Hub* 29 (September 1887): 415–16; "Made in Cleveland," 45, indicates the fire in June 1892, but contemporary trade press news has it as the previous summer: "Trade News," *The Hub* 33 (August 1891): 272; Sanborn Fire Insurance Map, Cleveland, 1886 (corrected to 1894), vol. 1, sheet 16.

51. "Cleveland Hardware Co.'s Factories," *The Hub* 42 (March 1901): 565; *75 Years of Progress;* Sanborn Fire Insurance Map, Cleveland, 1896 (corrected to 1910), vol. 1, sheet 58, and vol. 3, sheet 399; Cleveland Hardware Company, *Catalog* (Cleveland, [ca.1906]), incomplete catalog in company records; Cleveland Chamber of Commerce Exposition Committee, *Souvenir Book of the Cleveland Industrial Exposition, June 7–19, 1909* (Cleveland: J. B. Savage, 1909), 53; "Made in Cleveland," 45.

52. "Trade News," *The Hub* 31 (August 1889): 365; "Death Ends Work of Lee M'Bride," 16; "Associate Members Elect Officers," *The Carriage Monthly* 43 (October 1907): 219; Thomas A. Kinney, "Flowers on the Roof: Charles E. Adams and Industrial Reform at Cleveland Hardware Company," *Timeline* 18 (July/August 2001): 16–27. Avery, *History*, 24, and several other popular histories of Cleveland erroneously report Adams assuming the presidency of the firm in 1891 instead of 1909.

53. "Carriage and Wagon Materials," *Tenth Census, 1880*, 2:25–26; U.S. Bureau of the Census, "Carriage and Wagon Materials," *Census Reports: The Eleventh Census of the United States, 1890* (Washington, D.C., 1895), 6:150–51.

54. "Carriage Materials: Their Influence on the Carriage Industry," *The Hub* 43 (February 1902): 510.

55. Cleveland Hardware Company, *Carriage Hardware Catalogue* (1897–98), 10; Columbia Spring Company, *Catalog* (1893), 4–7.

56. *Industries of Cleveland* [1888], 126.

57. Henry C. Mercer, *Ancient Carpenters' Tools*, 5th ed. (Doylestown, Penn.: Bucks County Historical Society, 1975), 222; Salaman, *Dictionary*, s.v. "Lathe, Wheelwright's"; "No. 1 Patent Automatic Hub Turning Machine," *The Hub* 33 (January 1892): n.p., reprinted in Peloubet, *Wheelmaking*, 89; "An Extensive Hub Manufactory," *The New York Coach-Maker's Magazine* 2 (August 1859): 56.

58. "The Hayes Hub Boring and Mortising Machine," *The Coach-Makers' Illustrated Monthly Magazine* 4 [?] (February 1858): 17–18; "The Guard Wheel Machine," *The New York Coach-Maker's Magazine* 1 (November 1858): 120; advertisement for the Egan Company, *The Hub* 30 (November 1888): 651; advertisement for Defiance Machine Works, *The Hub* 29 (March 1888): 907; adver-

tisement for J. A. Fay & Company, *The Hub* 29 (April 1887): 53; "Heavy Automatic Double Chisel Hub Mortising Machine," *The Hub* 50 (August 1908): 175–76; M. Powis Bale, *Woodworking Machinery: Its Rise, Progress, and Construction* (London: Crosby, Lockwood & Co., 1880; reprint Lakewood, Colo.: Glen Moor, 1992), 201–2; Bale misnames the firm "Silver and Denning"; advertisement for the Silver Mfg. Co., *The Hub* 40 (September 1898): 468.

59. Carolyn C. Cooper, *Shaping Invention: Thomas Blanchard's Machinery and Patent Management in Nineteenth-Century America* (New York: Columbia University Press, 1991), 190–93, 232; Peter Haddon Smith, "The Industrial Archeology of the Wooden Wheel Industry in America," (Ph.D. diss., George Washington University, 1971), 26; "History of Machine-Made Wheels in America," *The Hub* 18 (November 1876): 315–16; "A Triumph of American Ingenuity," *The Hub* 33 (August 1891): n.p., reprinted in Peloubet, *Wheelmaking*, 100.

60. "Labor-Saving Machines," *The Coach-Makers' International Journal* 6 (December 1870): 43; "Spoke Driving Machine," *The Carriage Monthly* 9 (July 1873): n.p., "Improved Spoke Facing Machine," *The Carriage Monthly* 28 (March 1893): n.p., and "Hosler's Patent Spoke Driving Machine," *The Hub* 35 (June 1893): n.p., reprinted in Peloubet, *Wheelmaking*, 99, 102, 103; advertisement for the Silver Mfg. Co., *The Hub* 40 (September 1898): 468; advertisement for the Defiance Machine Works, *The Hub* 29 (March 1883): 907; advertisement for J. A. Fay & Co., *The Hub* 29 (April 1887): 53; Bale, *Woodworking Machinery*, 204.

61. Cooper, *Shaping Invention*, 211, 214–17; "Carriage Wood Bending," *The Carriage Monthly* 40 (April 1904): 114.

62. Cooper, *Shaping Invention*, 218–22; Salaman, *Dictionary*, s.v. "Wheelwright"; Smith, "Industrial Archeology," 12, 30–31; George Sturt, *The Wheelwright's Shop* (Cambridge: Cambridge University Press, 1923), 98–99; H. G. Shepard, "The Principles of Wood Bending," *The Hub* 24 (March 1883): 737; "Patent Automatic Rim and Felloe Bending Machine," *The Hub* 34 (June 1892): 111–12; advertisement for J. A. Fay & Co., *The Hub* 29 (April 1887): 53; "Carriage Wood Bending," 114–15.

63. "New Wheel-Tread Sanding and Equalizing Machine," *The Hub* 28 (January 1887): n.p., "A Brace of New and Indispensable Wheel Machines," *The Hub* 33 (September 1891): n.p., "New Sectional and Bent Felloe Borer," *The Hub* 35 (December 1893): n.p., and "Bending Machine," *The Hub* 37 (July 1895): n.p., reproduced in Peloubet, *Wheelmaking*, 122–23, 126–28; advertisement for J. A. Fay & Co., *The Hub* 29 (April 1887): 53; advertisement for the Defiance Machine Works, *The Hub* 29 (March 1888): 907, "An Improvement for Bending Tires," *The Carriage Monthly* 11 (April 1875): n.p., "Furnaces for Heating Tires," *The Hub* 24 (May 1882): n.p., and "Our Tire Heater," *The Carriage Monthly* 21 (June 1885): n.p., reproduced in Peloubet, *Wheelmaking*, 135, 145–47.

64. Advertisement for West's American Tire Setter," *The Coach-Makers' In-*

ternational Journal 8 (March 1873): n.p., reprinted in Peloubet, *Wheelmaking*, 165; George Worthington Company, *1909 Catalog* (Cleveland, 1909), partially reprinted in Towana Spivey, ed., *A Historical Guide to Wagon Hardware & Blacksmith Supplies* (Lawton, Okla.: Museum of the Great Plains, 1979), 188–89; Cray Bros., *Carriage and Wagon Materials* (Cleveland, 1902), 20–21; "A Perfected Cold Hydraulic Tire Setting Machine," *The Hub* 36 (June 1894): 220.

65. "The Wholesale Carriage Industry in the United States," *The Carriage Monthly* 40 (April 1904): 95; A. P. Daire, "Popular Carriages at Popular Prices" (part 1), *The Hub* 30 (December 1888): 673; "Rise and Development of the Carriage Building Industry in America," *The Carriage Monthly* 40 (April 1904): 100–101; "The Dealer and the Salesman," *The Carriage Monthly* 40 (April 1904): 200W, 200X; "The Supplanting of the Jobber," *The Carriage Monthly* 40 (April 1904): 93; Van Tassel and Grabowski, *Encyclopedia*, s.vv. "William Bingham," "W. Bingham Company."

66. Van Tassel and Grabowski, *Encyclopedia*, s.v. "George Worthington Company"; Worthington, *1909 Catalog; Industries of Cleveland* [1888], 62.

67. "Made in Cleveland," 51; Cray Bros., *Carriage and Wagon Materials* (1902), 1 andfollowing.

Chapter 5. An Empire of Taste

1. C. H. E. Redding, "A Fifty Year Old Association," *The Automotive Manufacturer* 64 (September 1922): 21. Chapter epigraphs: James Brewster, "Lectures by James Brewster to his Apprentices in 1826–7," 50, manuscript notebook in GCRL; "James Brewster: The Father of American Carriage Making," *Varnish* 4 (August 1891): 351.

2. James Brewster, "Autobiography of James Brewster," 1856, chaps. 1 and 2, typescript copy, NHCHS.

3. Ibid., chap. 2.

4. Ibid., chaps. 1–3; Brewster, "Lectures," 181–82; Ezra Stratton, *The World on Wheels; or, Carriages, with their Historical Associations from the Earliest to the Present Time* (New York, 1878; reprint, New York: Benjamin Blom, 1972), 424. In 1858 Ezra Stratton reported that he thought James Goold and Jason Clapp also served their apprenticeships in Chapman's shop, a speculation repeated in later trade press pieces: Ezra Stratton, "Biography of James Brewster," *The New York Coach-Maker's Magazine* 1 (June 1858): 1; "Pioneers of the Carriage Trade," *The Hub* 50 (October 1908): 237–39. But Clapp was six years older than Brewster and eight years older than Goold, the latter actually apprenticing in Clapp's own carriage shop in Pittsfield, Mass., in 1810. While Clapp was from Northampton, no evidence has come to light about his being apprenticed in Chapman's shop. See "Pioneers of the Carriage Trade," 238–39; and William D. Goold, "The Story of an Old House," typescript copy of speech to the Albany Rotary Club, 4 November 1921, and "Memorial of James Goold," typescript copy of address given at Goold's funeral, 3 October 1879, both in GCRL.

5. Brewster, "Autobiography," chaps. 3, 4, and 8.

6. Ibid., chap. 4.

7. Ibid., chaps. 5 and 6; J. Leander Bishop, *A History of American Manufactures* (Philadelphia: Edward Young, 1868; reprint, New York: Augustus M. Kelley, 1966), 3:427–28; Chauncey Jerome, *History of the American Clock Business* (New York: F. C. Dayton, 1860), 134–35; *Dana Collection Scrapbook* 127:14, in NHCHS; Walter A. Elwell, ed., *Evangelical Dictionary of Theology* (Grand Rapids, Mich.: Baker Book House, 1984), s.v. "Timothy Dwight."

8. Brewster, "Lectures," 182; Brewster, "Autobiography," chaps. 6, 7, and 9.

9. Brewster, "Autobiography," chap. 10.

10. Ibid., chap. 12; Bishop, *History of American Manufactures*, 3:427; advertisement for James Brewster, carriage manufacturer, in *The Connecticut Journal*, 26 June 1821, in *Dana Collection Scrapbook* 127:13.

11. Brewster, "Autobiography," chap. 13; Brewster advertisement in *The Connecticut Journal*, 8 June 1824, in *Dana Collection Scrapbook* 127:13; Stratton, *World on Wheels*, 427; "Obituary: Henry Brewster," *The Hub* 29 (October 1887): 479; Lawrence, Bradley & Pardee, *Illustrated Catalogue of Carriages* (New Haven, 1862; reprint, New York: Dover Publications, 1998), 5.

12. Brewster, "Autobiography," chap. 6; Brewster, "Lectures," preface, 72–73, 80–82, 160, 174–75, 182.

13. Brewster, "Autobiography," chap. 14.

14. Brewster advertisements in *Dana Collection Scrapbook* 127:13; Brewster, "Lectures," 49–50; typical carriage wheels had anywhere between twelve and sixteen spokes depending on the style and weight of the vehicle.

15. Brewster, "Autobiography," chap. 5; *Evangelical Dictionary of Theology*, s.vv. "Timothy Dwight," "New Haven Theology," "Scottish Realism."

16. Brewster, "Autobiography," chap. 5; *Evangelical Dictionary of Theology*, s.v. "Timothy Dwight"; Brewster, "Lectures," 72, 76–77, 79–80, 184.

17. Brewster, "Autobiography," chaps. 15–19.

18. Ibid., chap. 11; Brewster, "Lectures," preface, 5, 7–8.

19. Brewster, "Lectures," preface, index, 1–3, 9; James F. Babcock, who had known Brewster since his own boyhood in New Haven in the 1820s, noted Brewster's aversion to political strife as well as his enthusiastic temperament, but the connection I make between them is my own; James F. Babcock, "Address upon the Life and Character of the Late James Brewster" (New Haven, 1867), 15–16, 25.

20. Brewster, "Autobiography," chaps. 15–19; Brewster, "Lectures," preface, 1–3.

21. Letters of testimony from his former apprentices in the 1850s appear in handwritten original of Brewster, "Autobiography"; James Brewster, "An Address Delivered at Brewster's Hall," (New York: Isaac J. Oliver, 1857), remarks by E. W. Andrews, 28–37.

22. Stratton, *World on Wheels*, 413–15, 428–29; Brewster, "Autobiography," chap. 23.

23. Stratton, *World on Wheels*, 425–26, 428–29; Brewster, "Autobiography," chaps. 12 and 23; Lawrence, Bradley & Pardee, *Catalog* (1862), 5.

24. Brewster's involvement with the Mechanics' Institute bulks large in accounts of his life, but the exact details vary. Carl Bode's generally fine history of the lyceum movement, *The American Lyceum: Town Meeting of the Mind* (New York: Oxford, 1956), 51, traces the confusing name changes of the New Haven group but provides no documentation. Brewster himself gives better detail in his writings, which are largely substantiated by those of his apprentices and others involved with the group; see Brewster, "Autobiography," chaps. 6 and 21; Brewster, "Lectures," 194–98; Brewster, "Address," remarks by E. W. Andrews, 28, 37–38; Babcock, "Address," 22, 27–28. Brewster's characteristic modesty (or perhaps his political convictions) prevented him from making prominent notice of the Jackson carriage in his own writing, but others acknowledged it: "Death of James Brewster," *The New York Coach-Maker's Magazine* 8 (January 1867): 124; "Testimonial to Messrs. Brewster & Co., of Broome-Street, New-York," *The Hub* 21 (April 1879): 16.

25. Brewster, "Autobiography," chaps. 23–24; Brewster, "Lectures," 198–200; James Brewster, New Haven, to J. H. Conklin, n.p., 22 May 1861, in Brewster lawsuit copy book (BLCB), 70, manuscript copy book of materials related to James B. Brewster's 1869 lawsuit against Henry Brewster, James W. Lawrence, and John W. Britton, GCRL; "A Brewster & Collis Business-Card," *The Hub* 32 (March 1891): 977; Lawrence, Bradley & Pardee, *Catalog* (1862), 5.

26. Brewster, "Autobiography," chap. 24; "Regulations of This Establishment," *The Hub* 26 (February 1885): 767; Babcock, "Address," 11; a lithograph of Brewster's factory in 1860—long after he had vacated the premises—can be found in *Dana Collection Scrapbook* 127:11.

27. Brewster, "Autobiography," chap. 24.

28. Ibid., chaps. 26–27; Brewster, "Lectures," 202–6; "Father of American Carriage Making," 397.

29. Brewster, "Autobiography," chaps. 27–28; Brewster, "Lectures," 206.

30. Brewster, "Autobiography," chap. 28; Lawrence, Bradley & Pardee, *Catalog* (1862), 5.

31. Change of business announcement by James Brewster in 1837, 1838 advertisement for James Brewster & Son and James Brewster & Company, and 1840 advertisement for Brewster & Allen in *Dana Collection Scrapbook* 127:12; sworn statement of John W. Britton, October 1869, BLCB, 128–30; though incorrect in a few details (Britton incorrectly recalls Allen's first name as Matthew and admits to not remembering the exact firm name), the essential information agrees with other sources. Further research has yet to verify the details of Benjamin's flight and return, but the 1842 bankruptcy is well established.

A brief and highly sanitized account appears in the 1891 *Varnish* biography of Benjamin, which could only have been based on material provided by him and may even have been written by him. It goes beyond mere coyness to outright falsehood: the anecdote about Benjamin's alleged determination to repay creditors comes almost verbatim from chap. 29 of James Brewster's autobiography where he describes his brother Joseph's response to his own bankruptcy. Whoever recycled the material—and all evidence points to Benjamin—did so knowingly.

32. Brewster, "Autobiography," chap. 30; Britton's October 1869 statement, BLCB, 129–30; "John W. Britton" [obituary], *The Carriage Monthly* 28 (September 1886): 162.

33. Britton obituary, 161–62; "Relations between American Carriage Builders and their Workmen," *The Hub* 26 (July 1884): 272–73; Brewster & Company's 1860 statement, BLCB, 116; Britton's October 1869 statement, BLCB, 130.

34. John W. Britton, New Orleans, to James B. Brewster, New York, 31 March 1852, BLCB, 92–94; John W. Britton, New Orleans, to James B. Brewster, New York, 18 May 1852, BLCB, 99–101; John W. Britton, New Orleans, to James B. Brewster, New York, June 1852, BLCB,104–106; John W. Britton, New Orleans, to James B. Brewster, New York, 7 June 1852, BLCB, 106–108; Britton obituary, 162; "Biography of John C. Denman, Esq.," *The New York Coach-Maker's Magazine* 2 (July 1859): 21–22.

35. Britton to Brewster, 31 March 1852, 93; Britton obituary, 162; Phineas Taylor Barnum, *Barnum's Own Story: The Autobiography of P.T. Barnum, Combined and Condensed from the Various Editions Published during His Lifetime*, ed. Waldo R. Browne (New York: Viking, 1927), 265–66; Britton's October 1869 statement, 130–131.

36. Britton's October 1869 statement, 131–33; sworn statement of James W. Lawrence, 22 October 1869, BLCB, 39; sworn statement of Henry Brewster, 21 October 1869, BLCB, 41–42.

37. Henry Brewster's 21 October 1869 statement, 41–42; sworn statement of John W. Britton, 1 November 1869, BLCB, 60; James B. Brewster, New York, to Henry Brewster, Bridgeport [or possibly New York], 27 June 1856, BLCB, 61; Brewster & Company, New York, to James B. Brewster, New York, 12 January 1860, BLCB, 64.

38. Brewster & Company's 1860 statement, 127.

39. James Brewster, New Haven, to James B. Brewster, New York, 7 January 1856, BLCB, 86–87; James Brewster, New Haven, to James B. Brewster, New York, 27 October 1856, BLCB, 91–92; John L. H. Mosier, "Reminiscent: 1858–1908," *The Hub* 50 (October 1908): 242.

40. James Brewster, New Haven, to James B. Brewster, New York, 16 March 1857, BLCB, 90–91; James Brewster, New Haven, to James B. Brewster, New York, 1 April 1857, BLCB, 81; James B. Brewster, New York, to Henry Brewster, New York, 11 July 1857, BLCB, 139; James Brewster, New

Haven, to James B. Brewster, New York, 27 July 1857, BLCB, 84; James Brewster, New Haven, to Henry Brewster, New York, 15 October 1857, BLCB, 82; James Brewster, New Haven, to James B. Brewster, New York, 15 October 1857, BLCB, 87–88.

41. James Brewster, New Haven, to James B. Brewster, New York, 17 July 1858, BLCB, 84–86; Henry Brewster, New York, to James Brewster, New Haven, 28 September 1858, BLCB, 88.

42. Britton's October 1869 Statement, 133–34.

43. James Brewster, New Haven, to James B. Brewster, New York, 17 July 1858, BLCB, 84–85; Britton's October 1869 statement, 134–35; Brewster & Company's 1860 statement, 121–22; April 1859 advertisement for Brewster & Company, BLCB, 111; sworn statement of Isaac W. Britton, [ca. 1869], BLCB, 45–46.

44. James B. Brewster, New York, to Brewster & Company, New York, 19 January 1860, BLCB, 67; Britton's October 1869 statement, 132.

45. Brewster, "Autobiography," chaps. 32 and 36; Brewster, "Address," 3–7, 9; James Brewster, New Haven, to James B. Brewster, New York, 31 July 1857, BLCB, 96; Britton's October 1869 statement, 132.

46. Britton's October 1869 statement, 132–33; Babcock, "Address," 37, 39.

47. Britton's October 1869 statement," 129–30, 133; advertisement for Brewster & Company, 11 October 1869, BLCB, 112–13; the Long Island Museum (formerly the Museums at Stony Brook) has a J. B. Brewster coupe from this period having the "Old House of Brewster" label: see Museums at Stony Brook, *The Carriage Collection* (Stony Brook, N.Y., 1986), 56–57.

48. Affidavit by Thomas Williams, 21 October 1869, BLCB, 11; sworn statement of James B. Brewster, 21 October 1869, BLCB, 19.

49. Summary of suit James B. Brewster against Henry Brewster, James W. Lawrence, and John Britton, BLCB, 1; summons for relief, October 1869, BLCB, 3; complaint, October 1869, BLCB, 4–5; injunction by order, 13 October 1869, BLCB, 9.

50. Copy of disputed advertisement no. 1, BLCB, 6–7; copy of disputed advertisement no. 2, BLCB, 7–8.

51. Opinion of Morris S. Miller, Esq., Counsel, BLCB, 151–53.

52. Advertisement for Brewster & Company, April 1859, BLCB, 40–41.

53. Opinion of Morris S. Miller, 153–55; summary of suit, 1.

54. "A Discovery in New York," *The Chicago Republican*, 23 August 1869, BLCB, 27–34.

55. Ibid., 27–34.

56. Ibid., 27–34.

57. "New York City Gossip," *The New York Coach-Maker's Magazine* 9 (December 1867): 89; "Fifth Avenue Carriage Repository," *The New York Coach-Maker's Magazine* 9 (February 1868): 136–37; "Brewster & Co.'s Repository," *The New York Coach-Maker's Magazine* 9 (April 1868): 165; "Gleanings," *The Coach-Makers' International Journal* 6 (June 1870): 130. "Brewster Factory Re-

moves from New York City to Long Island," *The Carriage Monthly* 46 (November 1910): 34, contains some useful geographic information but misdates the erection of the Brewster & Company repository by several years.

58. Brewster & Company, New York, to Morris S. Miller, New York, [ca. October 1869], BLCB, 145; Britton's October 1869 statement, 136; Opinion of Morris S. Miller, 150; "James B. Brewster," *Rider and Driver* 15 March 1902, reprinted in *The Carriage Journal* 3 (autumn 1965): 67–68. This last source erroneously credits Benjamin with inventing the side bar concept itself, but he did originate its most popular variant; see "Changes in Forms and Suspensions of Carriages in Twenty-Five Years," *The Hub* 39 (October 1897): 432; Don H. Berkebile, ed., *Carriage Terminology: An Historical Dictionary* (Washington, D.C.: Smithsonian Institution Press, 1978), s.v. "Brewster Gear"; "The Spring and Axle Industry," *The Carriage Monthly* 40 (April 1904): 124; Mosier, "Reminiscent: 1858–1908," 242; and advertisement for the Brewster Patent Combination Cross Springs, *The Carriage Monthly* 16 (June 1880): viii.

59. "Brewster & Co.'s New Factory, New-York City," *The Hub* 16 (June 1874): 76–79; Hamilton S. Wicks, "The Manufacture of Pleasure Carriages," American Industries no. 4 *Scientific American* 40 (8 February 1879): 79–80; Betsy H. Bradley, *The Works: The Industrial Architecture of the United States* (New York: Oxford, 1999), 29–34.

60. "Brewster & Co.'s New Factory," 76–79; Wicks, "Manufacture of Pleasure Carriages," 79–80.

61. "Brewster & Co.'s New Factory," 76–79; Wicks, "Manufacture of Pleasure Carriages," 79–80; "What Is the Kolber Laundaulet, and Is It Public Property?" *The Hub* 24 (May 1882): 90.

62. "Brewster & Co.'s New Factory," 76–79; Wicks, "Manufacture of Pleasure Carriages," 79–80.

63. "Brewster & Co.'s New Factory," 76–79; Wicks, "Manufacture of Pleasure Carriages," 79–80.

64. "Portrait of Henry Brewster," *The Hub* 20 (January 1879): 498; "Testimonial to Messrs. Brewster & Co., of Broome-Street, New-York," *The Hub* 21 (April 1879): 15–16; "The Brewster & Co. Quarter-Century Celebration," *The Hub* 23(August 1881): 242; "James B. Brewster," *Carriage Journal*, 68; "In Memoriam: John W. Britton, of New-York," *The Hub* 28 (August 1886): 286a; Paul H. Downing and Harrison Kinney, "Builders for the Carriage Trade," *American Heritage* 7 (October 1956): 96.

65. "In Memoriam: John W. Britton," *The Hub*, 286–87a; "John W. Britton," *The Carriage Monthly*, 161–63; "Trade Gossip of the Past Month," *The Hub* 28 (May 1886): 101; "Obituary: Henry Brewster," *The Hub* 29 (October 1887): 479.

66. "Father of American Carriage Making," 398; Brewster & Co.'s 1860 Statement, BLCB, 116; Downing and Kinney, "Builders for the Carriage Trade," 96–97.

67. "News among the Dealers," *The Carriage Monthly* 34 (February 1899): 348; "Latest News in the Trade," *The Carriage Monthly* 38 (June 1902): 107; "James B. Brewster," *The Carriage Monthly* 38 (April 1902): 2–3; "J. B. Brewster & Co. Bankrupt," *The Carriage Monthly* 44 (June 1908): 88; this last erroneously lists Duncan as "A. M. Duncan" instead of Henry M. Duncan.

68. Downing and Kinney, "Builders for the Carriage Trade," 96–97; Brewster, "Address," 26.

69. James Brewster, New Haven, to James B. Brewster, New York, 27 October 1856, BLCB, 91; James J. Flink, *The Car Culture* (Cambridge, Mass.: MIT Press, 1975), 1, notes that there were six automobiles on display; Downing and Kinney, "Builders for the Carriage Trade," 96–97.

Chapter 6. A Wagon Every Six Minutes

1. DeWitt C. Goodrich, *An Illustrated History of the State of Indiana* (Indianapolis: Richard S. Peale, 1875): 435, 613–17. Chapter epigraphs: Charles Arthur Carlisle, "The Life of Clement Studebaker," ca. 1902–8, typescript, 160, Local History Collection, St. Joseph County Public Library, South Bend, Ind.; "The Farm Wagon Industry," *The Carriage Monthly* 40 (April 1904): 128.

2. The Studebaker Family National Association, *The Studebaker Family in America* (Tipp City, Ohio, 1976), 75–77, 386–87, 396–97; Carlisle, "Life of Clement Studebaker," 5. The first of these (and, to a lesser degree, the associated website www.studebakerfamily.org/history.html) remains the best source on this family's sometimes confusing history. The second informs parts of the SFNA history, as well as this chapter; despite the cryptic "author unknown" on extant copies, internal clues and additional research reveal Clement's son-in-law and prominent Studebaker employee Charles A. Carlisle as the author. Carlisle evidently wrote most of it prior to Clement's death, the latter having vetted at least one draft, though Carlisle or an assistant added material through ca. 1908. Besides relying on his own knowledge and that of Clement and his wife and their children, Carlisle drew heavily upon his subject's extensive business and personal correspondence. In 1969 a family member donated a copy to the St. Joseph Public Library, and at least one other (in the collection of the Studebaker Family National Association, Tipp City, Ohio) is in existence. Despite its mildly hagiographical tone and sometimes questionable interpretations, the account is factually reliable and preserves large amounts of otherwise unrecorded information.

3. *Studebaker Family in America*, 76–77; Carlisle, "Life of Clement Studebaker," introduction, 5–7; Walter A. Elwell, ed., *Evangelical Dictionary of Theology* (Grand Rapids, Mich.: Baker Book House, 1984): s.vv. "Radical Reformation," "Pacifism"; "General Schematic History of Christian Churches," appendix C in Joel F. Harrington, ed., *A Cloud of Witnesses: Readings in the History of Western Christianity* (Boston: Houghton Mifflin, 2001); Mark A. Noll et al., eds., *Eerdmans' Handbook to Christianity in America* (Grand Rapids: William

B. Eerdmans, 1983), 71–72, 74–75; Alice Felt Tyler, *Freedom's Ferment: Phases of American Social History from the Colonial Period to the Outbreak of the Civil War* (Minneapolis: University of Minnesota Press, 1944), 111–15.

4. *Studebaker Family in America*, 76–77, 396–97; Carlisle, "Life of Clement Studebaker," 5–8; "Outline History of Studebaker, 1851–1921," typescript, 23, archives of the Studebaker National Museum, South Bend, Ind. The latter is a detailed time line written by company secretary M. L. Milligan sometime prior to 1960. Based on the extensive company scrapbooks in the Studebaker archives, its value lies in its near-total reliance on contemporaneous sources (newspaper clippings, articles, company literature, letters, and more).

5. *Studebaker Family in America*, 76–77; 396–97; Carlisle, "Life of Clement Studebaker," 11–17; Stephen Longstreet, *A Century on Wheels: The Story of Studebaker* (New York: Henry Holt, 1952), 7. Some sources give 1835 as the date of the family's move, but Clement stated on numerous occasions that it took place in 1836; Carlisle, "Life of Clement Studebaker," 11–17; George W. Knepper, *Ohio and Its People* (Kent, Ohio: Kent State University Press, 1989), 113; Dumas Malone, ed., *Dictionary of American Biography* volume 9 (New York: Charles Scribner's Sons, 1935), s.v. "Clement Studebaker"; *Studebaker Family in America*, 78–80.

6. *Studebaker Family in America*, 78–80; Carlisle, "Life of Clement Studebaker," 15–17, 20; Malone, *Dictionary of American Biography*, s.v. "Clement Studebaker"; "The First Studebaker Wagon Works," *The Carriage Monthly* 39 (August 1903): 153, shows a version of the Ashland homestead, though the caption fails to identify it as such.

7. Carlisle, "Life of Clement Studebaker," 17–22.

8. Ibid., 22–24, 28–30, 32; *Studebaker Family in America*, 396–97.

9. Carlisle, "Life of Clement Studebaker," 33–34; *Studebaker Family in America*, 83–84.

10. Carlisle, "Life of Clement Studebaker," 36–42; Malone, *Dictionary of American Biography*, s.v. "Clement Studebaker"; Henry and Clement may have been employed by the same firm during 1851, but the surviving accounts do not specify.

11. Carlisle, "Life of Clement Studebaker," 34, 36–42; *Studebaker Family in America*, 84; Malone, *Dictionary of American Biography*, s.v. "Clement Studebaker."

12. "Outline History," 1–2; Longstreet, *Century on Wheels*, 2; *An Illustrated Historical Atlas of St. Joseph County, Indiana* (Chicago: Higgins, Belden & Company, 1875), 28; "John Moler [*sic*] Studebaker," *The Carriage Monthly* 38 (August 1902): 167; "Death of J. M. Studebaker," *The Spokesman* 33 (March 1917): 94.

13. "Outline History," 1–2; *Illustrated Historical Atlas of St. Joseph County*, 28, 52; "John Moler [*sic*] Studebaker," 167; "Death of J. M. Studebaker," 94; William Weber Johnson, *The Forty-Niners*, The Old West Series (New York: Time-Life Books, 1974), 116–17, 142; Albert Russel Erskine, *History of the*

Studebaker Corporation (South Bend, Ind.: Studebaker Corporation, 1918): 17, 19, 21; Hinds's name appears variously as "Hind," "Hines," "Hine," among others; all were Americanized versions of a German surname, probably Heinz.

14. "The Late John M. Studebaker, Sr.," *The Vehicle Monthly* 53 (April 1917): 90; Johnson, *Forty-Niners*, 222–23; Erskine, *History*, 21; Longstreet, *Century on Wheels*, 16- 18; both Erskine and Longstreet contain numerous errors about the early history of the family and firm, but their segments on Hangtown rely heavily on J. M.'s eyewitness account. Longstreet refers to Hinds as "Joe," a name that appears in some other sources, possibly a nickname.

15. Carlisle, "Life of Clement Studebaker," 42, 44; "Outline History," 1–2; Malone, *Dictionary of American Biography*, s.v. "Clement Studebaker"; Longstreet, *Century on Wheels*, 25; *Illustrated Historical Atlas*, 28. I have been unable to completely reconcile the different dates in the above sources for the new brick shop building, but it appears to have been completed sometime during 1856–58.

16. Carlisle, "Life of Clement Studebaker," 44, places J. M.'s return in 1857, but most other sources agree that it was 1858; John A. Garraty and Mark C. Carnes, *American National Biography* (New York: Oxford University Press, 1999), s.v. "John Mohler Studebaker"; "John Moler [*sic*] Studebaker," 167; "Death of J. M. Studebaker," 94; *Studebaker Family in America*, 90.

17. *Studebaker Family in America*, 85–86; "Outline History," 2; "John Moler [*sic*] Studebaker," 167; Garraty and Carnes, *American National Biography*, s.v. "John Mohler Studebaker." Curiously, Carlisle, "Life of Clement Studebaker," 44, puts J. M.'s return in 1857 and Henry's exit from the company prior to the subcontracting job. But J. M.'s own recollections clearly indicate 1858, and period advertisements suggest that Henry did not officially leave the partnership until that same year—well after the contract work was under way, if not actually completed. While this may simply be an error in an otherwise reliable manuscript, it could conceivably have been a deliberate alteration intended to shield Henry from posthumous criticism.

18. *Studebaker Family in America*, 86; "Outline History," 2.

19. "Outline History," 2.

20. Carlisle, "Life of Clement Studebaker," 43–44; "Outline History," 2; the early advertisement and 1860 company data is reprinted in Longstreet, *Century on Wheels*, 26, 29–30.

21. *Studebaker Family in America*, 92.

22. "Outline History," 3; Garraty and Carnes, *American National Biography*, s.v. "John Mohler Studebaker."

23. Frank Hastings Hamilton, *A Practical Treatise on Military Surgery* (New York: Bailliere Brothers, 1861), 129–134; John S. Haller Jr., *Farmcarts to Fords: A History of the Military Ambulance, 1790–1925* (Carbondale: Southern Illinois University Press, 1992), 46–48; Frederick Tilberg, *Gettysburg National Military Park* (Washington, D.C.: National Park Service, 1952), 34; Ulysses S. Grant, *Personal Memoirs of U.S. Grant* (New York: Charles L. Webster, 1894), 449, 451,

481, 520, 617; Fletcher Pratt, *Civil War in Pictures* (New York: Garden City Books, 1955), 201.

24. Huston Horn, *The Pioneers*, The Old West Series (New York: Time-Life Books, 1974), 22–23, 87; Francis Parkman, *The Oregon Trail: Sketches of Prairie and Rocky-Mountain Life* (1847; reprint, New York: Random House, 1949), 36–37; Mark L. Gardner, *Wagons for the Santa Fe Trade: Wheeled Vehicles and Their Makers, 1822–1880* (Albuquerque: University of New Mexico Press, 2000), 31–40; "Outline History," 3, 6; *Studebaker Family in America*, 92.

25. Carlisle, "Life of Clement Studebaker," 44–45, 111–12; "Outline History," 11; *Studebaker Family in America*, 93.

26. Carlisle, "Life of Clement Studebaker," 108–9, 380–81.

27. "Outline History," 3–4; *Turners' Guide from the Lakes to the Rocky Mountains* (South Bend, Ind.: T. G. & C. E. Turner, 1868), frontispiece, 4–5.

28. "Outline History," 6–7; Goodrich, *Illustrated History*, 614–16.

29. "Outline History," 4, 6–7; Goodrich, *Illustrated History*, 435, 614–17.

30. Goodrich, *Illustrated History*, 435, 616–17.

31. *Illustrated Historical Atlas*, 28; "Mr. Guiet's Report of the Centennial Carriage Display," *The Hub* 19 (November 1877): 355.

32. *Illustrated Historical Atlas*, 28; "Mr. Guiet's Report," 355.

33. *Illustrated Historical Atlas*, 28; David A. Hounshell, *From the American System to Mass Production, 1800–1932: The Development of Manufacturing Technology in the United States* (Baltimore: Johns Hopkins University Press, 1984), 149–51.

34. The assertion, usually made casually and without evidence, appears throughout the abundant secondary literature on Studebaker. The photograph in question appears in "History of the Studebaker Brothers Manufacturing Company and the Studebaker National Museum," *Driving Digest*, no. 54 (1989): 14, among other places.

35. Hounshell, *American System to Mass Production*, 149, briefly discusses Studebaker's innovative use of electric welding, as well as the firm's considerable investment in heavy metalworking equipment. Much of this machinery appears in the lavishly illustrated commemorative booklet Studebaker issued for the Columbian Exposition of 1893: *Studebaker Souvenir, 1893* (South Bend, Ind.: Studebaker Brothers Manufacturing Company, 1893), GCRL.

36. *Turners' Guide*, frontispiece, 5; Studebaker Brothers Manufacturing Company, *Catalogue and Price List, Studebaker Wagons, 1883* (South Bend, Ind., 1883), 7, 10–11,19, 45, 76; undated [ca. 1870s] farm wagon handbill, reproduced in *100 Years on the Road* (South Bend, Ind.: Studebaker Corporation, 1952), 6.

37. *Turners' Guide*, frontispiece, 5; *Catalogue and Price List, Studebaker Wagons, 1883*, 7, 10–13, 77; undated [ca. 1870s] farm wagon handbill, reproduced in *100 Years on the Road*, 6; Don H. Berkebile, ed., *Horse-Drawn Commercial Vehicles* (New York: Dover Publications, 1989), glossary.

38. *Turners' Guide*, frontispiece, 5; *Catalogue and Price List, Studebaker Wagons, 1883*, 7, 10–13; Berkebile, *Horse-Drawn Commercial Vehicles*, glossary.

39. *Catalogue and Price List, Studebaker Wagons, 1883*, 11, 13, 80.

40. *Turners' Guide*, 5; "Outline History," 3, 5, 8, 14; p. 3 of this document mentions a La Porte branch established in 1860, but I have been unable to obtain any corroborative information; it may have been a short-lived agency.

41. *Turners' Guide*, 5; "Outline History," 3, 5, 8, 14.

42. "Outline History," 6–7; "Old Wagons Factor in Winning of Wars," *New York Herald Tribune*, 22 February 1953, sec. 10:17.

43. Handwritten copy of 1867 advertisement reproduced in Erskine, *History*, 30; "Outline History," 5; 1868 advertisement reprinted in *100 Years on the Road*, 2.

44. *Catalogue and Price List, Studebaker Wagons, 1883;* "Outline History," 12; Frank Frost, "Studebaker Carried a Tune," *Driving Digest*, no. 77 (1993): 12–14; Frank Frost, "Studebaker Poster Stamps," *Driving Digest*, no. 79 (1993): 52–53.

45. The novelty poster is reprinted in *100 Years on the Road*, 6; "Who Speaks First?" *The Carriage Monthly* 38 (January 1903): 355.

46. All pertinent correspondence from this episode may be found in U.S. Congress, Senate, *Letter from the Secretary of the Interior, Transmitting, in response to a resolution of the Senate relative to the rejection of certain bids for wagons for the Indian service, a report of the Commissioner of Indian Affairs and accompanying correspondence*, S. Ex. Doc. 210, 46th Congress, 2d sess., 1880 (Serial 1886).

47. Ibid.

48. Ibid.

49. Ibid.; Gardner, *Wagons for the Santa Fe Trade*, 73.

50. "Outline History," 8–9, 11, 14; United States Centennial Commission, *Official Catalogue of the U.S. Centennial Exhibition, 1876, Complete in One Volume* (Philadelphia: John R. Nagle, 1876), 299; Carlisle, "Life of Clement Studebaker," 146–47.

51. "Outline History," 11; *Catalogue and Price List, Studebaker Wagons, 1883*.

52. "Outline History," 11; *Catalogue and Price List, Studebaker Wagons, 1883*.

53. "John Moler [*sic*] Studebaker," 167; Carlisle, "Life of Clement Studebaker," 195–96; "The Carriage-Builders' Convention. Positive Appointment of the First Meeting," *The Hub* 14 (September 1872): 145; "The Carriage-Builders' Convention," *The Hub* 14 (December 1872): 14; "History of the Carriage Builders' National Association, 1872–1913," *The Carriage Monthly* 50 (October 1914): 44.

54. Garraty and Carnes, *American National Biography*, s.v. "Clement Studebaker"; Carlisle, "Life of Clement Studebaker," 203–4; "Carriage-Makers' Societies," *The Hub* 28 (September 1886): 369.

55. Carlisle, "Life of Clement Studebaker," 115; "Mr. Guiet's Report," 355; Studebaker's average factory wage figures come from "Outline History," 6, 15–

16, while the national averages for 1890 come from U.S. Bureau of the Census, "Detailed Statement for Selected Industries," *Census Reports: The Eleventh Census of the United States, 1890* (Washington, D.C., 1895), 6:675–78.

56. "Outline History," 6, 15, 17; Carlisle, "Life of Clement Studebaker," 44–45, 114; "Detailed Statement," *Eleventh Census, 1890*, 6:678; U.S. Bureau of the Census, "The Carriage and Wagon Industry," *Census Reports: The Thirteenth Census of the United States, 1910* (Washington, D.C., 1913), 10:835–36.

57. Carlisle, "Life of Clement Studebaker," 104–105, 111; *Studebaker Family in America*, 77–78; Studebaker sleigh and winter carriage advertisement, *Country Life in America*, December 1905, 235; "Outline History," 17.

58. "Outline History," 15–17; a period illustration of the Aluminum Wagon appears in Berkebile, *Horse-Drawn Commercial Vehicles*, 33; *Studebaker Family in America*, 77–78, 396–97.

59. "Outline History," 18; Carlisle, "Life of Clement Studebaker," 324–25, 327–28; "Latest News in the Trade," *The Carriage Monthly* 35 (December 1899): 318, 37 (June 1901): 102, 37 (September 1901): 224.

60. Pre–World War I advertising poster reproduced in *100 Years on the Road*, 10–11.

61. Carlisle, "Life of Clement Studebaker," 127, 446–52.

62. *Studebaker Family in America*, 89–92; "Outline History," 16, 18; Erskine, *History*, 43, 45.

Chapter 7. From Craftsman to Assembler?

1. "Reflections by an Old-Time Wood Worker," *The Hub* 36 (February 1895): 806. Chapter epigraphs: George A. Thrupp, quoted in Joanne Abel Goldman, "From Carriage Shop to Carriage Factory: The Effect of Industrialization on the Process of Manufacturing Carriages," in *Nineteenth Century American Carriages: Their Manufacture, Decoration and Use* (Stony Brook, N.Y.: Museums at Stony Brook, 1987), 15; "Carriage Materials: Their Influence on the Carriage Industry," *The Hub* 43 (February 1902): 510.

2. Basic biographical information on Wright comes from *Webster's Biographical Dictionary* (Springfield: G. & C. Merriam & Co., 1943) and *Encyclopedia Britannica*, New American Supplement, ed. Otis Day Kellogg (New York: The Werner Co., 1903), vol. 29; U.S. Bureau of Labor, *Thirteenth Annual Report of the Commissioner of Labor, 1898: Hand and Machine Labor* (Washington, D.C., 1899), 1:6; "Carriage Materials," 510.

3. J. B. Hampton, *A Pocket Manual for the Practical Mechanic; or The Carriage Maker's Guide* (Indianapolis: Frank H. Smith, 1886), 7; W. J. Rorabaugh, *The Craft Apprentice: From Franklin to the Machine Age in America* (Oxford: Oxford University Press, 1986), 133–34.

4. *Thirteenth Annual Report*, 2:714–19. The dates of the hand and machine methods were not absolute; in many cases the machine methods surveyed had been in use for some time, while in others hand methods lasted well beyond 1865.

5. *Thirteenth Annual Report*, 2:714–19 (the table does not specify water wheel or turbine); F. R. Hutton, "Report on Machine Tools and Wood-Working Machinery," in U.S. Bureau of the Census, *Census Reports: Tenth Census of the United States, 1880* (Washington, D.C., 1888), 22:228–30, 237–40.

6. *Thirteenth Annual Report*, 2:718–21; advertisement for the Stow Mfg. Co., *The Hub* 40 (September, 1898): 468; "Machinery and Tools in Modern Carriage and Wagon Factories," *The Carriage Monthly* 37 (March 1902): 454.

7. *Thirteenth Annual Report*, 2:672–75. The tube cutter may have been a form of punch press.

8. *Thirteenth Annual Report*, 2:682–83, 690–91, 708–11. The saw was probably a large frame saw. The table on 682–83 inexplicably refers to the hand method of hub turning as involving the use of a "turning machine," while the machine method used a lathe. Both hand and machine methods used lathes, the former a hand-powered great wheel lathe, the latter a special-purpose, steam-powered hub lathe.

9. Robert B. Gordon, "Who Turned the Mechanical Ideal into Mechanical Reality?" *Technology and Culture* 29 (October 1988): 744–78; Robert B. Gordon and Patrick M. Malone, *The Texture of Industry: An Archaeological View of the Industrialization of North America* (Oxford: Oxford University Press, 1994), 372.

10. Gordon and Malone, *Texture of Industry*, 236.

11. Gordon and Malone, *Texture of Industry*, 236, 359; Brooke Hindle and Steven Lubar, *Engines of Change: The American Industrial Revolution, 1790–1860* (Washington: Smithsonian Institution Press, 1986), 153–54.

12. Gordon and Malone, *Texture of Industry*, 372–73; "Proportioning Wheels," *The Hub* 35 (July 1893): 299. Gordon and Malone cite the experiences of Smithsonian curators with the Howe pin machine as an example of the skill required to keep sophisticated equipment operating properly.

13. H. J. Habakkuk, *American and British Technology in the Nineteenth Century: The Search for Labor-Saving Inventions* (Cambridge: Cambridge University Press, 1962), chaps. 4- 6; Edward P. Duggan, "Machines, Markets, and Labor: The Carriage and Wagon Industry in Late-Nineteenth-Century Cincinnati," *Business History Review* 51 (Autumn 1977): 308–9, 312- 13, 321–22.

14. "The Subdivision of Labor in American Carriage Shops and Its Results," *The Carriage Monthly* 43 (August 1907): 130; Douglas Harper, *Working Knowledge: Skill and Community in a Small Shop* (Chicago: University of Chicago Press, 1987), 21.

15. *Thirteenth Annual Report*, 2:672–75, 702–703.

16. *Thirteenth Annual Report*, 2:715–20; "Hope Ahead for the Skilled Carriage-Painter," *The Hub* 30 (May 1888): 113; "Subdivision of Labor in Carriage-Building," *The Hub* 17 (July 1875): 122.

17. "The Wheelwright of the Past, Present and Future," *The Hub* 36 (March 1895): 886.

18. "Hope Ahead for the Skilled Carriage Painter," 113.

19. "Does Machinery Pay?" *The Hub* 17 (April 1875): 5–6.

20. "Castle Garden in the Trimming Shop," *The Coach-Makers' International Journal* 3 (June 1868): 199–200, 3 (August 1868): 248–49, 4(October 1868): 5–6, and 4 (November 1868): 21; David A. Hounshell, *From the American System to Mass Production, 1800–1932: The Development of Manufacturing Technology in the United States* (Baltimore: Johns Hopkins University Press, 1984), 49–50.

21. *Thirteenth Annual Report*, 1:20; Ben Riker, *Pony Wagon Town* (New York: Bobbs-Merrill, 1948), 113.

22. Charles F. Raun, "Use of Steam-Power in Carriage-Building," *The Hub* 14 (July 1872): 97; A. P. Daire, "Popular Carriages at Popular Prices" (part 2), *The Hub* 30 (January 1889): 752; "Subdivision of Labor in Carriage-Building," 122; "Hope Ahead for the Skilled Carriage Painter," 113; "Custom Builders," *The Carriage Monthly* 40 (April 1904): 96; "The Conflict between Custom-Made and Wholesale Work," *The Hub* 26 (January 1885): 704; "What Is to Become of the Small Carriage Shops?" *The Hub* 19 (February 1878): 509–10; "Condensed Review of the Carriage Trade: 1887," *The Hub* 29 (January 1888): 711.

23. "Condensed Review," 711; "Trade News," *The Hub* 28 (January 1887): 648; Ohio, Department of Inspection of Workshops, Factories and Public Buildings, *Third Annual Report, 1886* (Columbus: Westbote, 1887), 39, 47, and *Fifth Annual Report, 1888* (Columbus: Westbote, 1889), 45.

24. "Trade Gossip of the Past Month," *The Hub* 26 (May 1884): 109; Daire, "Popular Carriages" (part 2), 752; "Subdivision of Labor in American Carriage Shops," 30; "What Is to Become of the Small Carriage Shops?" 509–10; "Every Carriage Builder His Own Teacher," *The Hub* 26 (November 1884): 557; "The Conflict," 704; "Condensed Review," 711. The Selle Gear Company's gable-end advertising can still be read on the company building, which as of early 2002 still stood in the city of Akron.

25. "Trade News," *The Hub* 29 (September 1887): 416.

26. "The Small and Large Builder," *The Carriage Monthly* 42 (October 1906): 192; "Custom Builders," 96; "Trade News," *The Hub* 29 (September 1887): 416; Cleveland City Directory, 1901, 1345–46.

27. Cleveland City Directory, 1891, 725; "Custom Builders," 96.

28. Raun, "Use of Steam-Power," 97; U.S. Bureau of the Census, "Carriage and Wagon Works," *Census Reports: The Tenth Census of the United States, 1880* (Washington, D.C., 1886), 20:411; Duggan, "Machines, Markets, and Labor," 315.

29. Jacob Hoffman Wagon Company, sales brochure (Cleveland: J. B. Savage, [ca. 1885]), Western Reserve Historical Society, Cleveland.

30. "Every Carriage-Builder His Own Teacher," 557; "Trade Gossip of the Past Month," *The Hub* 26 (May 1884): 109; "Subdivision of Labor in American Carriage Shops," 130; Harper, *Working Knowledge*, 21–24.

31. Charles Eckhart, "How Carriages Were Built in Early Days," *The Spokesman* 26 (September 1910): 368.

32. *Encyclopedia Britannica*, vol. 27, s.v. "Labor Organizations," by Carroll D. Wright.

33. Some sources report the name as the International Association of Coachmakers, but the Coach Makers' International Union became standard. "Our Trip to New Haven," *The Coach-Makers' International Journal* 2 (September 1866): 3–4; "Reports from Subordinate Unions," *The Coach-Makers' International Journal* 2 (January 1867): 77. "The History of 'The Carriage Monthly,'" *The Carriage Monthly* 40 (April 1904): 200Y, written many years later by *The Carriage Monthly* editors, contains a generally accurate account, though it significantly downplays Ware's considerable early union enthusiasm. "From a Veteran Accessory Manufacturer," *The Carriage Monthly* 40 (April 1904): 174, written by H. G. Shepard, a long-time manufacturer of bent wood carriage parts, recounts more about the origins of the union and its journal but without latter-day apologetics.

34. "I. D. Ware as a Carriage Builder," *The Carriage Monthly* 40 (April 1904): 220P; "History of 'The Carriage Monthly,'" 200Z; "Subordinate Unions," *The Coach-Makers' International Journal* 2 (November 1866): 48; "Our Charter," *The Coach-Makers' International Journal* 2 (June 1867): 146; "The Convention," *The Coach-Makers' International Journal* 2 (September 1866): 2; "Admission of Members," *The Coach-Makers' International Journal* 2 (May 1867): 136; "What Is Our Duty?" *The Coach-Makers' International Journal* 2 (April 1867): 113.

35. "I. D. Ware as a Carriage Builder," 220P; "History of 'The Carriage Monthly,'" 200Y–Z; Philadelphia City Directory, 1866–67; "From a Veteran Accessory Manufacturer," 174; Charles B. Sherron, "Early Days of 'The Carriage Monthly,'" *The Carriage Monthly* 40 (April 1904): 200R; author's survey of vols. 2–4 in the collections of the Ohio Historical Society, Columbus.

36. Letter to the editor, *The Coach-Makers' International Journal* 2 (October 1866): 24–26; "Reports from Subordinate Unions," *The Coach-Makers' International Journal* 2 (September 1866): 16; author's survey of vols. 2–4.

37. "Reports from Subordinate Unions," 15–16; "President's Journey Continued," *The Coach-Makers' International Journal* 2 (December 1866): 50–51; "Admission of Members," 136. Unions' secretive membership practices, combined with the accompanying tendency to publish sometimes exaggerated overall numbers, renders estimates of the CMIU's size at any given time rather difficult, but the following sources provide an approximate figure for both the number of locals and total membership to approximately April 1867: "The Convention," 2; "Letter from the President," *The Coach-Makers' International Journal* 2 (October 1866): 22; "Our Progress," *The Coach-Makers' International Journal* 2 (January 1867): 65; "What Is Our Duty?" 113.

38. "A Coach-Maker on a Rail," *The Coach-Makers' International Journal* 2 (September 1866): 10; "Letter from the President," *The Coach-Makers' International Journal* 2 (October 1866): 22–23; "President's Journey Continued,"

The Coach-Makers' International Journal 2 (November 1866): 37–38, 2 (December 1866): 51.

39. "President's Journey Continued," *The Coach-Makers' International Journal* 2 (December 1866): 51, 2 (January 1867): 69–71, and 2 (February 1867): 84; John W. Gosling," *The Carriage Monthly* 40 (April 1904): 128; "Geo. C. Miller," *The Carriage Monthly* 40 (April 1904): 200CC; "Our Monthly Report," *The Coach-Makers' International Journal* 2 (January 1867): 68.

40. Henry C. McLear's name is misspelled as "McLeary" in Harding's account of the dispute: "Official," *The Coach-Makers' International Journal* 2 (September 1866): 9; "Difficulty Settled," *The Coach-Makers' International Journal* 2 (October 1866): 21; editorial note, *The Coach-Makers' International Journal* 2 (September 1866): 3; letter to the editor, *The Coach-Makers' International Journal* 2 (March 1867): 101; letter to the editor, *The Coach-Makers' International Journal* 2 (September 1866): 11–12.

41. William Harding, "Reducing Wages in New York," *The Coach-Makers' International Journal* 2 (May 1867): 132–33; editorial note and reprint of *New York Herald* piece, *The Coach-Makers' International Journal* 2 (March 1867): 98; "Trade News," *The New York Coach-Maker's Magazine* 9 (June 1867): 11.

42. "Reports from Subordinate Unions," *The Coach-Makers' International Journal* 2 (May 1867): 137; "Death of John C. Parker," *The New York Coach-Maker's Magazine* 8 (June 1867): 13, contains Ezra Stratton's version of the events relating to Parker, with a rebuttal of the same appearing in Ware's "Far-Fetched," *The Coach-Makers' International Journal* 2 (July 1867): 162. The firm J. C. Parker & Co. was named in 1863 as successor to Parker, Brewster, and Baldwin (est. 1860), but both were located in Parker's 25th Street shop; for background on Parker and his connection with Benjamin Brewster, see chap. 5.

43. "Death of John C. Parker," 13; "Far-Fetched," 162; "Reports from Subordinate Unions," 137; Harding, "Reducing Wages," 132–33.

44. "The Rights of Labor Recognized," *The Coach-Makers' International Journal* 2 (May 1867): 129; "Behind Time," *The Coach-Makers' International Journal* 2 (September 1866): 1; editorial note, *The Coach-Makers' International Journal* 2 (November 1866): 35, 2 (February 1867): 83.

45. "Our Monthly Report," *The Coach-Makers' International Journal* 2 (November 1866): 34; "Reports from Subordinate Unions," *The Coach-Makers' International Journal* 2 (January 1867): 76, 2 (March 1867): 107; "History of 'The Carriage Monthly,'" 200Y; "Executive Department," *The Coach-Makers' International Journal* 2 (February 1867): 83; letter to the editor, *The Coach-Makers' International Journal* 2 (March 1867): 105–6, 2 (April 1867): 120.

46. "The Cincinnati Convention," *The Coach-Makers' International Journal* 2 (May 1867): 129–30; "Convention of the Journeymen's International Union," *The New York Coach-Maker's Magazine* 9 (October 1867): 65–67; "A Coach-Maker on a Rail," *The Coach-Makers' International Journal* 3 (March 1868): 126–28; "Letters from the President," *The Coach-Makers' International Journal* 3 (November 1867): 29, 2 (December 1867): 53, 3 (February 1868): 103;

"Union Directory," *The Coach-Makers' International Journal* 3 (February 1868): 120; "A Change," *The Coach-Makers' International Journal* 3 (February 1868): 121; "The Wise Man of Gotham," *The Coach-Makers' International Journal* 4 (February 1869): 66; Philadelphia City Directory, 1865–69; "I. D. Ware as a Carriage Builder," 200P. "History of 'The Carriage Monthly,'" 200Y, 200CC, while otherwise correct, contains some bibliographic errors about the journal, including stating that the name change occurred with vol. 5 (it was vol. 8).

47. *Encyclopedia Britannica*, vol. 29, s.v. "Mutual Benefit Societies," by Carroll D. Wright; "Carriage-Makers' Societies," *The Hub* 28 (September 1886): 370.

48. "John W. Britton" [obituary], *The Carriage Monthly* 28 (September 1886): 162; Phineas Taylor Barnum, *Barnum's Own Story: The Autobiography of P.T. Barnum, Combined and Condensed from the Various Editions Published During his Lifetime*, ed. by Waldo R. Browne (New York: Viking, 1927), 265–66.

49. James Brewster, New Haven, to James Conklin, n.p., 22 May 1861, copy in BLCB, 69–71, GCRL; Britton obituary, 162–63; John W. Britton, New Orleans, to James Brewster, New Haven, 31 March 1852, BLCB, 93; sworn statement of John W. Britton, October 1869, BLCB, 130–31; "Co-operative Labor Associations," *The New York Coach-Maker's Magazine* 9 (June 1867): 11–12; "Co-operation," *The Coach-Makers' International Journal* 2 (July 1867): 161–62.

50. "Co-operative Labor Associations," 11; Britton obituary, 162- 63; "Relations between American Carriage Builders and Their Workmen," *The Hub* 25 (January 1884): 649–50; this and subsequent installments reproduced Britton's testimony on these events before an 1883 U.S. Senate labor and capital committee. Hughes and Ludlow were two of the founders of Christian socialism, a movement attempting to leaven laissez-faire capitalism with Christianity's collective social values. Son of an English Quaker and cotton mill owner, Bright also sought such a balance, and by 1868 he was head of his government's Board of Trade. Mill, the most influential of the four, had long since achieved fame as a reform-minded social theorist, philosopher, and economist, and he was just completing a Parliamentary term when the committee's letters arrived; *Webster's Biographical Dictionary* (Springfield, Mass.: G. & C. Merriam Co., 1943), s.vv. "John Bright," "Thomas Hughes," "John Ludlow," "John Stuart Mill."

51. Britton obituary, 162–63; Brewster & Company, *Proposal for an Industrial Partnership* (New York, 1869), 1–7 (this copy includes the constitution written up soon thereafter).

52. Brewster & Company, *Proposal*, 1–7, constitution, 3–8.

53. "Relations between American Carriage Builders," 650; Britton obituary, 162–63, provides a reasonably accurate summary of the strike, though it benefits from comparison with Britton's own recollections in "Relations between American Carriage Builders," 650.

54. Brewster & Company, *Proposal*, 1–7, constitution, 3–8; Britton obituary, 162–63; "Relations between American Carriage Builders," 650.

55. "Carriage Makers' Societies," 370–71.

56. The trade papers covered these strikes in great detail; for a representative sample, see "The Great Strike of New-Haven Carriage Mechanics," *The Hub* 28 (April 1886): 37–40 and sequels; "Strike of the Wagon and Truck Mechanics in New-York City and Brooklyn," *The Hub* 28 (June 1886): 177–78; and "General Strike of the Cincinnati Carriage Mechanics," *The Hub* 28 (June 1886): 178–81. Edward M. Miggins, "Introduction," *Twenty Year History of District Assembly 47, Knights of Labor* (Cleveland, 1902; reprint, Cleveland: Greater Cleveland Labor History Society and the Cleveland AFL-CIO Federation of Labor, 1987), 2; Jack W. Skeels, "Early Carriage and Auto Unions: The Impact of Industrialization and Rival Unionism," *Industrial and Labor Relations Review* 17 (July 1964): 566; Wright, "Labor Organizations."

57. Skeels, "Early Carriage and Auto Unions," 566–69; George W. Knepper, *Ohio and Its People* (Kent: Kent State University Press, 1989), 305; Wright, "Labor Organizations."

58. Skeels, "Early Carriage and Auto Unions," 566–72; Wright, "Labor Organizations."

Chapter 8. That Damned Horseless Carriage

1. Henry M. Meyer, *Memories of the Buggy Days* (Cincinnati, 1965), 4–5, 52–53. Chapter epigraphs: Ben Riker, *Pony Wagon Town* (New York: Bobbs-Merrill, 1948), 301; "Proceedings of the 43rd Annual Convention of the Carriage Builders' National Association," *The Carriage Monthly* 51 (October 1915): 38.

2. Riker, *Pony Wagon Town*, 298, passes on the CBNA's claim to have invented the annual trade convention, a distinction not borne out by reliable research. The census data combines "Carriages and Wagons, Including Custom Work and Repairing" and "Carriages and Wagons, Factory Product" in U.S. Bureau of the Census, *Census Reports: The Eleventh Census of the United States, 1890* (Washington, D.C., 1895), 6:150–53, 675–85; the accessory industry figures would add 539 firms, 10,928 employees, and $16,262,293 in value of product—also only approximations of what were likely much larger numbers.

3. "Origin and History of the Carriage Builders' National Association," *The Carriage Monthly* 40 (April 1904): 102–6; "History of the Carriage Builders' National Association, 1872- 1913," *The Carriage Monthly* 50 (October 1914): 42–47; "A Fifty Year Old Association," *The Automotive Manufacturer* 64 (September 1922): 21–24; Charles Arthur Carlisle, "The Life of Clement Studebaker," ca. 1902–8, typescript, 199–200, Local History Collection, St. Joseph County Public Library, South Bend, Ind.

4. "Origin and History," 102–6; "History of the Carriage Builders' National Association, 1872–1913," 42–47.

5. "History of the Carriage Builders' National Association, 1872–1913," 42, 44; "1895—Cleveland—1915," *The Hub* 58 (September 1915): 5; "The Carriage Builders' National Association," 94; "The Carriage-Builders' Con-

vention; Positive Appointment of the First Meeting," *The Hub* 14 (September 1872): 145.

6. "History of the Carriage Builders' National Association, 1872–1913," 43–44; "Official Notice to Exhibitors," *The Hub* 57 (May 1915): 13–14; "Carriage Builders' National Association; The Twenty-Third Annual Convention," *The Hub* 37 (November 1895): 588.

7. Cleveland City Directory, 1896, 1116, 1118, 1222; "Carriage Builders' National Association Convention and Exhibition of 1895," *The Hub* 37 (October 1895): 465–67; "1895—Cleveland—1915," 5; "Carriage Builders' National Association; The Twenty-third Annual Convention," 550–51, 558, 602–4.

8. "Carriage Builders' National Association; The Twenty-third Annual Convention," 550–51, 558, 602–4; "1895—Cleveland—1915," 5–6; Riker, *Pony Wagon Town*, 300, 306–7; "Origin and History," 102–6; "History of the Carriage Builders' National Association, 1872–1913," 42–47.

9. "Carriage Builders' National Association; The Twenty-third Annual Convention," 550; "Business Troubles," *The Hub* 32 (December 1890): 722, 32 (March 1891): 980; Carlisle, "Life of Clement Studebaker," 125.

10. "Carriage Builders' National Association; The Twenty-third Annual Convention," 551, 564; "W. N. FitzGerald," *The Carriage Monthly* 40 (April 1904): 200; "The Late W. N. FitzGerald," *The Carriage Monthly* 40 (June 1904): 239; "International Outrage at Newark, N.J." *The New York Coach-Maker's Magazine* 9 (August 1867): 42–43; "Newark Outrage," *The Coach-Makers' International Journal* 3 (November 1867): 31–33; "1895—Cleveland—1915," 6; James J. Flink, *The Automobile Age* (Cambridge, Mass.: MIT Press, 1988), 17.

11. "Carriage Builder's National Association; The Twenty-third Annual Convention," 564–66; "Salutory," *The Horseless Age* 1 (November 1895): 1; "Horseless Vehicle Items," *The Hub* 38 (September 1896): 408.

12. The word "automobile" originated in early 1880s France as a general reference to self-propelled devices such as trolleys and torpedoes; during the 1890s it began to designate a four-wheeled, self-propelled passenger vehicle. "Automobile" gradually replaced "horseless buggy," "motor carriage," and "motor car," and by the early twentieth century was understood to indicate a four-wheeled vehicle powered by an internal combustion engine. *Oxford English Dictionary*, 2d ed. (1989), s.v. "automobile"; James J. Flink, *The Car Culture* (Cambridge, Mass.: MIT Press, 1975), 5–6, 8–9, 15; Flink, *Automobile Age*, 1, 6–7; Richard Wagar, *Golden Wheels: The Story of the Automobiles Made in Cleveland and Northeastern Ohio, 1892–1932* (Cleveland: John T. Zubal and the Western Reserve Historical Society, 1986), xiii, 53–63; Bryan Bunch and Alexander Hellemans, *The Timetables of Technology: A Chronology of the Most Important People and Events in the History of Technology* (New York: Touchstone, 1993), 181, 197, 205; E. D. Kennedy, *The Automobile Industry: The Coming of Age of Capitalism's Favorite Child* (New York: Reynal & Hitchcock, 1941), 9;

John S. Bowman, ed., *The Cambridge Dictionary of American Biography* (Cambridge: Cambridge University Press, 1995), s.v. "Francis Edgar and Freelan O. Stanley"; David D. Van Tassel and John J. Grabowski, eds., *The Encyclopedia of Cleveland History* (Bloomington: Indiana University Press, 1989), s.vv. "White Motor Corporation," "Rollin Henry White."

13. Flink, *Automobile Age*, 7–8; Flink, *Car Culture*, 15–16; Wagar, *Golden Wheels*, 64, 205–6; Bunch and Hellemans, *Timetables*, 258, 288; Michael Brian Schiffer, *Taking Charge: The Electric Automobile in America* (Washington, D.C.: Smithsonian Institution Press, 1994), 30–33, 49–50; Schiffer tends to underestimate the electric's persistent technological shortcomings.

14. Bunch and Hellemans, *Timetables of Technology*, 258, 263; Flink, *Car Culture*, 9–14, 16; Flink, *Automobile Age*, 10–12, 23–24; Kennedy, *Automobile Industry*, 13–14; "Manufacture in New England," *Motor Age* 1 (12 September 1899): 4; J. T. Sullivan, "New England a 1900 Leader," *Motor Age* 19 (2 March 1911): 1–3.

15. Flink, *Car Culture*, 16, 45–46; Flink, *Automobile Age*, 24; Kennedy, *Automobile Industry*, 14–15.

16. Van Tassel and Grabowski, *Encyclopedia*, s.vv. "Automotive Industry," "Alexander Winton," "Winton Motor Carriage Company," "Walter C. Baker," "Baker Materials Handling Company," "Rollin Henry White"; Wagar, *Golden Wheels*, 3–12, 22, 53–69, 204–10; Schiffer, *Taking Charge*, 117.

17. Reference to each may be found by alphabetical heading in Wagar, *Golden Wheels;* the "over eighty" figure includes automobiles made by firms headquartered elsewhere but with branch plants in Cleveland; Van Tassel and Grabowski, *Encyclopedia*, s.v. "Automotive Industry"; William Ganson Rose, *Cleveland: The Making of a City* (Cleveland: World, 1950), 503; L. V. Spencer, "Detroit, the City Built by the Automobile Industry," *The Automobile* 28 (10 April 1913): 791–94; Flink, *Automobile Age*, 24–25.

18. "The Future of the Horse Vehicle," *The Carriage Monthly* 48 (December 1912): 53; "The Passing of the Horse?" *The Carriage Monthly* 42 (April 1906): 24; "The Horse-Drawn Vehicle Business—Its Present Condition and Its Future Prospects," *The Carriage Monthly* 51 (July 1915): 54.

19. "Forty Years from Now—What?" *The Carriage Monthly* 40 (April 1904): 200NN.

20. Riker, *Pony Wagon Town*, 301; "Why Not Some Industrial Progress?" *The Hub* 54 (January 1913): 321–22; "The C.B.N.A. Needs Not Revolution, but Evolution," *The Carriage Monthly* 35 (October 1899): 194; "First Exhibit of an Automobile Chassis at a C.B.N.A. Convention," *The Carriage Monthly* 46 (September 1910): 41.

21. "When the Big Carriage Builder Enters the Motor Car Field," *The Carriage Monthly* 45 (April 1910): 80–81; "Buggy Type Automobiles," *The Carriage Monthly* 43 (October 1907): 49–52; Edward V. Rickenbacker, *Rickenbacker* (Englewood Cliffs, N.J.: Prentice-Hall, 1967), 39–47.

22. Much late Brewster & Company history comes from an undated reprint

of an article in the periodical of the Rolls-Royce Owners' Club (Francis N. Howard, "Brewster & Company," *Flying Lady*, n.d., 27–36), GCRL.

23. Untitled news item, *The Carriage Monthly* 41 (July 1905): 128.

24. "Opinions of Leaders in the Carriage and Automobile Industry for 1904," *The Carriage Monthly* 39 (March 1904): 391; Arthur W. Soutter, *The American Rolls-Royce: A Comprehensive History of Rolls-Royce of America, Inc.* (Providence, R.I.: Mowbray Company, 1976), 58; "Selling Second-Hand Vehicles," *The Carriage Monthly* 40 (January 1905): 503; "Increasing Sale of Carriages," *The Carriage Monthly* 41 (August 1905): 146; "J. B. Brewster & Co. Secure Acme," *The Carriage Monthly* 43 (August 1907): 60; "Brewster & Co. [*sic*] Get New Acme," *The Carriage Monthly* 43 (March 1908): 56; "Motor News in Brief," *The Carriage Monthly* 44 (December 1908): 60; "J. B. Brewster & Co. Bankrupt," *The Carriage Monthly* 44 (June 1908): 88; Howard, "Brewster & Company," 29.

25. Soutter, *American Rolls-Royce*, 58–59; "The Brewster Building, Long Island City, N.Y., *The Carriage Monthly* 46 (October 1910): 44; "Brewster Factory Removes from New York City to Long Island," *The Carriage Monthly* 46 (November 1910): 34.

26. Soutter, *American Rolls-Royce*, 58–59; "Brewster Factory Removes," 34.

27. Soutter, *American Rolls-Royce*, 58–59.

28. Ibid., 59; "News in the Trade," *The Carriage Monthly* 47 (March 1912): 34; "Brewster & Co. Move," *The Hub* 53 (March 1912): 463; "The 'Brewster' Car," *The Hub* 57 (January 1916): 17–18; advertisement for Valentine & Company, *The Vehicle Monthly* 52 (March 1917): 49.

29. Soutter, *American Rolls-Royce*, 59; "The 'Brewster' Car," 17–18; advertisement for Valentine & Company, 49; Beverly Rae Kimes et al., *Standard Catalog of American Cars, 1805- 1942*, 3d ed. (Iola, Wis.: Krause Publications, 1996): 145–47. Downing and Kinney, "Builders for the Carriage Trade," 97, include a short list of William Brewster's auto body innovations, as does Soutter (60–61); even if Brewster did not invent the "crank up" automobile window, he and his firm continued to shape the American automobile body.

30. Albert Russel Erskine, *History of the Studebaker Corporation* (South Bend, Ind.: Studebaker Corporation, 1918), 39, 41, 43; John A. Garraty and Mark C. Carnes, eds., *American National Biography* (New York: Oxford, 1999), s.v. "John Mohler Studebaker"; The Studebaker Family National Association, *The Studebaker Family in America* (Tipp City, Ohio, 1976): 396–97.

31. Board minutes, 1897, reproduced in *Studebaker Centennial Report* (South Bend, Ind.: Studebaker Corporation, 1952), 10; "Outline History of Studebaker, 1851–1921," typescript, 18, archives of the Studebaker National Museum, South Bend, Ind.; Erskine, *History*, 41, 43, 45; Carlisle, "Life of Clement Studebaker," 151, 161B–161C; "Items," *The Carriage Monthly* 35 (June 1899): 95; "A New Departure for the Studebakers," *The Carriage Monthly* 35 (July 1899): 97.

32. *Studebaker Centennial Report*, 10.

33. Erskine, *History*, 43, 45; "The Studebaker Automobile," *The Carriage Monthly* 38 (September 1902): 196.

34. "Studebaker Automobile," 196; "Outline History," 18, 22; Richard M. Langworth, *Illustrated Studebaker Buyer's Guide* (Osceola, Wis.: Motorbooks International, 1991), 9–10.

35. "Studebaker Automobile," 196; "Outline History," 21.

36. Wagar, *Golden Wheels*, 204–6; *100 Years on the Road* (South Bend, Ind.: Studebaker Corporation, 1952), 13; Ronald W. Clark, *Edison: The Man Who Made the Future* (New York: G. P. Putnam's Sons, 1977), 192, 209–11; "Outline History," 26.

37. "Outline History," 21–22, 26; "Electric Victoria Phaeton," *The Carriage Monthly* 44 (May 1908): plate 354.

38. Wagar, *Golden Wheels*, 243–47; "Outline History," 23; Erskine, *History*, 45; Langworth, *Illustrated Studebaker Buyer's Guide*, 11. Most accounts of Studebaker automobile production credit the Garford collaboration as producing the first Studebaker gasoline automobile, but there is evidence of an earlier gasoline Studebaker that stemmed from a brief arrangement with a Cleveland inventor ca. 1903–4: see Wagar, *Golden Wheels*, 99–102.

39. Erskine, *History*, 45; Langworth, *Illustrated Studebaker Buyer's Guide*, 11–13; James M. Flammang, *Chronicle of the American Automobile: Over 100 Years of Auto History* (Lincolnwood, Ill.: Publications International, 1994), 18; "President Buys a Studebaker," *The Carriage Monthly* 43 (July 1907): 112; "Studebaker Cars at Chicago Show," *The Carriage Monthly* 43 (December 1907): 61.

40. Erskine, *History*, 45; Langworth, *Illustrated Studebaker Buyer's Guide*, 11–13; David A. Hounshell, *From the American System to Mass Production, 1800–1932: The Development of Manufacturing Technology in the United States* (Baltimore: Johns Hopkins University Press, 1984), 218, 220–23.

41. Erskine, *History*, 47, 55–59; "Outline History," 27–29; Langworth, *Illustrated Studebaker Buyer's Guide*, 14–15; "Studebakers Secure Control of the E-M-F," *The Carriage Monthly* 46 (April 1910): 81.

42. Erskine, *History*, 47, 55–59, 63; "Outline History," 27–29.

43. Erskine, *History*, 45; "Electrics May Be Charged at Home," *The Carriage Monthly* 43 (February 1908): 56; "Outline History," 29–30; Langworth, *Illustrated Studebaker Buyer's Guide*, 9–15; Wagar, *Golden Wheels*, 246–47.

44. "Outline History," 23, 27, 34.

45. In 1934 Rolls-Royce America became the Springfield Manufacturing Corporation, producing approximately 135 Brewster Town Cars (lower-priced bodies on Ford V-8 chassis), but this could not save the firm from liquidation in 1936. Downing and Kinney, "Builders for the Carriage Trade," 97; Kimes, *Standard Catalog*, 145–47; Soutter, *American Rolls-Royce*, 59–63, 105, 123–24.

46. "Why Not Some Industrial Progress?" 321; Riker, *Pony Wagon Town*, 111; Ernest I. Miller, "The Death of an Industry," *Bulletin of the Historical and*

Philosophical Society of Ohio 12 (January 1954): 25–26; "Carriage and Motor Industries Related," *The Carriage Monthly* 44 (May 1908): 52.

47. Numerous historians have made this point, especially David Hounshell, who in chap. 5 of his *American System to Mass Production* discusses in detail the bicycle as precursor to the automobile, as does Martha Moore Trescott, "The Bicycle, A Technical Precursor to the Automobile," *Business and Economic History* 2d ser., 5 (1976): 51–75; "Carriage and Motor Industries Related," 52.

48. *How Carriage Axles Are Made* (South Egremont, Mass.: Dalzell Axle Co., [ca. 1905]), 22–23; "Carriage Lamps," *The Hub* 42 (February 1901): 509–10; Col. James H. Sprague, "The Auto-Top and Front as a Factor in the Automobile Industry," *The Carriage Monthly* 42 (January 1907): 54–55; "Made in Cleveland," *The Carriage Monthly* 51 (September 1915): 51; assorted receipts (1920s-1930s) for the Love, Thompson Company, GCRL; Sears, Roebuck & Company, *Fall and Winter Catalogue, 1927* (Chicago, 1927; reprint, New York: Bounty Books, 1970), 455–97,1001–3.

49. "Made in Cleveland," 45; "The Commercial Car and the Wagon Builder," *The Spokesman* 26 (June 10, 1910): 213–14.

50. James J. Flink, "The Olympian Age," *Invention and Technology* 7 (Winter 1992): 54- 56; Bowman, *Cambridge Dictionary of American Biography*, s.v. "Frederick John Fisher"; *The Story of Fisher Body* (Detroit, 1963).

51. Bowman, *Cambridge Dictionary of American Biography*, s.v. "Frederick John Fisher"; *The Story of Fisher Body*, 2–5, 7–8, overstates the firm's role somewhat, but it did act as one of the major proponents of closed auto body architecture.

52. Flink, *Car Culture*, 172–73; Flink, "Olympian Age," 62–63; Kennedy, *Automobile Industry*, 164–65; *The Story of Fisher Body*, 4–5.

53. "The Wagon Builders' Opportunity," *The Carriage Monthly* 47 (June 1911): 28; "Ingenious Rocking Fifth Wheel Device," *The Hub* 57 (October 1915): 27–28; advertisement for the Ohio Trailer Company, *The Carriage Monthly* 51 (December 1915): 79; "Labor Saving Bodies for Motor Trucks," *The Hub* 57 (March 1916): 9–12; photograph of the Aug. C. Fruehauf shop in Charles Philip Fox, *Working Horses* (Whitewater, Wis.: Heart Prairie Press, 1990), 12; Van Tassel and Grabowski, *Encyclopedia*, s.v. "Schaefer Body, Inc."; Roger B. White, *Home on the Road: The Motor Home in America* (Washington, D.C.: Smithsonian Institution Press, 2000), 55, 65–67, 134.

54. Kennedy, *Automobile Industry*, 88; "Detroit: The City Built by the Automobile Industry," *The Automobile* 28 (April 1913): 791–92; "Proceedings of the 43rd Annual Convention," *The Carriage Monthly* 51 (October 1915): 37; "Carriage Builders' National Association Convention and Exhibition of 1895," 468; "Made in Cleveland," 38; Van Tassel and Grabowski, *Encyclopedia*, "Cleveland: A Historical Overview," xxxiii, and s.v. "Automotive Industry"; Rose, *Cleveland*, 509, 680, 698–99, 724 25; *Cleveland: Some Features of the Industry and Commerce of the City and a Collection of Pictures of Leading Business Institutions* (Cleveland: Walter E. Pagan, 1917), 9, 17–18.

55. "Made in Cleveland," 45; "1895—Cleveland—1915," 5; "Exhibition of Vehicle Parts and Materials," *The Carriage Monthly* 51 (October 1915): 58–59; "Proceedings of the 43rd Annual Convention," 36–40; "Charles O. Wrenn," *The Carriage Monthly* 40 (April 1904): 139; "The Forty-Third Annual Convention of the Carriage Builders' National Association," *The Hub* 57 (October 1915): 38.

56. "Program of the Forty-Third Annual Meeting of the C.B.N.A.," *The Hub* 57 (September 1915): 8; "Forty-Third Annual Convention," 41–46, 48, 50–51.

57. "Forty-Third Annual Convention," 39, 46–47.

58. Ibid., 54–55; "1895—Cleveland—1915," 5–6; "C.B.N.A. Convention Entertainment," *The Hub* 57 (October 1915): 18 (this source misprints the date as 23 instead of 21 September); Riker, *Pony Wagon Town*, 292–93, 308; "Meeting of the Associate Members of the C.B.N.A.," *The Hub* 57 (October 1915): 18.

59. "History of the Carriage Builders' National Association, 1872–1913," 47; "A Fifty Year Old Association," 24; "Carriage Builders' National Association," *The Hub* 61 (June 1919): 32; "Carriage Builders Optimistic over Future," *The Automotive Manufacturer* 61 (October 1919): 14; "Use of Horse to Be Urged at Session," *The Enquirer* (Cincinnati), 3 October 1923, 12.

60. "Forty-Third Annual Convention," 41; "Good Advertising for the Horse-drawn Vehicle," *The Hub* 57 (February 1916): 13; "Happiness Supreme," *The Hub* 58 (April 1916): 13.

61. "Here Is a Worth While Association, The N.A.A.H.I.," *The Hub* 55 (January 1914): 331–33; Bowman, *Cambridge Dictionary of American Biography*, s.v. "August Belmont"; "Horse Supreme on the Milk Route," leaflet no. 183 of the Horse Association of America (n.p., [ca. 1927]), "Milk Wagons" subject file, NMAH.

62. Ralph Douglas Fleming, "Labor Conditions and Wages in Street Railway, Motor and Wagon Transportation Services in Cleveland" (Ph.D. diss., University of Pennsylvania, 1916), 25; Don H. Berkebile, ed., *Horse-Drawn Commercial Vehicles: 255 Illustrations of Nineteenth-Century Stagecoaches, Delivery Wagons, Fire Engines, etc.* (New York: Dover Publications, 1989), vi, vii; "Here Is a Worth While Association," 332.

63. The January 1916 issue was the first under *The Vehicle Monthly* heading, while the December 1921 issue was the first as *The Motor Vehicle Monthly;* September 1919 marked the first appearance of *The Automobile Manufacturer;* "City Horse Is Holding His Own against the Motor Car," *The Spokesman and Harness World* 41 (October 1925): 17; "What Would Happen in This Country If Horses Were Eliminated," *The Spokesman and Harness World* 41 (December 1925): 17–18; "Buggy and Wagon Industry Goes On Despite Increase of Autos," *The Spokesman and Harness World* 42 (October 1926): 31–32; "Manufacture of Carriages, Wagons and Materials, 1921," *The Automotive Manufacturer* 64 (Febru-

ary 1923): 16; "Manufacture of Horse-Drawn Vehicles Increased 120 Percent in 1923," *The Automotive Manufacturer* 66 (December 1924): 26.

Epilogue. Hail and Farewell

1. Discovering the date of the final CBNA convention proved a daunting task due to the confusion generated by a misdated photograph. Henry W. Meyer's *Memories of the Buggy Days* (Cincinnati, 1965), 3–4, contains a photograph identified as the final CBNA meeting in 1927. But there is no mention of a CBNA convention in the trade press, or in the September, October, or November issues of the *Cincinnati Enquirer* for that year. The identified men in the photograph (one person is mislabeled as Henry McLear, who died in 1924) were present at the 1925 meeting, which was held unusually late (November), and the men are dressed for cold weather. This photograph appears to have been taken at the fifty-third annual CBNA convention, November 19, 1925, in Cincinnati. The final convention was the fifty-fourth, held on September 15, 1926. Meyer, *Memories*, 3–4, 147; "Henry C. MacLear [*sic*]," *The Automotive Manufacturer* 66 (October 1924): 26; "Carriage Builders Will Gather at Cincinnati for 53rd Reunion," *The Spokesman and Harness World* 41 (November 1925): 12; "Fifty-Third Reunion of Carriage Builders Was Gala Affair," *The Spokesman and Harness World* 41 (December 1925): 29–30; "Carriage Builders Hold Fifty-Fourth Reunion at Cincinnati," *The Spokesman and Harness World* 42 (October 1926): 27–28.

2. Untitled correspondence school announcement, *The Vehicle Monthly* 54 (February 1919): 8; "Census Figures on Carriage and Wagon Industry," *The Automotive Manufacturer* 63 (June 1921): 27; "The C.B.N.A. Convention and Anniversary," *The Automotive Manufacturer* 64 (September 1922): 20; "Executive Committee Meeting of C.B.N.A.," *The Automotive Manufacturer* 64 (December 1922): 20; "Golden Anniversary of the Carriage Builders' National Association," *The Automotive Manufacturer* 64 (October 1922): 23; "C.B.N.A. to Meet at West Baden," *The Automotive Manufacturer* 65 (December 1923): 28; "Our Cincinnati Letter," *Motor Vehicle Monthly* 60 (September 1924): 55; "Use of Horse to Be Urged at Session," *The Enquirer* (Cincinnati), 3 October 1923, 12; "Cincinnatian Is Winner," *The Enquirer*, 4 October 1923, 3; "Carriage Builders Will Gather at Cincinnati for 53rd Reunion,"12; "Fifty-Third Reunion of Carriage Builders Was Gala Affair," 29–30; "Carriage Men Hold Reunion," *The Spokesman and Harness World*, 42 (September 1926): 28; "Carriage Builders Hold Fifty-Fourth Reunion," 27. The 1924 cancellation had precedent: apathy killed what would have been the fourth convention in 1875 (1876 became the fourth), economic depression prevented the 1896 convention (an executive committee meeting counted in its stead), which also held true for 1924 when a July executive committee meeting appears to have been counted as the fifty-second convention.

Glossary

Every field has its language, but the long history and plethora of styles have given horse-drawn vehicles an especially complex vocabulary. The following is obviously not exhaustive, but it will acquaint readers with the most important terms used in the text. The standard reference work and the source of many of these definitions is Don H. Berkebile's *Carriage Terminology: An Historical Dictionary* (Washington, D.C.: Smithsonian Institution Press, 1978), supplemented by "Vehicle Glossary," *The Carriage Monthly* 40 (April 1904): 139–43.

Vehicle Types

Barouche. Heavy, open coach with a back seat sheltered by a folding top; eighteenth-century European in origin, it became popular in America during the early nineteenth century.

Brougham. Closed, glass-fronted, twin passenger carriage driven by a coachman on an open box; originated in late 1830s England and later common in the United States.

Buckboard. Durable, four-wheeled passenger carriage built on a springy floorboard serving in place of springs; developed during the first third of the nineteenth century and one of a few uniquely American vehicle types.

Buggy. Four-wheeled passenger carriage for one to two people, usually with a folding top, and produced in huge variety and quantities; easily the most common horse-drawn carriage in nineteenth-century America and another original American type.

Cabriolet. Low, four-wheeled, doorless vehicle with a folding top and open driver's seat; an expensive, European, coachman-driven design bearing a name originating in seventeenth-century France and Italy.

Caleche. Also calèche or calash; essentially a folding top coach like a barouche, which name it generally replaced in the United States after 1850.

Carriage. Originally referred to any wheeled vehicle, technically designated a vehicle's running gear (wheels, axles, suspension, and underpinnings), but commonly (in the United States and this study) any horse-drawn vehicle other than carts, wagons, or sleighs.

Cart. Simple, two-wheeled vehicle built in many varieties, originally for freight but later also for pleasure driving.

Chaise. Essentially a chair mounted on a sparse, two-wheeled gear, frequently with a folding top, and drawn by a single horse; an eighteenth-century English and American vehicle, replaced in the United States by the buggy after the mid-nineteenth century.

Coach. Ancient term for a heavy, ornate carriage used by European royalty; in America, where the term gradually faded from use, it denoted any of a number of heavy, enclosed, four-wheeled passenger carriages.

Coupe. Technically a cut-down coach; in America this French name usually denoted a small, low, glass-fronted closed coach similar to, but a little larger than, a brougham.

Cutter. Early-nineteenth-century American term used throughout the horse-drawn vehicle era to denote a light, twin passenger sleigh with a single seat and drawn by one or two horses; the Portland and the Albany were the most popular varieties.

Dog cart. Originally an eighteenth-century English bird-hunting rig, this open cart became a fashionable American gentleman's pleasure-driving vehicle during the late nineteenth century.

Dray. Late-nineteenth-century American usage usually denoted a heavy-duty, four-wheeled freight vehicle, but in the eighteenth and early nineteenth centuries it indicated a two-wheeled, skeletonized cart used for hauling heavy and awkward freight.

Freighter. Slang term for a freight wagon, a large, heavy wagon with a bowed canvas top; the Conestoga was the most popular early version, with the term eventually designating improved square-box wagons used in the American West.

Gig. Essentially a chaise with updated suspension; this name gradually replaced the former in nineteenth-century America.

Landau. An old design, possibly from sixteenth-century Germany, this coach had a collapsible leather top and drop-down glass windows permitting use as a fully closed or open vehicle; one of the last heavy carriages to pass from regular American use.

Omnibus. Long covered wagon with windows and a pair of seats running the length of the vehicle, used as a public conveyance; in decline after the advent of electric street cars.

Phaeton. Open carriage of eighteenth-century French origin, driven by the occupant rather than a coachman; made in a variety of styles and popular in the United States through the end of the nineteenth century.

Pony vehicle. Vehicles built specifically for use with ponies, which were not only smaller than horses but also had different proportions.

Rockaway. Essentially a passenger wagon equipped with springs, standing top, and other conveniences making it a popular family vehicle from the mid-nineteenth century; an American vehicle type evidently originating in the 1830s.

Sleigh. Any of a wide variety of runnered passenger vehicles, including cutters.

Stagecoach. Heavy coach for long-distance passenger transport; the standard American form became the curved body, thoroughbrace-suspended Concord coach, used in remote areas through the end of the nineteenth century.

Sulky. Lightweight two-wheeled vehicle for track use.

Surrey. A late-developed (1870s) American vehicle design, this multiple passenger family carriage came in many styles, including the fringed-top version made famous by a Broadway play.

Tally-ho. American term for a heavy coach pulled by four horses and designed for private use as a wealthy man's excursion vehicle during the last quarter of the nineteenth century; exclusive, expensive, and relatively uncommon.

Trap. Loosely referring to a number of open, four-wheeled carriages with shifting seats which accommodated two or four passengers; a popular fair-weather pleasure carriage in the late-nineteenth-century United States.

Truck. Robust four-wheeled wagon for heavy hauling; made in increasing variety after the mid-nineteenth century, when it largely replaced the older dray.

Van. Tall, boxlike wagon for moving furniture and other bulky, fragile goods.

Victoria. A style of phaeton popular in nineteenth-century Europe and named for Queen Victoria; this low-hung, four-wheeled carriage sported a curved body and folding top and was widely used for park driving in England and the United States.

Vis-a-Vis. A coach, usually open, with face-to-face seating.

Volante. Two-wheeled, ornate cart with a folding top, of Spanish origin and popular in Cuba.

Wagon. Ancient term for wheeled vehicles in general; in the United States it applied to a variety of pleasure and work vehicles but eventually indicated wheeled vehicles other than carriages (the usage observed in this study).

Vehicle Parts

Axle. Cross-member on which vehicle wheels turn, also called an axle tree; while many were made out of wood, iron axles became common on light carriages.

Axle, patent. Any of a number of friction-reducing metal vehicle axles, such as the mail and Collinge.

Axle, thimble-skein. Axle type having hollow metal arms over the tapered wooden axle arms.

Axle arm. The tapered end of an axle which fit into the wheel hub.

Axle bed. Wooden support to which iron axles were affixed by clips; also that portion of a wooden or iron axle between the arms.

Axle box. Iron cone in the hub's center to protect the wood from wear.

Axle clip. Iron strap with threaded ends, used to clamp iron axles to wooden axle beds.

Axle skein. Iron strip sunk into the top and bottom of a wooden axle arm to protect it from wear; also called a clout.

Band. Circular metal hoops on hubs to prevent splitting.

Bobs. New England term for paired runners to be used alone (as in logging) or in place of wheels on carriages.

Bolster. Heavy timber on top of and parallel to a wagon's axles, designed to support the box.

Bow. Narrow wooden strips supporting a folding carriage top; wider versions supported canvas wagon tops.

Bow socket. Metal sockets receiving bow ends and forming a carriage top's pivot point.

Coach box. Elevated driver's seat on a coach.

Coach lace. Intricately woven fabric used to trim fine carriage interiors.

Coupling. Hinged metal fasteners attaching pole or shafts to a vehicle.

C-spring. Curved "C" shaped springs used on heavy coaches and carriages.

Cut-Under. Portion of a vehicle body cut away to allow front wheels a tighter turning radius.

Dash. Leather-covered iron frame at a carriage's front to protect occupants from hoof splatter; called a dashboard when made of wood.

Dish. Outward splay of spokes forming a "hollow" wheel to better resist lateral thrust.

Doubletree. Same as evener.

Draft. Amount of animal power necessary to move a vehicle; moving loads by pulling; a team and load.

Evener. Moveable front crossbar of a pair-horse vehicle to which the singletrees are attached; also called a doubletree.

Felloe. Wheel rim segment, usually sawn; largely supplanted by steam-bent rims in all but the largest wheels by the late nineteenth century.

Felloe plate. Metal plate used to bind felloes together to form a wheel rim.

Fifth wheel. Horizontal metal circle or circular segment between the front axle and body allowing axle to pivot smoothly.

Gear. Wheels, axles, and springs (a carriage's underparts) were collectively referred to as a gear, or gearing.

Hound. One of a pair of crooked timbers framed through the front axle to receive the pole and whose back ends support the fifth wheel.

King-bolt. Bolt connecting the front axle to the rest of the gear and on which the front axle pivots.

Leaf spring. General term for any wagon or carriage suspension composed of graduated length, flat steel leaves bound at the center.

Panel. Thin boards, usually whitewood, which fit between and to the framing to form a carriage body's outer skin.

Perch. A wooden or metal bar connecting the front and rear axles of a heavy carriage or coach, also called a reach; often eliminated in later designs by improved springs and suspension.

Pillar. Any upright frame member in a carriage body.

Platform spring. A combination of springs designed to create a lower mounted body, usually a pair of side springs perpendicular to the axle and joined by a cross-spring.

Pole. Long timber connecting a team to the vehicle; also called a tongue.

Rim. A wheel's outer framework, composed either of a pair of bent rims or a series of sawn felloes.

Rub iron. Also called a wear iron, this metal plate prevented the front wheels from injuring the wooden body during sharp turns.

Shaft. One of a pair of poles used to attach a single horse to a vehicle; also called a thill.

Sill. Main, horizontal body-framing timbers supporting the pillars.

Singletree. Also called a swingletree, one of a pair of short bars (one behind each horse in a team) attached to an evener and from whose ends the traces run up to each horse.

Slat iron. Thin iron plates used to connect the wooden top bows and pivot; generally replaced on carriage tops by bow sockets.

Stake. Upright timber at the ends of wagon bolsters used to position and contain the box; also a removable upright timber on the side of a freight vehicle.

Thoroughbraces. Heavy leather straps used on early coaches to suspend the bodies; used throughout the nineteenth century on stagecoaches

Top. Roof of a carriage, whether permanent (standing; usually wood) or collapsible (folding; cloth or leather).

Track. Width between wheels, often measured from the center of each tire on the ground; never truly standardized, though fifty-six inches and sixty-two inches became especially common.

Essay on Sources

In the course of nearly a decade's work on this topic I consulted everything from handwritten journals to a mouse-ravaged copy of the 1903 *Encyclopedia Britannica*. Rather than a complete list, what follows is an annotated guide to the most important sources informing this study. I have paid special attention not only to such predictably significant manuscript items as letters and photographs but also to the extensive government documentary record, voluminous trade literature, and trade catalogs that form the vast majority of the primary sources on the American wagon and carriage industry.

Manuscript Collections

Perhaps the single most complete repository of primary sources on American horse-drawn vehicles, the Gerstenberg Carriage Reference Library at the Long Island Museum of American Art, History & Carriages (formerly the Museums at Stony Brook) houses a tremendous amount of material on nearly every aspect of the trade. My examination of the Brewster carriage dynasty is anchored by a handful of splendid manuscript sources here. At least two copies exist of James Brewster's 1856 "Autobiography" (one of the handwritten originals resides here, while a convenient typescript can be found in the New Haven Colony Historical Society, New Haven, Conn.). This remains the best source of information on James Brewster's life, trade activities, outside business ventures, and especially his religious and philanthropic work. More of his world view can be obtained in his "Lectures by James Brewster to his Apprentices in 1826–7," another manuscript notebook in this collection in which Brewster recorded his religious and philosophical convictions. These two sources inform the majority of the secondary trade journal and newspaper articles. Two published sources contain additional information insufficiently explored or missing from the manuscript material: James Brewster, "An Address Delivered at Brewster's Hall," (New York: Isaac J. Oliver, 1857), which consists of Brewster's speech to his former apprentices, and James F. Babcock, "Address upon the Life and Character of the Late James Brewster," (New Haven, 1867), an informative posthumous assessment of Brewster's life.

In 1996 the musuem acquired a previously unknown manuscript of tremendous importance, what I refer to as the Brewster lawsuit copybook. This is a manuscript copybook of materials related to James B. Brewster's 1869 lawsuit against Henry Brewster, James W. Lawrence, and John W. Britton. The contents consist of carefully handwritten copies of dozens of letters, sworn statements, and advertisements, all relating to the businesses of the Brewsters and their assorted partners during the 1850s and 1860s. Internal evidence indicates compilation by or for Henry Brewster, perhaps for future use in the event of further litigation. It is likely that the body of letters was limited to what the copyist had on hand, but it may also mean that ones damaging to Henry were deliberately excluded. Still, there is enough material from each point of view—including letters, ads, and affidavits—to give a reasonably balanced picture of both sides of the dispute, and it is a remarkable record of the trade during the extremely competitive and rapidly changing 1850s and 1860s.

The Studebaker Brothers Manufacturing Company stands alone as the only major American wagon manufacturer to have a sizeable museum and archive devoted to its long history. South Bend, Indiana's Studebaker National Museum contains the majority of the company's extant records, as well as photographs, catalogs, and vehicles. The present study only skims the surface of this material; in particular I relied on M. L. Milligan's "Outline History of Studebaker, 1851–1921," compiled by a late company secretary and based on the firm's extensive scrapbook collection. Though not without minor error and omission, it forms an excellent supplement to other sources and a means of verifying key dates, people, and events. A highly underutilized source on the family, the wagon business, and the company's first and longest-running president is Charles Arthur Carlisle, "The Life of Clement Studebaker" (ca. 1902–08), typescript in the Local History Collection of the St. Joseph County Public Library, South Bend, Ind. Carlisle knew his father-in-law intimately and also made use of family members and his subject's extensive correspondence. While this source benefits from careful reading and verification, it nevertheless preserves a tremendous amount of otherwise unknown information. Much anecdotal material in secondary works originates here.

Back in Connecticut, the New Haven Colony Historical Society's *Dana Collection Scrapbook* (volume 127) preserves useful early information on James Brewster's shop and business from a variety of sources, while in Massachusetts the Amesbury Public Library's Jacob R. Huntington files contain a handful of informative sources on this wholesale pioneer, particularly "Origin of Amesbury Industry: Founder of Vehicle Business and His Early Struggles," *Newburyport Herald*, January 1903, n.p., and Margaret S. Rice," "Brief History of the Carriage Industry in Amesbury, Massachusetts," an undated typescript by a local historian.

Government Documents

The U.S. Census abounds in detailed information on the American wagon and carriage industry but has been astonishingly neglected by historians. The best general introduction to the manufacturing data is Meyer H. Fishbein, "The

Census of Manufactures, 1810–1890," *National Archives Accessions* 57 (Washington, D.C.: National Archives, June 1963): 1–20, while a prime entry point for those interested specifically in the wagon and carriage industry would be James K. Davis, "Carriages and Wagons," in U.S. Bureau of the Census, *Census Reports: The Twelfth Census of the United States, 1900* (Washington, D.C., 1902), vol. 10, part 4:291–322. Dawes summarizes the entire body of nineteenth-century data on the trade and gives invaluable pointers on the considerable limitations of the earliest material. Severely limited data renders the 1810 and 1820 reports of little use at all, while shifting categorization of the 1840–70 material (itself an indication of profound structural change within the trade) restricts it to only the most generalized comparisons. The post-1870 reports contain far more detailed and better organized figures, and while the categories continued to shift somewhat (reflective of the Census Bureau's ongoing attempt to pin down a rapidly changing industry), they allow a wider range of comparisons. The 1880, 1890, and 1900 reports provide exceptionally detailed information on the industry at its peak, while 1910 and 1920, through their diminished content as well as the figures themselves, accurately reflect the decline of the industry in the face of the automobile.

U.S. commissioner of labor Carroll D. Wright's interest in "the great machinery question" created a pivotal source for anyone interested in the intersection of skill and machinery in nineteenth-century industry: see U.S. Bureau of Labor, *Thirteenth Annual Report of the Commission of Labor, 1898: Hand and Machine Labor*, 2 vols. (Washington, D.C., 1899). Wright's research on the nexus between mechanization and skill and his incisive exploration of hand versus machine labor in both the wagon and carriage and the accessory industries provides data and interpretation of enormous utility. F. R. Hutton, "Report on Power and Machinery Employed in Manufactures," in U.S. Bureau of the Census, *Census Reports: Tenth Census of the United States, 1880* (Washington, D.C., 1888), vol. 22, part 2:178–294, is a heavily illustrated report on woodworking machinery which is essential for identifying and understanding its application in wagon and carriage shops. U.S. Bureau of Labor, *Thirteenth Annual Report of the Commissioner of Labor, 1898: Hand and Machine Labor* 2 vols. (Washington, D.C., 1899) contains industry-specific listings of the tools and machinery, which are also vital to any discussion of work practices before and after industrialization.

Two additional federal sources supplied me with insight into the wagon-making portion of the trade and the trade's presence at the Centennial Exposition. A government contract debate containing a comparative assessment of Studebaker product quality can be found in U.S. Congress, Senate, *Letter from the Secretary of the Interior, Transmitting, in response to a resolution of the Senate relative to the rejection of certain bids for wagons for the Indian service, a report of the Commissioner of Indian Affairs and accompanying correspondence* S. Ex. Doc. 210, 46th Congress, 2d sess., 1880 (Serial 1886), which also serves as a counterweight to the company's typically florid advertising claims. Two publications from the United States Centennial Commission, *Grounds and Buildings of the Centennial Exhibition, Philadelphia, 1876* (Philadelphia: J. B. Lippincott & Company, 1878) and *Official Catalogue of the U.S. Centennial Exhibition, 1876, Complete in One Volume* (Philadelphia: John R. Nagle & Company, 1876), contain general information about the exposition's architecture as well as a comprehensive list of exhibitors.

State-level government documents contain additional valuable material, though I drew most heavily on those for Ohio, particularly that state's factory inspection reports. These are variously titled, for example: Chief State Inspector of Workshops and Factories, *Second Annual Report of the Chief State Inspector of Workshops and Factories to the General Assembly of the State of Ohio, for the Year 1885* (Columbus: Westbote, 1885). On the municipal level, city directories provided me with information on individuals and businesses alike. Beyond the actual name listings, the city maps furnished spatial context, this often supplemented by advertisements typically included in the volumes. Like federal government documents, these sources remain surprising underutilized by historians.

Trade Catalogs and Company Publications

By the mid-nineteenth century advances in printing technology and the growing importance of advertising led most wagon and carriage manufacturers to issue catalogs and other promotional literature. Some of these are readily available in reprint editions. *G. & D. Cook & Co.'s Illustrated Catalogue of Carriages and Special Business Advertiser* (New Haven, 1860; reprint, New York: Dover Publications, 1970) is one of the best early illustrated carriage catalogs; likewise Lawrence, Bradley & Pardee, *Illustrated Catalogue of Carriages* (New Haven, 1862; reprint, New York: Dover Publications, 1998), contains information about Brewster's early business partner.

Later trade catalogs supply greater amounts of manufacturing detail, particularly Alliance Carriage Company, *Successful Salesman, "On the Road for the Alliance Carriage Co."* (Cincinnati, 1895), and Ohio Carriage Manufacturing Company, *Those Good Split Hickory Buggies and How They Are Made* (Columbus, 1913), both in the files of the Division of Transportation, National Museum of American History, Smithsonian Institution, Washington, D.C. (hereafter NMAH); Studebaker Brothers Manufacturing Company, *Catalogue and Price List, Studebaker Wagons* (South Bend, Ind., 1883), in the Gerstenberg Carriage Reference Library, The Long Island Museum of American Art, History, & Carriages, Stony Brook, N.Y. (hereafter GCRL); and Ohio Carriage Manufacturing Company, *Portfolio and Book of Split Hickory Vehicles for 1910* (Columbus, 1910; reprint, Hampshire, U.K.: John Thompson, n.d.). Brewster & Company, *Proposal for an Industrial Partnership* (New York, 1869), in GCRL contains important details about the firm's groundbreaking experiment in profit sharing.

While some of the earliest Studebaker advertisements are reproduced in later company literature (see below), I made extensive use of materials such as the *1876 Price List, Studebaker Bros. Manufacturing Company* (South Bend, Ind., 1876) in the collection of Mark L. Gardner, and *Catalogue and Price List, Studebaker Wagons, 1883* (South Bend, Ind., 1883) in GCRL. Studebaker advertising ephemera remains popular with collectors, so I am indebted to Frank Frost for sharing with me selected portions of his extensive collection, in particular segments of the lavishly illustrated *Studebaker Souvenir, 1893* (South Bend, Ind., 1893). The company's commemorative *100 Years on the Road* (South Bend, Ind., 1952) is valuable for its reproductions of early advertising and promotional materials, but it also includes a crudely altered photograph of the first shop (3) that has been carelessly reproduced elsewhere; secondary material and late company publications must be used with caution.

The Dalzell Axle Company's *How Carriage Axles are Made* (South Egremont, Mass., [ca. 1905]) contains important manufacturing information, while the Eberhard Manufacturing Company's *Saddlery Hardware, Catalogue and Price List no. 8* (Cleveland, 1887) and *Eberhard Manufacturing Co.'s Production in Malleable Iron Casting of Carriage and Wagon Hardware . . . Catalog no. 8* (Cleveland 1915), both in Division of Transportation, NMAH, supply abundant evidence of the importance of malleable iron castings. *Wagon Hardware Catalogue* and *Carriage Hardware Catalogue* (Cleveland, 1897–98), as well as *75 Years of Progress: Cleveland Hardware & Forging Company* (Cleveland, [ca. 1956]), in company records, Cleveland Hardware & Forging Company, Cleveland, Ohio, provided information on Eberhard's primary competitor.

A partial reprint of the 1909 catalog of the George Worthington Company showcases similar products in a general hardware context: see Towana Spivey, ed., *A Historical Guide to Wagon Hardware & Blacksmith Supplies* (Lawton, Okla.: Museum of the Great Plains, 1979). Sears, Roebuck & Co., *Consumer's Guide*, catalog no. 104 (Chicago, 1897; reprint, New York: Chelsea House, 1968), is one of numerous reprints of various Sears catalogs, most of which include at least a selection of the firm's many horse-drawn vehicle offerings.

Trade Periodicals

The American wagon and carriage trade press forms a particularly rich example of nineteenth-century technical literature. Generally speaking, *The Hub* and *The Carriage Monthly* led the others in content, length, and prestige; Ben Riker's widely repeated comparison of them in his *Pony Wagon Town* (chap. 21) is highly subjective and based on limited familiarity with the publications. Both are useful and within the limitations of the genre reliably accurate, and they each published a commemorative issue of particular importance to the historian. *The Carriage Monthly* 40 (April 1904) is an unparalleled compilation of brief company histories, portraits of "leading carriage men," and articles on various aspects of the trade, and *The Hub* 50 (October 1908) is similarly valuable, if less comprehensive. Based on filed and submitted material, both have their flaws, but aside from the odd factual error and the absence of an overall organizational scheme, they preserve a great deal of otherwise unobtainable information.

My account of the Coach-Makers' International Union comes from the pages of the early issues of *The Coach-Makers' International Journal;* gaps remain in the story, but regular features such as the editorial column and "Reports from Subordinate Unions" give a wealth of information on the locals and the opinions of at least some of the membership, while union president Harding's recruitment efforts can be profitably traced through his "President's Journey Continued" and "Letters from the President" in 1866 and 1867. Latter-day accounts such as "I. D. Ware as a Carriage Builder," *The Carriage Monthly* 40 (April 1904): 220P, "The History of 'The Carriage Monthly,'" *The Carriage Monthly* 40 (April 1904): 200Y–200CC, and H. G. Shepard's "From a Veteran Accessory Manufacturer," *The Carriage Monthly* 40 (April 1904): 174, contribute additional details, though some tend to downplay Isaac D. Ware's enthusiasm for trade unionism. My account of the spring 1867 New York strike comes not only from the obviously pro-union journal but also from the avowedly hostile commentary of Ezra M. Stratton in his *New York Coach-Maker's Magazine.* Indeed, much of the early history of the CMIU reads best when contrasted by the criticism of Stratton, who knew the trade well and whose commentary tends to balance some of Ware's more utopian utterances.

The best sources of information on the Carriage Builders National Association are the extensive annual convention accounts published in *The Hub* and *The Carriage Monthly* in the late fall or early winter of each year, though the final few received better coverage in *The Spokesman and Harness World.* There is, so far as I have been able to ascertain, no comprehensive published list of all CBNA conventions, but the following sources in conjunction with supplementary trade journal and newspaper searches have finally yielded a complete list: "Origin and History of the Carriage Builders' National Association," *The Carriage Monthly* 40 (April 1904): 102–6, and "History of the Carriage Builders' National Association, 1872–1913," *The Carriage Monthly* 50 (October 1914): 42–47. "A Fifty Year Old Association," *The Automotive Manufacturer* 64 (September 1922): 21–24, contains some additional information but provides only sketchy details on the post-1887 conventions. By the early 1920s the association had outlived most of the trade journals. My search for an account of the final meeting began with Henry Meyer's *Memories of the Buggy Days*, which contains a photograph identified as taken at the final CBNA convention in 1927. Careful research into the one remaining journal willing to devote space to such matters (*The Spokesman*), as well as Cincinnati newspapers, revealed the error in the caption and the information on the real final convention, which took place in 1926. The additional late history of the CBNA can be found in the sources mentioned in the Epilogue's endnotes.

Detailed, firsthand accounts of life and work in the preindustrial carriage trade are exceedingly rare, but one of the best comes from the early trade press. Ezra M. Stratton's "Autobiography of Caleb Snug, of Snugtown, Carriage-Maker," serially published in *The New York Coach-Maker's Magazine* 2 (July 1859) and 3 (November 1860) is a thinly disguised history of his own youth and an unmatched source of detailed information on early-nineteenth-century carriage apprenticeship. Besides the literally hundreds of instructional articles and thousands of notices in the long-running "Trade News" columns, the trade press also covered individual firms and proprietors via short biographies and portraits (especially *The Carriage Monthly*'s "Leading Men in the Trade" series), as well as in obituaries. The wagon and carriage trade press had always been conscious of the trade's long history, and feature articles on the history of leading firms, selected vehicle styles, and even some of the more prominent draftsmen and mechanics round out a usefully historical journalistic literature. Categorically arranged technical pieces, themselves excellent guides to shop and factory work practice, appear in Isaac D. Ware's *The Coach-Makers' Illustrated Hand-Book*, 2d ed. (Philadelphia, 1875; reprint, Mendham, N.J.: Astragal Press, 1995) from *The Carriage Monthly* and its predecessor *The Coach-Makers' International Journal*, and M. T. Richardson's *Practical Carriage Building*, 2 vols. (New York, 1892; reprint, Mendham, N.J.: Astragal Press, n.d.), a similar collection from *The Blacksmith and Wheelwright*.

Newspapers, Periodicals, and Journals

While the majority of period press coverage for this study comes from the industry's own journals, some outside sources supplied additional useful information. Hamilton S. Wicks, "The Manufacture of Pleasure Carriages," American Industries no. 4, *Scientific American* 40 (8 February 1879): 79–80 provides a rare detailed glimpse of the interior of Brewster & Company's factory, while "A Discovery in New York," *The Chicago Republican*, 23 August 1869, describes the inner workings of Benjamin Brewster's factory. The wagon and carriage industry has received exceedingly scant coverage in most academic and popular historical journals, with a few exceptions: Richard E. Powell Jr., "Coachmaking in Philadelphia: George and William Hunter's Factory of the Early Federal Period," *Winterthur Portfolio* 28 (winter 1993): 247–77, is the single best source on the eighteenth-century carriage trade. The widely cited piece by Paul H. Downing and Harrison Kinney, "Builders for the Carriage Trade," *American Heritage* 7 (October 1956): 90–97, contains several inaccuracies but still stands as a useful general introduction to the Brewsters. Though Edward P. Duggan's "Machines, Markets, and Labor: The Carriage and Wagon Industry in Late-Nineteenth-Century Cincinnati," *Business History Review* 51 (autumn 1977): 308–25 contains substantial erroneous generalizations, I concur with his observation that mechanization stemmed from owners' desire to increase production rather than to eliminate skilled labor.

In the realm of American manufacturing history, Philip Scranton, "Diversity in Diversity: Flexible Production and American Industrialization, 1880–1930," *Business History Review* 65 (spring 1991): 27–90, provides an expansive treatment of custom/batch versus bulk/mass production. John K. Brown, "Design Plans, Working Drawings, National Styles: Engineering Practice in Great Britain and the United States, 1775–1945," *Technology and Culture* 41 (April

2000): 195–238, insightfully contextualizes improved draftsmanship as a means of enhanced and centralized managerial control, while Robert B. Gordon's "Who Turned the Mechanical Ideal into Mechanical Reality?" *Technology and Culture* 29 (October 1988): 744–78, gauges the effects of this machinery upon the skill quotient of its operators. Martha Moore Trescott "The Bicycle, A Technical Precursor to the Automobile," *Business and Economic History* 2d ser., 5 (1976): 51–75, addresses the contributions of the bicycle to the nascent automobile industry.

Turning to workers' collective responses to industrialization, U.S. commissioner of labor Carroll D. Wright's "Labor Organizations" in Day Otis Kellogg, ed., *Encyclopedia Britannica*, New American Supplement, vol. 27 (New York: The Werner Company, 1903), remains a useful period survey of the most prevalent and influential nineteenth-century American trade unions, and Wright also penned an equally useful overview of worker's mutual benefit groups for the same source ("Mutual Benefit Societies," vol. 29).

Monographs, Collection Catalogues, and General Works

Monographs devoted specifically to the American craftsman and craft system remain surprisingly scarce. A sense of the system's ancient roots can be gleaned from L. F. Salzman's classic *English Industries of the Middle Ages* (Oxford: Clarendon Press, 1923) and David S. Landes, *The Unbound Prometheus: Technological Change and Industrial Development in Western Europe from 1750 to the Present* (Cambridge: Cambridge University Press, 1969), contains useful contextual material. Though dated and vastly oversimplified, Carl Bridenbaugh's *The Colonial Craftsman* (New York: New York University Press, 1950; reprint, New York: Dover Publications, 1990) remains a passable general introduction to the topic, while Ian M. G. Quimby's "Apprenticeship in Colonial Philadelphia," (M.A. thesis, University of Delaware, 1963; reprint New York: Garland, 1985) is a much more detailed single-city study. The latter's *The Craftsman in Early America* (New York: W. W. Norton, for the Henry Francis du Pont Winterthur Museum, 1984), is a much-needed expansion featuring more comprehensive research, specific examples, and an accurate portrayal of the system in eighteenth- and early-nineteenth-century America, but the single best survey is W. J. Rorabaugh, *The Craft Apprentice: From Franklin to the Machine Age in America* (Oxford: Oxford University Press, 1986). Charles F. Hummel's *With Hammer in Hand: The Dominy Craftsmen of East Hampton, New York* (Charlottesville: University Press of Virginia, 1968) and Edward S. Cooke Jr.'s *Making Furniture in Preindustrial America: The Social Economy of Newtown and Woodbury, Connecticut* (Baltimore: Johns Hopkins University Press, 1996) tackle the system as it existed in the woodworking trades, while Hummel's "The Business of Woodworking," in *Tools and Technologies: America's Wooden Age*, ed. Paul B. Kebabian and William C. Lipke (Burlington: Robert Hull Fleming Museum and the University of Vermont, 1979), pays particular attention to the business side of the small woodworking shop.

Ezra M. Stratton's *The World on Wheels; or, Carriages, with their Historical Associations from the Earliest to the Present Time* (New York, 1878) is one of the best detailed, nineteenth-century histories of the carriage worldwide. Ben Riker's *Pony Wagon Town* (New York: Bobbs-Merrill, 1948) is an enchantingly well-written account of life and work in a small Ohio carriage factory, and though potentially misleading in some respects, it records much about the ideas and attitudes of late-nineteenth-century members of the trade. Henry W. Meyer was a late carriage industry draftsman and catalog illustrator who spent a long retirement collecting information about the wagon and carriage industry. The self-published fruit of his efforts, *Memories of the Buggy Days* (Cincinnati, 1965), consists of letters, photographs, and anecdotes dealing primarily with the 1890s and later. While containing some erroneous information and lacking organization, the book rewards careful reading with an abundance of information difficult to find elsewhere. J. B. Hampton, *A Pocket Manual for the Practical Mechanic; or The Carriage Maker's Guide* (Indianapolis: Frank H. Smith, 1886), provides a rare perspective on drafting and many other subjects from the eyes of a carriage workman in a large firm.

Easily the most comprehensive source ever compiled on the subject, Don H. Berkebile's *Carriage Terminology: An Historical Dictionary* (Washington, D.C.: Smithsonian Institution Press, 1978), defines every imaginable cart, wagon, carriage, and component and provides the single best source on the decidedly nonstandardized realm of wagon and carriage nomenclature. His compilations of trade journal illustrations differ from most others in their selection and inclusion of useful interpretive commentary: *American Carriages, Sleighs, Sulkies, and Carts: 168 Illustrations from Victorian Sources* (New York: Dover Publications, 1977) and *Horse-Drawn Commercial Vehicles: 255 Illustrations of Nineteenth-Century Stagecoaches, Delivery Wagons, Fire Engines, etc.* (New York: Dover Publications, 1989).

For something so fundamental to the histories of Europe, America, and technology in general, the literature on the process of industrialization is also frustratingly thin. While numberless works use the term in passing, some give it a more central role in studies of craftsmanship, working-class consciousness, and village and city life. I have already made reference to the old but still useful work by David Landes, which covers the overall contours of the British Industrial Revolution. Maxine Berg's *The Age of Manufactures: Industry, Innovation and Work in Britain, 1700–1820* (Oxford: Oxford University Press, 1986) discusses early-eighteenth-century and nonfactory industry, two understudied aspects of British industrialization. Anthony F. C. Wallace gives an excellent picture of the antebellum industrial transformation of a small Pennsylvania village in *Rockdale: The Growth of an American Village in the Early Industrial Revolution* (New York: W. W. Norton, 1972).

The historian treating industrialization and the worker is still forced to rely on a literature that remains stubbornly wedded to Marxist and neo-Marxist presuppositions. This approach invariably forces the widely varied but near-universal experience of "work" into the far more restricted and ideologically contested category of "labor." Besides producing scholarship convincing primarily to those of the same persuasion, it continues to ignore or distort the experiences of the vast majority of people whose working lives had little to do with class consciousness, organized labor, or socioeconomic protest. Unionization is a legitimate and important part of the history of work, but its hegemony in the field of labor history stands in stark contrast to its far more diminished role in the lives of millions of working people. Despite this major weakness, the careful reader can garner a great deal of useful information from the better monographs without endorsing the underlying world view. In terms of general context I used Bruce Laurie's *Artisans into Workers: Labor in Nineteenth-Century America* (New York: Noonday, 1989), supplemented by three citywide case studies: Susan E. Hirsch, *Roots of the American Working Class: The Industrialization of Crafts in Newark, 1800–1860* (Philadelphia: University of Pennsylvania Press, 1978), Sean Wilentz, *Chants Democratic: New York City and the Rise of the American Working Class, 1788–1850* (Oxford: Oxford University Press, 1984), and Steven J. Ross, *Workers on the Edge: Work, Leisure, and Politics in Industrializing Cincinnati, 1788–1890* (New York: Columbia University Press, 1985). Monographs that wrestle with the basic meaning and overall impact of industrialization as a main focus, rather than ancillary to a study of urbanization or worker identity, remain surprisingly few in number; the best (and with a bibliographic slant) is Walter Licht's *Industrializing America: The Nineteenth Century* (Baltimore: Johns Hopkins University Press, 1995).

Robert B. Gordon and Patrick M. Malone, *The Texture of Industry: An Archaeological View of the Industrialization of North America* (Oxford: Oxford University Press, 1994), present their insights into industrialization from the material and mechanical point of view, particularly in their characterization of general machine types. Likewise Carolyn C. Cooper's *Shaping Invention: Thomas Blanchard's Machinery and Patent Management in Nineteenth-Century America* (New York: Columbia University Press, 1991) conveys information about the mechanization of woodworking that goes beyond the immediate scope of her story. Betsy Hunter Bradley's *The Works: The Industrial Architecture of the United States* (New York: Oxford University Press, 1999) is probably the single most comprehensive survey of factory architecture while Lindy Biggs, *The Rational Factory: Architecture, Technology, and Work in America's Age of Mass Production* (Baltimore: Johns Hopkins University Press, 1996), discusses the same from the perspective of rationalization.

Brooke Hindle and Steven Lubar's *Engines of Change: The American Industrial Revolution, 1790–1860* (Washington, D.C.: Smithsonian Institution Press, 1986) remains a useful survey of pre-1860 American industrialization while David A. Hounshell's *From the American System to Mass Production, 1800–1932: The Development of Manufacturing Technology in the United States* (Baltimore: Johns Hopkins University Press, 1984) is an indispensable scholarly survey of the evolution of manufacturing methodologies from the Springfield Armory to

Ford Motor Company. Philip Scranton's scholarship provides the best counterweight to overemphasis on large, vertically integrated corporations in American industry and has already spawned a substantial literature examining smaller and proprietary organizations that played an enormous part in American industrialization. His *Proprietary Capitalism: The Textile Manufacture at Philadelphia, 1800–1885* (Cambridge: Cambridge University Press, 1983) explores the Philadelphia textile industry, which was so different from Lowell's, while *Endless Novelty: Specialty Production and American Industrialization, 1865–1925* (Princeton: Princeton University Press, 1997) provides the most comprehensive treatment of the subject across the American industrial landscape. An accessible, brief overview of industrialization in the carriage trade can be found in Joanne Abel Goldman's "From Carriage Shop to Carriage Factory: The Effect of Industrialization on the Process of Manufacturing Carriages," in *Nineteenth Century American Carriages: Their Manufacture, Decoration and Use* (Stony Brook, N.Y.: The Museums at Stony Brook, 1987), 9–33.

The obvious starting place for understanding the power sources used in American industry is Louis C. Hunter's magisterial *History of Industrial Power in the United States, 1780- 1930*, vol. 1: *Waterpower in the Century of the Steam Engine* (Charlottesville: University Press of Virginia, for the Hagley Museum and Library, 1979) and vol. 2: *Steam Power* (Charlottesville: University Press of Virginia, for the Hagley Museum and Library, 1985), while Carroll W. Pursell Jr., *Early Stationary Steam Engines in America: A Study in the Migration of a Technology* (Washington, D.C.: Smithsonian Institution Press, 1969), provides a shorter and more accessible overview of steam. Polly Anne Earl, "Craftsmen and Machines: The Nineteenth-Century Furniture Industry," in *Technological Innovation and the Decorative Arts*, ed. Ian M. G. Quimby and Polly Anne Earl (Charlottesville: University Press of Virginia, 1974), discusses mechanization in a woodworking trade, and Paul A. Kube's "A Study of the Gruber Wagon Works at Mt. Pleasant, Pennsylvania" (M.Ed. thesis, Millersville State College, 1968) contains a site-specific study of the machinery and work practices in a Pennsylvania wagon shop.

In explaining the manufacture of metal vehicle hardware, particularly the origins and development of the malleable iron casting, I found useful W. David Lewis's study, *Iron and Steel in America* (Greenville, Del.: Hagley Museum, 1976), while the single best source on the iron industry in particular is Robert B. Gordon's *American Iron, 1607–1900* (Baltimore: Johns Hopkins University Press, 1996). H. A. Schwartz, *American Malleable Cast Iron* (Cleveland: Penton, 1922), presents a wealth of specific information on the origins and development of the malleable iron casting industry.

Mark L. Gardner's *Wagons for the Santa Fe Trade: Wheeled Vehicles and Their Makers, 1822–1880* (Albuquerque: University of New Mexico Press, 2000) is a refreshingly well researched monograph on wagons used in the American West, while William Louis Gannon, "Carriage, Coach, and Wagon: The Design and Decoration of American Horse-Drawn Vehicles," (Ph.D. dissertation,

Iowa State University, 1960), adds much about carriage design. There are numerous published works showcasing major collections of American horse-drawn vehicles, but the best (given its scope as well as its unparalleled degree and quality of interpretation) is easily The Museums at Stony Brook, *The Carriage Collection* (Stony Brook, N.Y., 1986). A related publication provides additional interpretation: *Nineteenth Century American Carriages: Their Manufacture, Decoration and Use* (Stony Brook, N.Y., 1987).

Studebaker is also unique among horse-drawn vehicle manufacturers for having an extensive secondary literature, but here one must proceed with caution: even recently published monographs rely heavily on a handful of well-meaning but misleading company publications, which contain incomplete, vague, or incorrect information. While it serves as a useful general outline of the managerial and financial aspects of the company, the source of some of the most common inaccuracies is Albert Russel Erskine's *History of the Studebaker Corporation* (South Bend, Ind.: The Studebaker Corporation, 1918); likewise Stephen Longstreet's *A Century on Wheels: The Story of Studebaker* (New York: Henry Holt, 1952). Still, both contain accurate information on J. M.'s California years through their reliance on the latter's first-hand accounts. There are a number of more recent monographs on Studebaker, but they share some of these same flaws. The definitive nineteenth-century history of the company, particularly from a manufacturing standpoint, remains to be written.

James J. Flink's *The Car Culture* (Cambridge, Mass.: MIT Press, 1975) and better-organized and less polemical *Automobile Age* (Cambridge, Mass.: MIT Press, 1988), as well as the dated but still useful E. D. Kennedy, *The Automobile Industry: The Coming of Age of Capitalism's Favorite Child* (New York: Reynal & Hitchcock, 1941), provided vital facts and contextual detail about the early automobile industry. Michael Brian Schiffer, *Taking Charge: The Electric Automobile in America* (Washington, D.C.: Smithsonian Institution Press, 1994), is one of several recent retrospectives on the early-twentieth-century electric automobile, while Richard Wagar, *Golden Wheels: The Story of the Automobiles Made in Cleveland and Northeastern Ohio, 1892–1932* (Cleveland: John T. Zubal and the Western Reserve Historical Society, 1986), remains the standard work on Cleveland automobiles and their makers. Arthur W. Soutter was a long time Rolls-Royce America employee (and its last upon its 1936 liquidation), and his *The American Rolls-Royce: A Comprehensive History of Rolls-Royce of America, Inc.* (Providence, R.I.: Mowbray Company, 1976), provides a marvelous insider's view of the rise and fall of that famed marque's American division, including a wealth of information on Brewster & Company's own brand of automobile, as well as its final phase as a Rolls-Royce division. Intimately familiar with both Rolls-Royce America and Brewster & Company finances, Soutter's account takes a hard-headed look at the massive indebtedness William Brewster piled on Brewster & Company during its ten years of automotive production and points to the great harm it did to both firms. Soutter provides a counterweight to some popular accounts that tend to portray Brewster & Company's demise as owing solely to its unflinching commitment to product quality and traditional craftsmanship. Finally, of the various works dealing with Studebaker automobiles I found Richard M. Langworth's *Illustrated Studebaker Buyer's Guide* (Osceola, Wis.: Motorbooks International, 1991) to be of the greatest use. This contains a surprisingly detailed and well-organized amount of information on each of the major vehicles but is especially useful for anyone attempting to disentangle the confusing progression of Studebaker-Garford models.

Index